Biotechnology principles and applications

Biotechnology

EDITED BY I. J. HIGGINS, D. J. BEST & J. JONES

principles and applications

BLACKWELL SCIENTIFIC PUBLICATIONS

OXFORD LONDON EDINBURGH

BOSTON PALO ALTO MELBOURNE

© 1985 by
Blackwell Scientific Publications
Editorial offices:
Osney Mead, Oxford, OX2 0EL
8 John Street, London, WC1N 2ES
23 Ainslie Place, Edinburgh, EH3 6AJ
52 Beacon Street, Boston
  Massachusetts 02108, USA
667 Lytton Avenue, Palo Alto
  California 94301, USA
107 Barry Street, Carlton
  Victoria 3053, Australia

First published 1985

Printed and bound in
Great Britain by
R. J. Acford, Chichester, Sussex

DISTRIBUTORS

USA and Canada
  Blackwell Scientific Publications Inc
  PO Box 50009, Palo Alto
  California 94303

Australia
  Blackwell Scientific Publications
  (Australia) Pty Ltd
  107 Barry Street, Carlton
  Victoria 3053

British Library
Cataloguing in Publication Data

Higgins, I.J.
  Biotechnology principles and
  applications.—(Studies in
  microbiology, ISSN 0267–3134; 3)
  1. Biotechnology
  I. Title   II. Best, D.J.   III. Jones, J.
  IV. Series
  660′.62   TP248.3

  ISBN 0–632–01029–0
  ISBN 0–632–01034–7 Pbk

Contents

Contributors

Beech G.A. RHM Research Limited, The Lord Rank Research Centre, Lincoln Road, High Wycombe, Buckinghamshire HP12 3QR

Best D. J. Biotechnology Centre, Cranfield Institute of Technology, Cranfield, Bedfordshire MK43 0AL

Brierley C.L. New Mexico Bureau of Mines and Mineral Resources, Socorro, NM 87801, USA

Coombs J. Department of Plant Sciences, School of Biological Sciences, University of London, 68, Half Moon Lane, London SE24 9JF

Hall D.O. Department of Plant Sciences, School of Biological Sciences, University of London, 68, Half Moon Lane, London SE24 9JF

Hamer G. Institute for Aquatic Sciences, Swiss Federal Institute of Technology, Ueberlandstrasse 133, CH-8600 Dubendorf, Switzerland

Hardy K.G. Biogen S.A., Route de Troinex 3, 1227 Carouge, Geneva, Switzerland

Higgins I.J. Biotechnology Centre, Cranfield Institute of Technology, Cranfield, Bedfordshire MK43 0AL

Jones J. Biotechnology Centre, Cranfield Institute of Technology, Cranfield, Bedfordshire MK43 0AL

Kelly D.P. Department of Environmental Sciences, University of Warwick, Coventry CV4 7AL

Melvin M.A. RHM Research Limited, The Lord Rank Research Centre, Lincoln Road, High Wycombe, Buckinghamshire HP12 3QR

Oliver S.G. Department of Biochemistry and Applied Molecular Biology, University of Manchester Institute of Science and Technology, PO Box 88, Manchester M60 1QD

Pickup J.C. Unit for Metabolic Medicine, Department of Medicine, Hunts House, Guy's Hospital Medical School, London Bridge, SE1 9RT

Seal K.J. Biotechnology Centre, Cranfield Institute of Technology, Cranfield, Bedfordshire MK43 0AL

Skinner F.A. 5, Carrisbrooke Road, Harpenden, Hertfordshire AL5 5QS

Stafford D. Cardiff Laboratories for Energy and Resources Limited, Lewis Road, East Moors, Cardiff CF1 5EG

Taggart J. RHM Research Limited, The Lord Rank Research Centre, Lincoln Road, High Wycombe, Buckinghamshire HP12 3QR

Preface and acknowledgements

The purpose of this volume is to present a broad, informed view of contemporary biotechnology, a technology with ancient roots and enormous future potential. We hope that it will prove valuable both to students of the subject and practitioners of research in specific aspects seeking a broader view. Clearly, it is not intended to be an advanced treatise but we believe that it is an authoritative introduction to most major aspects of biotechnology written by leading researchers. A substantial collection of references has been included after each chapter for those wishing to pursue more specialist studies.

The fact that this volume was first conceived by one of the editors some four years prior to publication is a measure of the pressure of time on many practising biotechnologists in recent years rather than of their slothfulness. During that period the number of editors grew from one to three and some manuscripts are still awaited! Perhaps they will make the next edition.

The editors are most grateful for the extreme forebearance shown by some of the contributors to this volume who originally offered manuscripts some three years prior to publication. We would also like to thank all contributors and the publishers for their efforts to make a success of a difficult project. We are delighted to have eventually 'got the act together' and sincerely hope that our readership will derive both knowledge and pleasure from this book.

The authors of Chapter 3 wish to thank Dr J. Edelman, Director of Research, RHM PLC, for permission to publish, and the management of the Frish Dytes Partnership for the release of information on MycoProtein. The authors of Chapters 4 and 6 acknowledge the assistance of Mrs D.P. Fowler in the preparation of diagrams.

The author of Chapter 9 is greatly indebted to the following colleagues at Rothamsted Experimental Station for photographs to illustrate the text of this chapter: Dr Barbara Mosse for Figs. 9.1 and 9.2; Dr D.S. Hayman for Figs 9.3–9.6; Dr M.G.K. Jones for Figs. 9.7 and 9.8; and Dr R. Nelson for Figs. 9.9 and 9.10.

Biotechnology Centre   I. J. Higgins
Cranfield Institute of Technology   D. J. Best
September 1984   J. Jones

1   What is Biotechnology?

I. J. HIGGINS

1.1   Introduction

Biotechnology is not simply, as some sceptics would have it, a latter-day buzz-word for man's oldest industry, but it is a word, the acceptance of which into common parlance is highly symbolic. This acceptance reflects the widely, but not universally held view that many facets of industry and society will be revolutionized over the next ten to fifty years by the application of biological principles and materials. There is no doubt that there has been an extraordinary increase in interest and activity in the subject in the last few years. This has manifested itself in many ways, from the development of countless small entrepreneurial biotechnology companies to the setting up of committees by national governments to examine the potential for the subject and the wide-spread introduction of university courses in biotechnology. The governments of most advanced nations and many less advanced ones have already committed substantial funds to help catalyse the development of biotechnology, although there are substantial differences in the level of commitment and efficiency of use of available funds. It is a commonly held view amongst professionals concerned with the development of biotechnology that, in general, for a nation to be successful in this endeavour, a commitment from central government is essential to facilitate the evolution of a technology of such a complex and uniquely multi-disciplinary nature. Many aspects of biotechnology involve long lead times from ideas to product and only a few nations, most notably the USA, have fully adequate, free-enterprise financial mechanisms to enable optimum development of such technology, largely independent of government.

Biotechnology is perhaps best defined as the industrial exploitation of biological systems or processes and it is largely based upon the expertise of biological systems in recognition and catalysis. This is reflected in the ability to recognize other biological systems and specific chemicals and an extraordinary ability of enzymes to catalyse a vast range of specific chemical reactions under moderate conditions. Even today, the catalytic chemist cannot compete with the efficiency and specificity of biological catalysis and our understanding of the processes of enzymic catalysis is very limited. In spite of our achievements in chemistry, upon which so much of our industry is based, we are still quite naive in our understanding of catalysis.

Clearly, therefore, man has exploited biotechnology for thousands of years in such activities as brewing, wine-making, bread-making, food preservation and modification by fermentation (e.g. cheese, vinegar and soy sauce), the manufacture of soap from fats, primitive medications

1

and waste treatment. It is, however, the discovery of genetic engineering techniques via recombinant DNA technology (see Chapter 7) which is responsible for the current 'biotechnology boom'. Not only do these techniques offer the prospect of improving existing processes and products, but they are enabling us to develop totally new products which were not previously possible, and facilitate the realization of other processes. The development of genetic engineering is a dramatic example of the difficulty of predicting those areas of basic science which are most likely to lead to important applications. The discovery of recombinant DNA technology is a consequence of the large amount of support given to research into molecular biology for more than forty years. Yet even as recently as the late sixties, a common criticism of giving relatively so much support to this 'glamour' area of chemistry and biology was that nothing directly useful had come from it. Now, however, it is clear that it has led to discoveries that will radically affect humanity.

Whilst recombinant DNA technology has doubtless been the main cause of much of the recent publicity for biotechnology, it must be emphasized that there have also been important recent developments in other areas of activity which are essential for the development of the technology. The most important subjects in this regard are shown in Table 1.1 and their respective contributions to the various different areas of biotechnology are discussed in the following chapters in this book.

In no area of scientific endeavour is the following quotation more appropriate: 'No a thousand times no, there does not exist a category of science to which one can give the name, applied science. There are science and the applications of science, bound together as the fruit to the tree which bears it' (Pasteur, 1871, translated from *Revue Scientifi-*

Table 1.1. Subjects in which there have been recent advances of importance to biotechnology

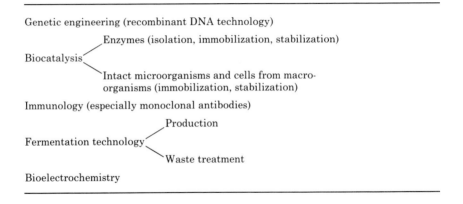

Genetic engineering (recombinant DNA technology)

Biocatalysis
— Enzymes (isolation, immobilization, stabilization)
— Intact microorganisms and cells from macro-organisms (immobilization, stabilization)

Immunology (especially monoclonal antibodies)

Fermentation technology
— Production
— Waste treatment

Bioelectrochemistry

que). In biotechnology the applications of the science are intimately linked with fundamental work and are following closely behind the frontiers of the science. The extraordinary multi-disciplinary nature of biotechnology alluded to above is emphasized in Fig. 1.1. Not all products or processes involve all the disciplines depicted but, in any one case, several are involved.

Another interesting aspect of the current stage of development of the subject is the need in many promising areas for close international

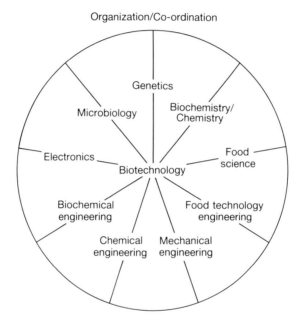

Fig. 1.1. The interdisciplinary nature of biotechnology.

co-operation between the scientists and technologists involved. This is because there are often only a few research groups in the world having the appropriate expertise. This phenomenon has already been exemplified most obviously by the highly multinational nature of several of the larger entrepreneurial biotechnology companies established over the last few years.

The remainder of this chapter is devoted to a brief survey of the development and future potential of the major areas of biotechnology, together with economic considerations, as a prelude to the more detailed discussions of the various areas in subsequent chapters.

1.2   Historical perspectives

Until the recent general acceptance of the all-embracing term, biotechnology, most biologically based technology was encompassed by terms

such as applied microbiology, applied biochemistry, enzyme technology, bioengineering, applied genetics and applied biology. Apart
from the manufacture of soap, the first of these 'technologies' to develop
was the forerunner of applied microbiology. Although ignorant of the
processes involved and, therefore, practising an art rather than a
science, our ancestors have for thousands of years exploited microbial
fermentations for food preservation (e.g. cheese, vinegar), flavour
enhancement (e.g. bread, soy sauce) and alcoholic beverage production.
Indeed, brewing still remains today in monetary terms the most
important biotechnology industry, with a world annual production of
about 10^{11} l, worth approximately M£100 000. All these processes
depend upon the mechanisms used by some microorganisms to grow and
reproduce under anaerobic conditions. It was Pasteur who, in the late
nineteenth century, really formed the basis for future developments in
applied (industrial) microbiology and, hence, much of biotechnology.
He realized that microbes were responsible for fermentation and
showed that different types gave different products. His work laid the
foundations for the subsequent development of industrial processes for
fermentative production of organic solvents (e.g. acetone, ethanol,
butandiol, butanol, isopropanol) and other chemicals by different
species in the late nineteenth century and early twentieth century.
These all involved the conversion of plant carbohydrates to useful
chemical products by microbes in the absence of oxygen. Under these
conditions, the microbe uses the entropy change in the interconversion
as the source of energy for growth. This is distinct from aerobic
processes, in which living organisms get far more energy from the
controlled oxidation of chemicals to carbon dioxide and water.

Such processes using biomass, i.e. renewable resource feedstocks,
for making chemicals constituted the first phase of modern biotechnology. Many were replaced by chemical processes, however, as the
petrochemical industry developed. Some chemicals, e.g. citrate, vinegar
acetate and itaconate, particularly where used extensively in food
processing, have continued to be produced by fermentation, which is the
most economic route. Even industrial alcohol in some parts of the world
(e.g. Italy) continued to be made by fermentation and now, of course, as
a result of the energy crisis, ethyl alcohol production from plant
material by fermentation is increasingly important in the USA and
Brazil and may well become so in the Far East.

There is likely to be a return to the manufacture of organics by
fermentation, but in Europe it may well be limited to higher value
chemicals. An obvious exception is the conversion of the EEC wine
lakes to industrial alcohol. This is a strange case of politics, biotechno-

logy and, to say the least, unconventional economics offering unfair competition to the chemical industry.

Fig. 1.2 depicts the biochemical basis of an early example of 'modern' biotechnology, of particular interest since it involves 'engineering' a modification of the normal metabolism involved in the yeast ethanolic fermentation. Addition of bisulphite to the fermentation broth prevents acetaldehyde from acting as an acceptor for reducing equivalents derived from the Embden–Meyerhof pathway. Dihydroxyacetone phosphate then acts as the acceptor, with the consequent generation of glycerol in place of ethanol. This process was particularly important for meeting the demands for glycerol for explosive manufacture during World War I.

The next major milestone in biotechnological generation of valuable products was the development of the antibiotic industry, arising initially from the discovery of the chemotherapeutic properties of penicillin by Fleming, Florey & Chain in 1940. Worldwide, the turnover of this industry is now about $£2 \times 10^9$/a.

Both fermentative production of chemicals and food additives and antibiotic synthesis have always involved aseptic operation of plant, although some more recent processes (e.g. single cell protein (SCP) production) are more demanding in this respect. This is an interdisciplinary problem for the chemical engineer and microbiologist and is discussed in Chapter 10. In contrast, the use of microorganisms in waste

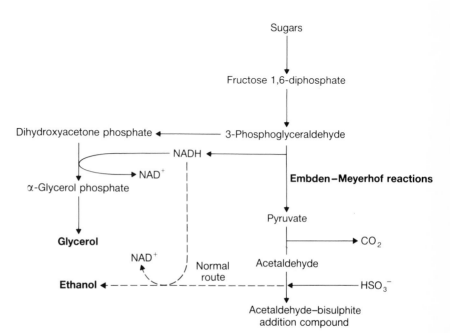

Fig. 1.2. Modification of the yeast ethanolic fermentation to generate glycerol.

treatment processes (Chapter 6) does not involve aseptic operation (in general, the more species present, the better) but the design and manufacture of waste treatment plants presents the chemical engineer and microbiologist team with a different set of problems. The activated sludge process for mineralizing organic waste was first developed in 1914 and since then it has developed dramatically in size and sophistication and is exploited worldwide for treatment of sewage.

The treatment of wastes by anaerobic digestion by mixed microflora, eventually generating biogas (mainly methane and CO_2), has become an increasingly important process throughout this century. It is energetically highly efficient in terms of conserving and concentrating the energy available in the waste (over 80% of the free energy is recovered in the gas). In some parts of the world, particularly in rural communities, the process represents a substantial part of overall energy consumption. There are, for example, over eighteen million biogas generators in rural China. In advanced nations, with high energy consumption, the conversion of wastes to biogas could only meet a few per cent of the energy demands. Nevertheless, in large municipal waste treatment plants, biogas is often burnt in heat engines which in turn drive electricity generators. Small systems for treatment of agricultural wastes have also been developed in recent years.

1.3   Developments in the biotechnology industry since World War II

In addition to continuing improvements in processes discussed above, there have been many other important practical developments over the last forty years, some of which are shown in Table 1.2. The examples given are discussed in subsequent chapters but four particularly interesting examples—amino acids, SCP (single cell protein) steroid transformations and the cultivation of plant and animal cells—are briefly examined further in this section.

During the last thirty years, the production of amino acids by aerobic microbiological processes has expanded rapidly. The two produced in the largest amounts are monosodium glutamate (annual world production of about 150 000 t), used primarily as a flavour enhancer, and lysine (annual world production of about 15 000 t) as a food supplement. The total annual world amino acid sales represent about £1 × 10^9, with Japanese companies fulfilling most of the demand. The rather special position of Japan in some aspects of biotechnology is discussed later in this chapter under 'Economic considerations'.

Microorganisms have the ability to upgrade low protein plant

Table 1.2. Some recent important products and services based on biotechnology

Industry	Examples
Agriculture	Strain selection, plant and animal breeding techniques (including cloning)
Chemicals	Organic acids (e.g. citric, itaconic), use of enzymes in detergent formulations
Energy	Increasing use of biogas, large-scale production of ethanol as a liquid fuel
Environment	Improved test and monitoring procedures. Prediction of the fate of xenobiotic chemicals via increasing understanding of microbial biochemistry. Improvement in waste treatment techniques, especially for industrial wastes
Food	New methods of food treatment and preservation. Food additives (e.g. microbial polymers, microbiologically derived amino acids), use of enzymes in food processing. Single cell protein
Materials	Mineral extraction, improved knowledge and control of microbial biodeterioration
Medicine	Improved diagnosis using enzymes, enzyme sensors, use of microorganisms and enzymes in manufacture of complex drugs (e.g. steroids), new antibiotics, use of enzymes in therapy

materials to high protein food, and the large-scale industrial exploitation of this phenomenon to grow *Saccharomyces cerevisiae* for human consumption was exploited in Germany during World War I. The yeast was incorporated mainly into sausages and soups and in this way about 60% of the country's pre-war food imports were replaced. Similar processes using the food yeasts, *Candida arborea* and *Candida utilis* were used during World War II. A number of oil and chemical companies began research and development work during the 1960s aimed at developing new processes for making SCP for animal or human consumption (see Chapter 3). This was, in part, a response to the world shortage of protein foodstuff. Substrates include petroleum, methane, methanol and starch, and most products so far have been developed for animal feed. In general, methanol and starch processs have proved most viable. In western countries, ICI have developed the largest plant, making about 70 000 t/a of feedstuff (Pruteen) from methanol in a single fermenter using the methanol-utilizing bacterium, *Methylophilus methylotrophus*. Modifications of the nitrogen assimilating mechanism of this bacterium by recombinant DNA technology, leading to an improved yield, was one of the earliest practical demonstrations of the potential industrial value of genetic engineering (Chapter 7). In the USSR, over 1 Mt/a of SCP is produced, largely from hydrocarbons and vegetable waste. One of the few examples of a high quality SCP product for direct human consumption is being test marketed in Britain by Rank Hovis McDougall and is derived from a fungus grown on carbohydrate feedstocks (Chapter 3).

Whilst there has been a steady increase in the use of the enzymes in industrial processing, particularly over the last twenty years, growth of the enzyme industry has been disappointingly slow. Their use is mainly limited to a small number of enzymes used primarily in food processing. However, the range of enzymes used in medicine, mainly for diagnosis, has been increasing quite rapidly over the last few years. The total enzyme market nevertheless remains relatively small, about £250 × 10⁶/a. The main reasons for this rather slow development are lack of stability, product recovery and problems of cofactor supply or replacement. In some cases, however, these problems have been circumvented by exploiting the specificity of enzymes in intact microorganisms. This technique became extremely important during the 1950s for the large-scale production of steroid drugs. It became established that many microorganisms are capable of hydroxylating the complex steroid nucleus with great regio- and stereo-specificity. An early and particularly important finding was that the mould, *Rhizopus arrhizus*, could hydroxylate the female sex hormone, progesterone, stereospecifically in the 11-position. This greatly simplified the procedure for manufacturing cortisone, used for treating arthritis. This compound had previously been manufactured by a 37-step chemical process with 0·02% yield, leading to a cost of about \$200/g. The introduction of the microbial biotransformation step reduced the cost to about 68 c/g by facilitating a much simpler synthetic procedure. Subsequently, a range of microorganisms were isolated capable of specific hydroxylation of the other ring carbon atoms and, more recently, microbial systems have been used for converting phytosterols to the smaller C-19 steroid hormones, used, amongst other things, in oral contraceptives.

There is currently considerable interest in the chemical industry, particularly in the USA and Japan, in the possibility of devising analogous processes for the production of lower cost chemicals, exploiting microbial oxidative biocatalysis as a basis for more economic processes than the conventional chemical one (see Chapter 4).

Our ability to cultivate plant and animal cells on a large scale has already had important industrial consequences, e.g. in the mass production of vaccines. Techniques for fusing different cell lines have also been developed, allowing, for example, the cloning of oil-palms with improved characteristics both in terms of yield and quality by Unilever scientists.

1.4   Future prospects for the development of biotechnology

Biotechnology will, in the future, make an important contribution to the quality of life via a range of goods and services, summarized in Table 1.3. In the short term, the most important contributions from biotechnology will, in general, be medical but in the longer term (ten years plus) most informed commentators are of the view that applications

Table 1.3. Goods and services that biotechnology can provide

Analytical chemical tools
Biosynthetic and biodegradation processes
Carbon chemical feedstocks
Chemical (downstream) processing
Consumer chemicals: adhesives, detergents, dyes, fibres, flavours, gelling and thickening agents, gums, mucilages, perfumes, pigments, plastics, surfactants, waxes, etc.
Energy sources
Environment control in air, water and soil
Food and drink (agriculture and processing)
Health care: (diagnosis, treatment) plant and animal care
Mineral extraction from land or sea

in agriculture and the chemical industry will be more important in economic terms. Some of the main developments anticipated in relation to medicine, energy, food and drink, chemistry, materials, the environment and agriculture are discussed briefly in the following sections. Many of these aspects are discussed in more detail in later chapters.

1.4.1   *Medicine*

In medicine, we have already witnessed some important advances resulting from recombinant DNA technology. Many organizations are well advanced in developing processes for the large-scale production of human interferon, having cloned human genes into microorganisms. However, there are admittedly still some doubts as to the efficacy of interferons as antiviral and antitumour agents. In addition, genes for human insulin and growth hormone have been cloned and expressed in bacteria, and the genes for many other human proteins of value for diagnosis or treatment are being cloned to facilitate their large-scale production. At the time of writing, microbially derived human insulin is already being marketed and used for therapy in diabetics. The development of techniques for producing monoclonal antibodies, which are discussed in more detail in Chapter 8, is of great importance and some applications are shown in Table 1.4.

There will, no doubt, be a steady increase in the use of enzymes in diagnosis, therapy and in tissue and cell transplantations.

An expanding area of biotechnology, still in its infancy, that is likely to have major effects on medicine as well as industry involves the interaction between biological and electrical and electronic system, i.e. bioelectronics and bioelectrochemistry. There have been significant advances in this area in recent years, both in electronic engineering, leading to, for example, infusion pumps, and in effecting connections between biochemicals, especially enzymes and antibodies, and various types of electrode. A range of specific sensing devices based on these principles have already been devised, e.g. glucose monitors, particularly for medical purposes, and nerve gas sensors for military use. Most sensors developed to date are based upon detecting the results of enzymic activity by a conventional electrode upon which the biological system is immobilized. New approaches, however, which are leading to more sensitive and effective devices offering a wide range of possibilities, include direct electron transfer between electrodes and protein redox centres and the accommodation of proteins on the gate of semiconductor devices.

In the short term, quantitatively the most important enzyme-based medical sensor is the glucose sensor, of which there are a variety of configurations. There will be a range of products based upon this; in particular, cheap, accurate, reliable devices for *in vivo* sensing are imminent. These are expected to have a major effect on improving the regulation of blood sugar levels in diabetics. Inadequate fine control is generally believed to be the cause of the long-term, life-threatening secondary effects of diabetes. Such sensors will allow closing of the control loop in artificial pancreas devices.

Table 1.4. Some possible applications of monoclonal antibodies

Medical field	Applications
Analysis	Structural probes to identify specific features on cell surface
Diagnosis	Pregnancy testing kits. Detection of oestrogen receptors associated with some breast cancers
Immunoassay	Precise assay of specific antigens
Immune purification	Purification of antigens, e.g. interferon
Therapeutics	Directing toxins to tumour cells, deactivating poisons, passive immunization, treatment of autoimmune diseases.

The commercialization of other sensors measuring blood chemicals, including bioelectronic immunosensors, in some cases based upon field effect transistors, is likely within the next decade. These could form the basis of relatively inexpensive systems capable of measuring and monitoring a wide range of substances in body fluids, which could revolutionize diagnosis.

Whilst the exploitation of the specific *catalytic* properties of biological systems has not been as rapid and widespread as many predicted in the early 1960s, these new methods of exploiting their *recognition* properties are likely to see more rapid and widespread application both in medicine and industry as the new technology is perfected. This is because in many practical situations chemical catalysis, in spite of its inadequacies, still remains economically superior to biocatalysis. Chemical approaches to specific recognition, however, are relatively naive and when some of the remaining problems of interfacing and stability are solved, biological approaches are likely generally to be superior. Rapid progress is currently being made in the development of a range of sensors based on the interaction between immobilized, stabilized microorganisms and electrodes.

1.4.2   *Energy*

A range of efficient energy interconversions have been perfected by biological systems during the course of evolution. Fig. 1.3 summarizes the major interconversions which are known, some of which are currently exploited in a variety of ways, especially reactions 1,2 and 9.

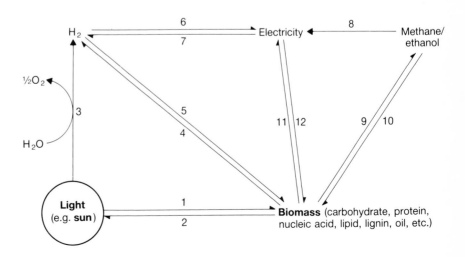

Fig. 1.3. Basic bioenergy interconversions.

These are discussed in more detail in Chapter 2. Biomass is, of course, currently converted to electricity via non-biological equivalents of reactions 8 and 11.

It is more difficult to make predictions about future contributions of biotechnology to man's energy requirements than in most other areas. Much has been said about the 'energy crisis' in recent years and the finite reserves of fossil fuels available to a world with an increasing population and increasing *per capita* energy demand. The nuclear energy debate, of course, continues. The unequal distribution of fossil energy reserves, together with complex political and economic factors, make predictions particularly difficult. A fundamentally important fact, however, is that about 99.4%, some 1.7×10^{23} cal/a, of non-nuclear energy available on the earth comes from the sun and and some of this is continually being converted to biomass, albeit with a relatively low efficiency of 1 or 2%. Biomass, therefore, represents a continuously renewable source of chemical energy which can be burnt or readily converted to convenient liquid or gaseous fuels (methane, ethanol or hydrogen) by microorganisms. However, biomass is also used for other purposes: it is man's food supply and a source of many of his basic materials. What is more, it is likely that biomass will become increasingly important as a source of chemical feedstocks via biotechnological processes, e.g. lignin conversion to aromatic intermediates. In addition to competition for available biomass, the area of the earth's surface available for its production will shrink due to population increase. The global future for conventional biomass-derived energy is, therefore, uncertain, although because of the shortage of fossil fuels in some parts of the world (e.g. the Americas) fermentative production of ethanol is becoming increasingly important, particularly as a supplement or replacement for petroleum for transport purposes. When considered in isolation, the economics of this process are highly questionable but they are for some countries (e.g. Brazil, USA) acceptable for political reasons.

In recent years, there has been a revival of interest in the development of biofuel cells which can effect the efficient and low temperature conversion of a range of fuels or waste biomass to electricity. Whilst these devices are already finding applications, e.g. as specific sensors, there are, in most cases, difficulties in obtaining adequate current densities at the electrodes for use as large-scale energy transducers, although they may find specialist energy applications quite soon. One exception is the hybrid hydrogen biofuel cell, in which hydrogen generated by fermentation is used in a conventional hydrogen–oxygen cell. The pure biofuel cell does, nevertheless, in the medium to long term, offer a potentially cheap method of generating electricity from wastes, sewage and carbon monoxide.

Given the massive amount of energy available in sunlight and the competition for use of available biomass, some biotechnologists working in the energy field are applying themselves to two main approaches to using biological systems to harness this energy efficiently: first, improving the practical efficiency of conversion of sunlight to biomass, e.g. by growing algae under high carbon dioxide conditions and light limitation in carefully controlled bioreactors; second, generating hydrogen by using the photosystems of photosynthetic organisms to cleave water, i.e. biophotolysis. It is technically easier to generate hydrogen using intact blue-green algae or by fermentation. However, for biotechnology to have a major world impact on energy production in the future, many believe it necessary to solve the considerable technical problems implicit in the development of a stable, biophotolytic reactor containing reconstituted, stable biophotosystems.

In the short term, biotechnology will have an increasingly important role to play in oil recovery. As the price of oil rises, recovery of the more recalcitrant reserves becomes increasingly economic. Microorganisms can assist in several ways: first, some microbial polymers, especially xanthan gums, can be used to provide aqueous displacing fluids with the appropriate rheological properties for secondary oil recovery (see Chapter 5); second, microbial surfactants are being exploited in the oil industry. The economics of biosurfactant production would be particularly promising if the surfactant was produced during the microbiological treatment of oily waste. In general, the economics of a biotechnology process are favoured by combining a waste treatment with the generation of useful product.

Some groups are working on longer-term, more ambitious programmes involving introduction of appropriate microorganisms directly into the oil reserves in order to assist recovery from porous rocks.

1.4.3   *Food and drink*

The development over several thousand years of the exploitation of microbial fermentation in the production of beers, wine and fermented foods has been a tradition and more of an art than a technology until relatively recently. Our understanding of the science of microbiology has led to control of food spoilage, greater reliability and reproducibility in fermentative processes and most recently to some new products, e.g. SCP and flavourings. Whilst it is difficult to make detailed predictions of the forthcoming effects of biotechnology in this area, the overall prognosis is clear. There will be two main effects, which are linked. First, traditional methods of producing food will be slowly replaced by bioreactors containing animal and plant cells or microorganisms. This is because the productivity of fermenters or reactors can be substantially greater than conventional agriculture and hence the process is more intensive. Increasing competition for available land resources will encourage these developments. Second, genetic engineering techniques will make this alternative technology increasingly more attractive via modified strains of cells and microorganisms for use in these processes.

This trend to food production via intensified biotechnology processes is already occurring for high value food additives; a particularly good example, quoted in Chapter 3, is citric acid, which is now largely produced by a microbiological process rather than by extraction of citrus fruits. An increase in the use of microbial proteins in human foodstuff is imminent.

There is considerable potential for the development of new types of food processing. For example, beef is expensive and in short supply in Japan. Hence, there is a large potential market there for a good beef substitute made from other protein sources.

1.4.4   *Chemicals*

In spite of the theoretical advantages of the use of biological systems for producing chemicals, there are currently few such biotechnological processes. They are restricted to the relatively low cost but high tonnage fuel chemicals, ethanol and methane, and a few important rather high value products for use in food or medicine, e.g. citric acid, itaconic acid, amino acids, steroids and antibiotics.

In general, biocatalytic chemical production has the following potential advantages: specificity, controllability, low temperature operation, environmental acceptability, simplicity. The first two properties are especially important. For example, much of the present organic chemical industry is based upon petroleum and most of the chemical intermediates generated are partial oxidation products. Specific, controlled, partial oxidation is difficult to effect by conventional catalysis, but microorganisms are masters of this type of reaction.

There are several reasons why biotechnology has yet to make a major impact on the chemical industry despite the large amount of

information on potentially interesting biological systems. One major
problem is that a biotechnology-based chemical industry will involve
substantially different technology. The reactions are generally low
temperature and low pressure, often in an aqueous phase, although a
number of experimental systems operate in an organic phase. In some
cases, rates are low or the catalyst shows poor stability properties.
Perhaps most important, however, is the fact that most of the technical
influence in the chemical industry comes, not unnaturally, from
chemists and, compared to other areas of technology, biotechnological
catalysis has until recently received relatively little research and
development investment.

There is currently something of a paradox in the status of biocatalysis in the chemical industry. On the one hand, the advent of genetic
engineering has reawakened interest in the area since it offers the
possibility of dramatically improving the performance of biocatalysts,
particularly intact microorganisms. On the other hand, the lead times
for development of such new processes are long (about ten to twenty
years) and in addition to the associated high costs of research and
development, substantial investment in the new plant (10–25% of the
total process cost) is required. This new opportunity to develop
larger-scale biocatalysis in the chemical industry has arisen during a
period when the industry worldwide has been beset with economic
problems.

In view of these considerations and whilst much publicity has been
given recently to experimental bioprocesses for high tonnage chemicals, e.g. propene oxide, it is likely that we will see exploitation of
biocatalysis in the short-to-medium term, primarily for the synthesis of
relatively low tonnage, high added value chemicals. Lessons learnt
from these processes will subsequently be used to develop systems for
high tonnage chemical production.

There are basically three approaches to chemical production by
biocatalysis: the use of plant or animal cell culture to generate
expensive products; the use of microorganisms (if necessary after
genetic engineering) to biosynthesize or interconvert chemicals; and
the use of genetically engineered microorganisms as vehicles for the
expression of plant or animal genes in order to intensify the production
of specific higher organism chemical products.

The combination of chemical, biological and electronic technology,
whilst in its infancy, has tremendous potential and will lead to a whole
range of specific sensing devices for process control in the chemical and
food industries, medical diagnosis, monitoring and control and environmental control. An especially exciting aspect of these developments is

the possibility of devising semiconductor biosensors based on micro-chips. Once perfected, such devices could be manufactured cheaply and would offer the possibility of simultaneous monitoring of several parameters by a single, minute bioelectronic device. The timescale for these developments is five to twenty years. Simpler sensors based on semiconductor or thin-layer manufacturing technology will be developed within the next five years.

The development of microbial plastics, emulsifying agents and thickening agents is discussed in the next section.

1.4.5   *Materials*

Biotechnology will affect the availability and performance of materials in three main ways: first, it will assist in the recovery of resources used in manufacturing, e.g. minerals and oil; second, microbial products will be used increasingly for making biodegradable plastics and as emulsifiers and thickening agents; finally, improvements will be made in our ability to control adverse effects of microorganisms on materials, i.e. biodeterioration.

Prospects for the further development of microbial mineral leaching processes and the potential for use of microbial polymers and surfactants in oil recovery are discussed in Chapter 5. The most promising material to serve as a basis for a bioplastic is a storage polymer poly-β-hydroxybutyrate, PHB. A considerable amount of industrial research and development is currently going into this material and its derivatives.

Biodeterioration is not a subject normally discussed under the heading 'Biotechnology' but of course it is ubiquitous, a considerable technical challenge and has substantial economic consequences. It is an inevitable consequence of the essential role of microorganisms in the cycling of elements on earth. Biodeterioration problems range from the obvious one of food spoilage to contamination of cutting oils and fuel systems, destruction of concrete and stimulation of electrochemical corrosion processes by microorganisms. Biotechnology will lead to improved control of biodeterioration as a consequence of advances in our understanding of the underlying processes involved, from which other biotechnological processes may evolve, e.g. the use of enzymes in the food industry.

The development of new types of monitoring devices to assess levels of microbial contamination of materials based on biotechnological concepts will also occur.

1.4.6   *The environment*

As the world population and overall industrial activity increase, so also do the problems associated with maintaining the quality of the environment. Biotechnology will play an increasingly important role, particularly via the development of improved or novel waste treatment procedures. Some improved procedures and systems will be developed largely as a result of chemical engineering work. However, microbial physiologists, biochemists and geneticists will also play an important role by conscripting organisms with newly discovered or 'engineered' catabolic activities into novel treatment processes for specific wastes. Both these approaches are discussed further in Chapter 6.

The environment is, of course, a common denominator for all the activities discussed in this book as they obviously have effects upon the environment. For example, increasing use of biotechnology in the chemical industry should lead to the evolution of an industry which is more compatible with the environment. Bioengineering offers a similar promise. The development of biosensors will assist in environmental monitoring and control. The environment and agriculture are clearly inseparable.

1.4.7   *Agriculture*

Biotechnology and agriculture interact in many different ways. Agricultural produce may provide feedstocks for industry, e.g. the manufacture of industrial ethanol from surplus low grade wine. This concept has been taken a step further to purpose grown crops to yield the same product. Much modern agricultural produce constitutes the raw material for a massive food processing industry.

Agricultural wastes provide another form of feedstock, and much effort is being directed into the production of fuel gas from manure while retaining the fertilizer value. A better understanding of the rates of degradation of different substances and identification of the microbes involved is required for process improvement in this area.

Biotechnology provides vaccines and antibiotics for veterinary use. If vaccines from genetically manipulated microorganisms live up to expectation, we will witness the eventual eradication of decimating diseases such as foot-and-mouth and sleeping sickness. Growth hormones may find application in the increase of meat yields. Modern biotechnology also provides animal feedstocks such as SCP but this product has yet to prove its economical viability, at least in the Western World.

Biotechnology can provide routes to the improvement of whole crops, both in terms of quality and yield. It can provide supplements or alternatives to expensive chemical fertilizers and pesticides. For example, nitrogen may be provided in the form of symbiotic biological nitrogen fixation technology and phosphate nutrition effected via mycorrhizal relationships. A longer term goal is the introduction of the ability to fix nitrogen directly into crop plants by the insertion of the nitrogenase gene, thus allowing the plants to produce the enzyme catalysing the fixation of nitrogen. This would allow considerable savings of energy currently used in the chemical synthesis of ammonia. There are also prospects for replacing pesticides with biological control methods, both in the traditional sense and also through the discovery and application of naturally occurring substances. For example, bacteria can be used to control plant disease such as crown gall disease; *Bacillus thuringiensis* can be used to control lepidopteran pests. Developments of this type will not only be of benefit to agriculture, but will be important in maintaining and improving the quality of the environment.

There is a concensus, however, that the greatest benefit biotechnology will confer on agriculture is through improvements in plants themselves, via genetic manipulation and plant protoplast technology. Little success has been achieved in the regeneration of an entire plant from an individual cell with legumes or cereals, but this has been effected with alfalfa, and is indicative that work with legumes will progress with increasing knowledge of suitable conditions of cultivation. An example of the application of this technology is the production of cereal proteins containing essential amino acids which they currently lack.

1.5   Economic and business aspects of biotechnology

Biotechnology, in its broadest sense, is of course already of the greatest economic and social importance. The main purpose of this section is to examine the likely economic impact and mechanisms of the new biotechnology with its rapidly developing, new, enabling techniques. However, in order to give an indication of the current size of the biotechnology industry, total UK fermentation industry sales alone are about M£9000/a, the world antibiotic market is about M£2000, and the amino acid and enzyme markets are about M£1000 and M£250 respectively.

Whilst it is relatively easy to assess market sizes for individual new biotechnology products (e.g. the potential Western World market for

cheap, reliable, specific, disposable enzyme glucose sensors is about
M£800/a), predicting global figures for all products and services
resulting from the new biotechnology is more difficult. Nevertheless, a
reasonable estimate gleaned from many sources is shown in Table 1.5.
The projected figure of about M£56 000/a represents a growth industry
worldwide within the next twenty years, equivalent to more than six
times the size of the existing UK fermentation industry. Prediction
becomes even more difficult thereafter but it is likely that there will
continue to be a high rate of expansion of the biotechnology industry
over the next century.

A particularly interesting feature of Table 1.5 is that, in spite of the
bulk of publicity going to medical applications of biotechnology, it is

	M£1000
Chemicals	9
Energy	12
Food	12
Medicine	8
Miscellaneous (e.g. enhanced oil recovery, pollution monitoring and control, bioplastics, mineral leaching)	15
Total	56

Table 1.5. Estimates of likely world markets for products of the 'New Biotechnology', by the year 2000

likely that this will, in fact, be the smallest of the markets by the turn of
the century. Nevertheless, it may well be the most profitable and, in the
short term, the most rapid to develop. For example, human insulin is
already available. Widespread use of interferon is likely for cancer
therapy by 1987 and then as an antiviral and anti-inflammatory agent by
1988 or 1989. Hepatitis B vaccine and human growth hormone are
expected to be widely available by 1990.

1.5.1   *Logistics of biotechnology development*

The mechanisms behind the practical development of biotechnology
differ substantially between different parts of the world and depend
upon political and economic factors characteristic of East and West,
North and South, Third World or advanced nations, capitalist or
socialist. The nature of the technology also varies in that Third World
countries, in general, are developing alternative technologies. For
example, small-scale local biogas generators can and do, in parts of the

Far East, supply a substantial proportion of the energy requirements of a rural community; such devices are of much less importance to advanced economies. In the USSR and other communist countries, the development of biotechnology is generally governed by central policy. This has led in the USSR, in particular, to the development of a major SCP programme with an estimated current output of well over 1×10^6 t/a produced from hydrocarbon or vegetable waste.

In Japan, the government is playing a major role in co-ordinating the development of biotechnology in the universities and research institutes whilst, in the main, development and commercialization of ideas is effected by the major bioindustry companies. The governments of most western democracies have announced programmes in the last few years to assist the development of biotechnology. This is necessary because of the complex interdisciplinary nature of the technology and because of the long lead times for many of the products. In addition, because many of the important developments are at or near to the 'frontiers of science', many of the key scientists involved in these countries are employed in the public sector, and particularly in institutions of higher education. It is especially important for governments to invest in appropriate research and development in the area and to assist the private (and public) sector to effect commercialization in spite of the long lead times. The adverse economic climate in recent years makes such catalysis by governments all the more important because of the cuts in research and development spending in parts of the private sector.

The USA is a special case. Although the government there has been involved in substantial investment in various aspects of biotechnology development and the major companies in the clinical, bio- and energy industries have substantial biotechnology programmes, the availability of high risk venture capital and a 'business-aware' scientific community has led to another major biotechnology sector in that country: the entrepreneurial biotechnology company. Such small, new enterprises are not entirely restricted to the USA, there being a few established in Western Europe. Some of these European companies are 'quasi-entrepreneurial' in that some have substantial government or traditional financial institution involvement. There are, therefore, basically four categories of companies involved in biotechnology development: (1) specialist biotechnology companies referred to above, often based mainly upon recombinant DNA technology; (2) large, usually multinational companies involved in the oil, food, chemical, agricultural, materials or pharmaceutical industries who devote part of their research and development effort to biotechnology; (3) smaller, special-

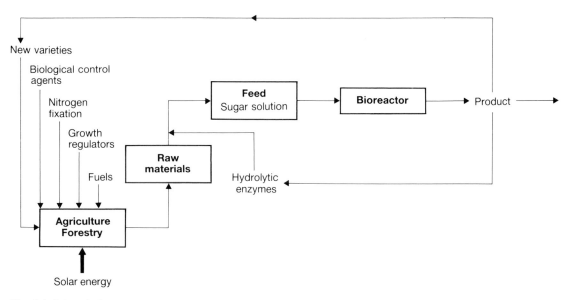

Fig. 2.1. Interrelations between biomass and biotechnology.

production and conversion in photosynthesis, in farming and in fermentation.

2.2   Biomass as a solar energy technology

Solar energy is an immense resource: the surface of the earth receives about 3×10^{24} J/a; since the estimated proven reserves of oil, natural gas, coal, and uranium are only about 2.5×10^{22} J (8×10^{11} t coal equivalent), we see that in less than one week the earth receives from the sun an amount of energy equal to the total reserves of non-renewable energy. Put in another way, if the earth were covered to an extent of only 0·1% of its surface with collectors converting solar energy at only 10% efficiency, then all of the current world annual needs for energy (3×10^{20} J/a) would be met.

However, solar energy has two major drawbacks—it is intermittent and diffuse. The first imposes some form of storage system so that the energy can be made available when it is needed and the second requires large areas of collectors. Both of these factors place severe but not insurmountable economic constraints on solar energy systems. Photosynthetic production of biomass overcomes both of these problems: first, the collectors can be grown from seeds; and, second, the product is available in a stable, storable form. However, the production and use of biomass for energy is not without its own specific problems, offset in turn by potential advantages in terms of world-wide availability, renewability and environmental and ecological appeal.

The advantage of the biomass route of solar energy utilization lies in the fact that the energy is trapped in the form of organic material and hence may be stored or transported over time and distance (Table 2.1). Disadvantages include the low efficiency (often less than 1% and seldom over 2%) of trapping solar energy by photosynthesis into the agricultural product, the diffuse and often seasonal nature of production and the high moisture content on a weight basis. Thus, to provide a high-grade, energy-rich product requires collection, transport, removal of water, concentration or chemical/biological conversion and packaging in order to upgrade the biological raw materials. However, in converting biomass to a premium fuel the objectives are not only to remove water and increase the energy density but to form a product which is compatible with the technology with which it is to be used.

The major use of plant material as a fuel in the past on a worldwide basis, and at present in many developing countries, is direct combustion—mainly burning wood and, to a lesser extent, crop residues and dung. Currently, a number of thermal conversion systems are in various stages of development. These include pyrolysis, gasification and hydrogenation. However, all these processes require feedstock of relatively low water content and operate at high temperatures. The advantage of the biological conversion processes is that they can handle feedstocks of high water content and operate at temperatures in the range of 25–65 °C. The range of suggested potential biomass raw materials is shown in Table 2.2. At present, the major feedstocks used are sugar-cane maize, wood and wastes, manure and domestic rubbish, as well as crop or process residues such as bagasse, molasses and by-products of paper or pulp production.

Table 2.1. Some advantages and problems foreseen in biomass for energy schemes

Advantages	Problems
Stores energy	Land and water use competition
Renewable	Land areas required
Versatile conversion and products; some products with high energy content	Supply uncertainty in initial phases
	Costs often uncertain
Dependent on technology already available with minimum capital input; available to all income levels	Fertilizer, soil and water requirements
	Existing agricultural, forestry and social practices
Can be developed with present manpower and material resources	Bulky resource; transport and storage can be a problem
Large biological and engineering development potential	Subject to climatic variability
	Low conversion efficiencies
Creates employment and develops skills	Seasonal
Reasonably priced in many instances	
Ecologically inoffensive and safe	
Does not increase atmospheric CO_2	

Table 2.2. Sources of biomass for conversion to fuels

Wastes	Land crops	Aquatic plants
Manures	Ligno-cellulose	Algae
Slurry	Trees	Unicellular
Domestic rubbish	Eucalyptus	*Chlorella*
Food wastes	Poplar	*Scenedesmus*
Sewage	Firs, pines	*Navicula*
Residues	*Leucaena, Casuarina*	Multicellular
		Kelp
Wood residues	Starch crops	
Cane tops		Waterweeds
	Maize	
Straw	Cassava	Water hyacinth
Husks		Water reeds/rushes
Citrus peel	Sugar crops	
Bagasse	Cane	
Molasses	Beet	

2.2.1   *The resource base*

The primary resource base for the production of biomass fuels is agriculture and forestry. In attempting to assess the present potential the assets may be considered in terms of available land, nature and yields of the present major sugar and starch crops, and the people employed in agricultural production. On a world-wide basis, the annual increment of biomass is estimated to be in the order of 2×10^{11} t (Table 2.3). This includes about 1.2×10^{11} t of dry matter in the form of wood. About 60% of the felled wood is used as fuel. If the regeneration and consumption of wood are kept in balance, the annual availability of wood, not used at present, is in the order of 3×10^{10} t, equivalent to some 10^{10} t of cellulose.

The distribution of these forest reserves on a global basis is shown in

Table 2.3. Annual biomass production

	Tonnes
Net primary production (organic matter)	2×10^{11}
Forest production (dry matter)	9×10^{10}
Cereals (as harvested)	1.5×10^9
as starch	1×10^9
Root crops (as harvested)	5.7×10^8
as starch	2.2×10^8
Sugar crops (as harvested)	1×10^9
as sugar	9×10^7

Table 2.4, which also indicates the land areas devoted to various types of agricultural practice and the areas of irrigated land.

Table 2.5 indicates the areas of closed forests and size of the growing stock and, in addition, the volume of wood available on a per capita basis. This table highlights two points: first, the tremendous wood-based resource which exists in Canada, the USSR and S. America (mainly Brazil and tropical S. America); and second, the scarcity which now exists in Asia due to a rapidly expanding population relying mainly on biomass fuels used at very low efficiency, without an established policy of replanting.

The present levels of sugar and starch production by the major crops, on a global basis, are shown in Table 2.6. These figures serve to emphasize what is fairly well known, that overall crop yields are generally considerably lower in the developing countries, with the

Table 2.4. Available land and current use on a global scale

	Land area (10^6 ha)	Use %				
		Permanent crops	Arable	Pasture	Forests	Irrigation
World	13078	0·7	10	23	32	1.5
Developed countries	5484	0.4	12	23	34	0.9
Developing countries	7593	0.9	10	23	30	1.9
USA	936	0.2	20	26	31	18.4
Canada	922	—	5	3	36	0.05
Europe	472	3.2	30	19	32	2.8
USSR	2227	0.2	10	17	41	0.7
Asia	2757	1.0	16	20	22	4.6
S. America	1753	1.2	65	25	53	0.4
Africa	2964	0.5	7	27	22	0.3
Oceania	842	0.1	5	56	22	0.2

Table 2.5. Areas of closed forest and size of the growing stock

	Closed forest (10^6 ha)	Growing stock of closed forests		
		Total (10^6 m^3)	m^3/ha	m^3/caput
USA	220	20 200	92.1	95.3
Canada	250	17 800	71.2	809.1
Europe	144	14 900	103.5	29.1
USSR	770	81 800	106.2	324.6
Asia and Oceania	448	38 700	72.9	18.7
S. America	631	92 000	145.8	362.2
Africa	188	35 200	187.2	90.0

	(i) Sugar				(ii) Starch: grains								(iii) Starch: roots and tubers							
	Cane		Beet		Cereals		Maize		Rice		Wheat		Roots and Tubers		Potatoes		Sweet Potatoes		Cassava	
	Total	Yield	Total	Yield	Total	Yield	Total	Yield	Total	Yield	Total	Yield	Total	Yield	Total	Yield	Total	Yield	Total	Yield
World	737	56	290	32	1459	1.9	349	3.0	366	2.6	386	1.7	570	11.0	272	15.0	138	9.6	110	8.8
Developed countries	70	79	263	32	765	2.5	236	4.7	26	5.4	261	1.9	225	15.7	222	15.7	2	14.7	—	—
Developing countries	667	54	26	30	687	1.6	113	1.7	340	2.5	125	1.3	345	9.2	70	10.4	136	9.5	110	8.8
USA	25	82	23	46	273	4.1	161	5.7	5	4.9	55	2.0	17	28.0	16	29.2	1	12.4	—	—
Canada	—	—	1	39	42	2.3	48	5.9	—	—	19	1.9	3	22.4	3	22.4	—	—	—	—
Europe	0.3	63	143	38	249	3.5	49	4.2	1.5	3.8	82	3.3	114	18.7	0.1	10.3	—	—	—	—
USSR	—	—	93	25	187	1.5	11	3.2	2.2	4.0	92	1.5	83	11.8	83	11.8	—	—	—	—
Asia	305	52	25	31	603	1.8	54	2.0	335	2.6	108	1.3	224	10.3	61	11.2	128	9.9	33	11.2
S. America	187	57	—	—	64	1.7	31	1.8	13	1.9	9	1.1	44	11.0	9	9.5	3	10.4	32	11.6
Africa	60	64	16	32	66	0.9	26	1.3	8	1.8	8	0.9	73	6.8	4	7.8	5	6.3	44	6.6
Oceania	26	75	—	—	16	1.0	0.4	4.5	0.5	5.3	10	0.9	2	10.3	1	23.0	0.6	5.4	2	11.0

These figures are on a crop basis, excluding the considerable amounts of crop wastes, straw, etc. The cereal figures exclude such crops grown for forage silage.

Table 2.6. Total production (10^6 t) and yields (t/ha) of the major sugar and starch crops

highest yields in North America and Europe, poorer yields in the USSR, and particularly low yields from African and many Asian countries.

These figures highlight the many conflicts in a biomass energy policy in countries where wood or food shortages exist, which have led to questions of food or fuel production. However, if agricultural and forestry production can be increased both goals can be met. The application of agroforestry systems will alleviate many of such problems. This requires the identification of limits to production in terms of photosynthetic efficiency, limits to crop productivity and, in particular, the dependence on and response to external energy inputs into the farming system. In addition, it may be possible to select or breed new, high yielding biomass species capable of being produced in arid, marshy, saline or aquatic conditions, where the production does not compete with conventional agriculture and forestry. These aspects will be covered in more detail.

2.3   Photosynthesis

Photosynthesis is the key process in life and, as performed by plants, can be simply represented as

$$H_2O + CO_2 \xrightarrow[\text{solar energy}]{\text{plants}} \text{organic materials} + O_2$$

In addition to carbon, hydrogen and oxygen, the plants also incorporate nitrogen and sulphur into the organic material via light-dependent reactions.

The basic processes of photosynthesis are now fairly well understood. Photosynthesis takes place in the chloroplasts (Fig. 2.2), which
assimilate CO_2 entering the plant along a diffusion gradient. The
primary carboxylation process takes place in the stroma (the region of
the chloroplast containing few membranes) and is catalysed by ribulose-
bis-phosphate carboxylase (RBPC). This reaction results in the formation of two molecules of a three-carbon acid (phosphoglycerate, PGA),
which are then reduced to a three-carbon sugar, triose phosphate. This
serves as the precursor for the synthesis of starch in the chloroplast or
may be exported to the cytoplasm and used in the synthesis of sucrose.

Fig. 2.2. Higher plant
chloroplasts.

Part of the fixed carbon is recycled through the photosynthetic carbon reduction (PCR) cycle in order to regenerate the CO_2 acceptor, ribulose-bis-phosphate.

The PCR cycle is driven in the direction of net synthesis by utilization of energy derived from the hydrolysis of ATP and the oxidation of NADPH. ATP and NADPH are generated in the so-called light reactions of photosynthesis associated with the membrane (thylakoid) component of the chloroplasts. A minimum of four electrons are needed to reduce CO_2 to (CH_2O). Evidence supports the 'Z-scheme' of photosynthesis, which involves the co-operative interaction of two parallel photoreactions (PS I and PS II; Fig. 2.3) linked by an

Fig. 2.3. Photosynthetic electron transport system.

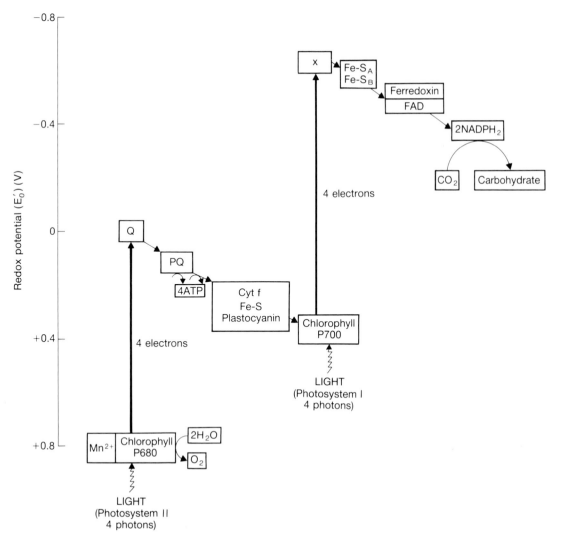

intermediate electron transport chain. Reduction of NADP to NADPH occurs at PS I through an iron/sulphur protein, ferredoxin.

Light is harvested by the antennae chlorophyll–protein complexes and channelled to the PS I reaction centre (P_{700}) which is a specific chlorophyll a–protein complex. Here, charge separation results in the reduction of ferredoxin. Electrons are donated to P_{700} through the intermediate electron transport chain from PS II and are derived from water. During this PS II-linked reaction, O_2 is evolved from water.

The electron carriers are arranged within the thylakoid membrane in such a way that an electrochemical gradient is formed across the membrane associated with the electron flow. This electro-chemical potential is used to drive the synthesis of ATP. The protein involved in this process is known as coupling factor and has reverse ATPase activity.

The wavelengths absorbed by the antennae chlorophylls and other photosynthetic pigments reflect the properties of these molecules and determine the proportion of the solar spectrum which can be utilized by the plants (photosynthetically active radiation, PAR). This comprises between 45% and 50% of the incident solar spectrum on an energy basis. The light-harvesting characteristics represent the first limit to the overall efficiency of photosynthesis if expressed in terms of incident radiation.

At low light intensities, photosynthesis is limited by the activity of the light reactions. At higher intensities or lower temperature, photosynthesis will become limited by the capacity of dark carbon metabolism. For many temperate (C_3) plants, light saturation occurs at around $100–150\,J\,m^{-2}\,s^{-1}$, or at about 50% of full sunlight. At high light intensities and temperatures, water stress is more likely and may lead to stomatal closure and again limit carbon metabolism. Hence, in the field situation, interaction of light and dark reactions and their regulation through dissipation of excess light energy becomes important.

In active chloroplasts each electron transport chain may turn over every 15 ms. Under bright light, the average chlorophyll molecule absorbs a photon once every 100 ms. Under low or difuse light, this rate is slower. However, since each reaction centre is associated with several hundred light-harvesting pigment molecules, the rate of electron flow is increased. The time from excitation to reaching the trap can be measured in picoseconds. However, the electron transport reactions are slower, with the rate limiting step associated with ATP synthesis being about 20 ms.

As shown in Fig. 2.3, during the photosynthetic process electrons are raised from the potential of water ($+0.8$ V) to that of (CH_2O), -0.43 V.

This potential difference is equivalent in energy terms to around 480 kJ. Hence, the PCR would appear to work at a high efficiency. However, in practice this is not the case.

Problems arise from the fact that the enzyme RBPC will also catalyse an oxygenase reaction betwen RBP and O_2, to give PGA plus phosphoglycollate. This oxygenase reaction is competitive with that of the carboxylation to 2(PGA), and results in the diversion of carbon to the C-2 pathway of photorespiration, during which CO_2 is generated and lost from the plant.

As a result of oxygenase activity and photorespiration in temperate (C_3) plants the efficiency of photosynthesis is reduced by a factor of between 15 and 50%, depending on species and environmental conditions. This inhibition is the result of one or more of the following effects: (1) the carboxylation substrate is lost; (2) carbon is lost from the PCR cycle; (3) CO_2 is lost from the plant; (4) the diffusion gradient into the plant is reduced in size.

The rate of photosynthesis depends on the flux of CO_2 into the leaf, which in turn depends on the gradient of CO_2 concentration and a number of resistances which reflect the physical and biological barriers to diffusion and assimilation, i.e.

$$P = K(\Delta CO_2)/(r_1 + r_2 + r_3)$$

where K is a diffusion coefficient and r_x represents the various resistances. The variable resistance in the physical diffusion pathway is r_2—the stomatal pore resistance. The size of this resistance is controlled by light, water and CO_2 concentration. In particular, the aperture will decrease under water stress.

The resistance r_3 (mesophyll resistance) reflects the enzyme components with the ability to assimilate CO_2. The affinity of the enzyme RBPC can be expressed in terms of its Michaelis constant (K_m). Most recent estimates of the K_m for CO_2 of RBPC are about 10–20 μM, equivalent to the atmospheric level of CO_2 in equilibrium with water. In other words, unless the cells contain a CO_2-concentrating mechanism, this enzyme will function at about half its optimum rate.

A CO_2 concentration mechanism exists in some algae and in some higher plants which possess the C_4 or the CAM pathway of photosynthesis. The C_4 plants contain a number of genera, most of which are tropical plants. These include the important commercial crops, sugar cane, maize and sorghum, as well as the tropical elephant grass (*Pennisetum*), for which the highest rates of dry matter production have been recorded. These plants have a secondary carboxylation cycle (the C_4 pathway) which functions as a pump to move CO_2 from the atmosphere to the site of reductive reassimilation through the PCR cycle in the bundle sheath cells. Atmospheric CO_2 is first assimilated into four-carbon acids (malate and aspartate) in an outer (mesophyll) layer of photosynthetic tissue. These acids are transported to the bundle sheath layer where the CO_2 is released and refixed. As a result of this mechanism, the effects of photorespiration are decreased in C_4 plants, leading to higher rates of photosynthesis and higher productivities. However, although these plants are more efficient in terms of carbon assimilation, they require additional energy to drive the C_4 pathway and are thus less efficient in terms of light use efficiency. The C_4 plants also show a greater efficiency of water use and are less susceptible to high temperature stress. However, few C_4 species are able to grow well in temperate climates.

2.3.1   *Photosynthetic efficiency*

As far as biomass production is concerned, photosynthetic efficiency may be defined as the percentage of the total solar radiation, falling over a given area in a given time, which is converted into harvestable organic material. The productivity of the system may be defined in terms of the amount of harvestable organic dry matter produced per unit land area per annum, expressed in terms of weight (kg) or energy content (MJ) of material produced per hectare per annum.

The harvested biomass is thus a function of the size of the solar collector available at any time of the year (leaf area) and the number of days during which a full active canopy is available to carry out photosynthesis at the maximum rate times the efficiency of the overall photosynthetic process.

Estimates of the percentage of the solar radiation available to plants (PAR), coupled with knowledge of the basic photochemical and biochemical process and its thermodynamic efficiency, enable calculation of possible maximum rates of production of organic material in the form of carbohydrate.

Plants use radiation between 400 and 700 nm; this PAR comprises about 50% of the total sunlight which, on the earth's surface, has an average normal-to-sun daytime intensity of about 800–1000 W/m^2. The overall practical maximum efficiency of photosynthetic energy conversion is approximately 5–6%. This is derived from knowledge of the process of CO_2 fixation and the physiological and physical losses involved. Fixed CO_2 in the form of carbohydrate has an energy content of 0.47 MJ/mol of CO_2 and the energy of a mole quantum of red light at 680 nm (the least energetic light able to perform photosynthesis efficiently) is 0.176 MJ. Thus the minimum number of mole quanta of red light required to fix 1 mol of CO_2 is $0.47/0.176 = 2.7$. However, since at least eight quanta of light are required to transfer the four electrons from water to fix one CO_2 the theoretical CO_2 fixation efficiency of light is $2.7/8 = 33\%$. This is for red light and, obviously, for a white light it will be correspondingly less. Under the best field conditions, values of 3% conversion can be achieved by plants; however, often these values are for short-term growth periods, and when averaged over the whole year they fall to between 1 and 3%.

In practice, photosynthetic conversion efficiencies in temperate areas are typically between 0.5 and 1.3% of the total radiation when averaged over the whole year, while values for sub-tropical crops are between 0.5 and 2.5%. The yields which can be expected under various sunlight intensities at different photosynthetic efficiencies can be easily calculated from graphical data (Fig. 2.4).

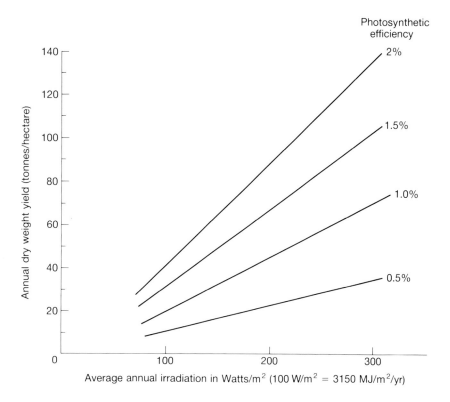

Fig. 2.4. Expected (good) annual plant yields as a function of annual solar irradiations at various photosynthetic efficiencies (from Hall, 1982).

2.3.2   *Improving photosynthesis through biotechnology*

Two distinct opportunities exist for modification of existing plant systems in order to increase annual biomass yields. One way is to increase the absolute net rate of photosynthesis which can be achieved under optimal conditions and the other is to increase the time over which optimal photosynthesis is carried out.

To achieve such objectives requires an assessment of the relative importance of various limitations to photosynthesis, dependent on both inherent photobiological and physiological limitations on the one hand, and on those environmental factors which limit expression of this potential on the other. Important factors are as follows: harvest index, light, CO_2, water, temperature, nutrients, pests/disease, oxygen effects/ photorespiration, dark respiration, rate limitations in electron transport, level of carboxylation enzymes, light-harvesting pigments, energy dissipation in wasteful reactions and rate of transport of material from the chloroplast.

In other words, the major limitations in the short term may be overcome by energy-dependent management inputs or may represent environmental factors which lie outside the control of man. Hence high biomass production will require careful selection and breeding of species best able to exploit the environmental conditions with low inputs.

Major improvements have been made in agricultural yields by a combination of breeding advances and management techniques. However, in general, breeding has been aimed at increasing the harvest index of crops rather than increasing the total amount of biomass· produced.

Increases brought about by management techniques largely reflect increased energy inputs in the form of tractor fuel, fertilizer and plant protection chemicals derived from fossil fuel. As a result, the yield per unit area has increased. However, the yield per unit of energy put into the system has decreased. Such changes are relevant to the production of the food crop but not to the production of an energy crop where high biomass yields at low energy inputs are needed. However, it is clear that effective biomass production systems will need energy-dependent inputs. The extent of such inputs and the benefits to be gained in terms of increased energy production have yet to be established.

Some significant improvements in yield have been made by breeding for leaf shape, erectness and position for improved canopy structure and light interception. In theory, photosynthetic changes might be introduced in one or more of the following partial processes of photosynthesis:

(1) light harvesting and electron transport;
(2) carbon dioxide assimilation and regulation of central carbon metabolism, including dark respiration and photorespiration;

(3) membrane characteristics with an emphasis on lipid structure and response to temperature;

(4) anatomy in relation to CO_2/water use economy.

Once existing varieties of potential biomass crops have been evaluated, possibilities exist for the application of new techniques in breeding, in analysis of photosynthetic potential and in the propagation of plant material by unconventional means. In the longer term, new methods of recombinant DNA technology may result in the possibilities of such improvements. To enable such advances, information is needed in the following areas:

(1) methods for recognition of beneficial photosynthetic characteristics, particularly extrapolation from sophisticated laboratory techniques to field use;

(2) methods of introducing genetic variation;

(3) an understanding of the plant genome and chloroplast chromosome and control of their expression;

(4) definition of the types of changes which might be introduced;

(5) breeding methods for speeding up propagation and genetic stabilization.

Methods of introducing genetic variation include a variety of techniques, such as the use of tissue culture, protoplast fusion, single gene transfer, haploids, radiation-treated pollen transfer, chemical mutagens, transmission of mitochondrial or chloroplast genomes, etc. Specific transfer of identifiable characteristics includes the techniques of genetic engineering, such as questions of vector selection, gene incorporation and expression of the new characteristic within the complex total plant regulation system at both the genetic and metabolic level.

Hence the opportunities for increasing productivity in plants grown for energy (biomass) using *in vitro* techniques depend on the following aspects:

(1) establishment of genetic limitations to productivity;

(2) identification of individual genes responsible for such traits;

(3) selection of DNA or RNA which code for these traits;

(4) incorporation of this material, using restriction enzymes and ligases, into a suitable vector for amplification in a bacterial system;

(5) insertion of the amplified material into a vector suitable for transfer of the material to the desired plant;

(6) stabilization of the new DNA in the genome of the new host in such a way that it is expressed as a dominant characteristic and inherited according to Mendelian patterns;

(7) development of screening procedures in order to recognize such
improved plants;

(8) an alternative approach, using less specific methods of introducing variability by use of mutagens (radiation or chemicals) on tissue
cultures or protoplasts, or using the natural variability which could
be screened for improved photosynthesis.

In spite of dramatic advances in knowledge of the nuclear genome
and chloroplast plastome and the regulation of gene expression in the
synthesis of chloroplast and other proteins, practical use of such
engineered improvements are still several decades away. However, it is
important that a continued emphasis is placed on basic photobiology in
relation to productivity in order that in time such developments will
become possible. In the meantime, the production of biomass fuels must
rely on conventional agriculture and forestry.

2.4   Agriculture and forestry

2.4.1   *Energy ratios*

In any agricultural system it is important to consider how much energy
is derived from the system compared to how much is put in to operate it
(energy output/input ratio). In biomass production systems this is even
more important. Table 2.7 shows some ratios which highlight the
problem of energy intensive agriculture, such as greenhouse production, and the low energy conversion efficiencies of animal systems. Both
greenhouses and animals convert less than a tenth of the input energy
into usable energy. Both greenhouse and animal products are important
components of our modern agriculture. However, a small decrease in
the rearing of animals for food would release large amounts of organic
material for food, fuel, etc., if this was considered desirable. Calculations in the US on energy output:input ratios in the production of maize
grain have shown that this ratio has fallen from 3.7 in 1945 to 2.8 in 1970,
i.e. a doubling of yield has been achieved (most important) by a trebling
of energy input, mostly as a result of increased fertilization. In the UK,
where maize production is mostly for forage and the whole plant is
considered, the output:input energy ratios are between 5 and 9. If the
great use of nitrogen fertilizers (which often contribute 50% of the
energy input) could be decreased without lowering yields, e.g. by
nitrogen fixation or manure or sludge from digesters, considerable
savings in energy could result. Even though there has been some
criticism of the excessive use of nitrogen fertilizers in the past, it should

Table 2.7. Energy output/input ratios for the UK

Whole farms (average)		Grass	
Specialist dairy	0.38	Low efficiency (grazing)	9.1
Mainly dairy	0.55	Low efficiency (grazing and hay)	5.6
Cattle and sheep	0.59	High efficiency (hay)	2.4
Sheep	0.25	High efficiency (silage)	2.4
Pig and poultry	0.32	*Peas* (fresh)	0.9
Cereal	1.9	(canned)	0.1
Barley (and oats)	2.4	*Carrots*	1.1
Maize (grain)	2.3	*Brussels sprouts*	0.2
Wheat (grain)	3.4	*Greenhouse lettuce*	0.002
Potatoes	1.6	*Poultry* (meat)	0.1
Sugar-beet		(eggs)	0.1
Sugar	3.6	*Fish*	0.05
Gross output	4.2	*Milk*	0.4

be pointed out that for every joule of nitrogen fertilizer energy added to the plant about 4–6 J of plant energy is produced—photosynthesis is the catalyst which 'increases' this energy and the plant cannot operate efficiently unless it has optimum amounts of nitrogen (and other minerals like phosphorus, potassium and trace elements).

In many agricultural systems a significant proportion of the net biomass produced is left in or on the soil at harvest. The problem with integrated fuel/food systems is that, since all parts of the plants can be used, all parts could be removed—such practice cannot be permitted. The removal of mineral nutrients through harvesting forest plantations can be offset by adding inorganic fertilizers. However, this on its own is often not sufficient, particularly in the tropics where soils are generally more fragile, because a most important component in soil sustainability is the organic matter. This serves both as a source of nutrients for various microorganisms responsible for the release of mineral nutrients, and the fixation of atmospheric nitrogen, as well as a physical agent to improve the crumb structure of the soil and its water-holding capacity. in general, an otherwise unsupported ecosystem can afford to export only 10–20% of the annual net dry matter increment . However, this applies in a situation where the canopy is kept closed throughout most of the year. Where soil is laid bare the rate of loss of organic matter will be increased. If agroforestry is to provide both fuel and food the complex of trees and field crops must be managed in such a way that the system also results in sufficient retention, or return, of organic matter.

2.4.2   *Forests and trees as a source of biomass fuels*

Trees have a number of attractions as potential biomass fuel feedstock. Eventually, yields per hectare are very high; there is more total biomass available as trees than from any other source, and agricultural inputs need to be much lower than for other crops. The disadvantages lie in the long period of growth to maturity and the fact that the main component (lignocellulose) is very intractable. At present, logging residues provide the greatest promise as a short-term source although, in the longer term, fuel plantations will be of increasing importance. Since the major costs are related to ground clearance and planting, particular attention is being given to fast-growing coppiced hardwoods, probably managed on the basis of one crop every five years. It has been suggested that a silviculture energy farm, at harvest, could average 10–30 t ha^{-1} a^{-1} of biomass.

Relatively large-scale (8000 ha) *Eucalyptus* energy farms dedicated to charcoal production for steel mills have operated since the early 1950s in Argentina and Brazil and plantations exist elsewhere (South Africa and Australia, for instance) for the production of pulp for cellulose acetate fibre, and widely for the production of paper pulp and timber. Studies have been made on *Salix* and *Populus* in temperate regions, *Leucaena* and *Casuarina* in the tropics but, in general, it would appear that the highest yields will be obtained from *Eucalyptus* in the places where it will grow.

Eucalyptus is a genus of evergreen trees of which there are over 450 identifiable species, varieties and hybrids. Mainly Australian in origin, where they constitute 95% of the forest in both wet and dry sites on a wide variety of soil types, they have now spread throughout most of the world, although they do not thrive in regions where frosts are common or extremes of rainfall occur. Under optimum conditions they will outgrow softwoods such as *Pinus radiata*, producing annual yields of 10 t/ha in ten years, whereas such a yield would not be reached in *P. radiata* for about twenty-five years. A recent development has been at the Aracruz pulp and fuelwood plantation (500 km NE of Rio de Janeiro) where 50 000 ha of selected *Eucalyptus* clones yield up to 45 t ha^{-1} a^{-1} under the prevailing excellent weather conditions; all propagation is by vegetative production of plants from leaf cuttings.

However, problems in use of wood or lignocellulose as feedstock for liquid energy production lie in its intractable nature, since for fermentations to fuels such as ethanol, or chemicals such as butanol, a fairly pure solution of suitable sugars is required.

2.4.3   *Algae and aquatic plants*

The potential yield of biomass from freshwater and marine plants is great. However, the extremely high water content of many of these plants as harvested and the difficulty in drying them in the sun precludes their use as a fuel by direct combustion. Hence, anaerobic fermentation of aquatic plants and wet agricultural wastes appears to be a most appropriate technology for processing into fuel, fertilizer, and feed. Waterweeds thrive on sewage; they clean the water effectively, and grow rapidly in the process. Thus, they may serve a dual role of improving the environment and providing a significant source of energy.

The production of biogas from water hyacinth has been carried out in a number of countries. They were selected as a feedstock because of their prolific growth rate and because they are floating plants which can be easily harvested. Alternatively, algal ponds which incorporate the use of organic wastes (often polluting and/or costly to dispose of) seem promising in many of the sunnier parts of the world where there are often problems with liquid waste disposal.

Thoughts of using algae and bacteria in biological solar systems are not new but have recently been receiving more attention. One advantage of microbial systems is that they can be technologically sophisticated or simple, depending on local conditions. The choice of the most suitable species will also depend on local occurrences and preferences, e.g. taking into account salinity and temperature; the species selected can then be easily fitted into the environmental requirements.

Many liquid and semi-solid wastes are ideal for growth of photosynthetic algae and bacteria. Under good conditions, rapid growth with about 3.5% solar conversion efficiency can be obtained to give 50–80 t $ha^{-1} a^{-1}$. The harvested algae may be fed directly to animals, fermented to produce methane, or burnt to produce electricity. Simultaneously, waste can be disposed of and water purified. It is estimated that such algal systems are 0.5–0.75 times as expensive as conventional waste disposal systems in California. The main economic problem is harvesting costs but the development of new techniques and using different, easily harvested species of algae is proving important. Two-stage algal ponds for complete liquid waste treatment are being tested. Algae that can be harvested by straining are grown in the first pond while nitrogen-fixing blue-green algae (also easily harvested) grow in the second pond, deriving their nutrients from the first treatment ponds. The use of wastes from industry, including CO_2, can also increase productivity. The harvested biomass can be fermented to methane (equivalent to 1.1 MJ/kg algae) while the residues would contain virtually all the nitrogen and phosphorus of the algal biomas, providing a good agricultural fertilizer. One hectare of algal ponds could supply the fertilizer required by 10–50 ha of agriculture. By optimization of yields and including energy inputs and conversion losses, a net production of 200 GJ $ha^{-1} a^{-1}$ of methane seems feasible. At a 30° latitude this would represent a 1.5% annual photosynthetic conversion efficiency.

In California, average yields of algae in excess of 100 kg dry wt. $ha^{-1} d^{-1}$ are obtained, with peak production in summer reaching three times this figure. Yields of 50–60 t dry wt. $ha^{-1} a^{-1}$ would produce 74 000 kWh of electricity. Algal ponds of 10^6 l have been constructed which give a 2–3% photosynthetic efficiency on a steady-state basis. Large feeding systems for cattle and chickens have now been provided with algal ponds where the animal waste is fed directly into the ponds. About 40% of the nitrogen is recovered in the algae, which is subsequently re-fed to the animals. The green algae presently grown have 50–60% protein but blue-green algae now being tried contain 60–70% of extractable protein. Algal ponds for oxidation of sewage are operating in at least ten countries and the interest in these systems as possible net energy and fertilizer producers and as water purifiers is increasing.

In addition, algae may be used as a source of specific hydrocarbons and speciality chemicals. For instance, *Botryococcus braunii* is a green alga which grows widely and can produce hydrocarbons up to 75% of its dry weight, depending on growth conditions and the strain used. The hydrocarbons range from C_{17} to C_{34} chain lengths and are produced as an intracellular product, causing the algae with higher proportions of hydrocarbon to float. Once the algae have been removed from the water, the hydrocarbons are easily separated by solvent extraction or destructive distillation (which can produce fractions similar to diesel oil and kerosene). *B. braunii* occurs widely in nature, from Australian brackish lakes (where the dried remains of the algae, known as Coorongite, were responsible for early oil rushes) to water butts in London. Similar materials found in other parts of the world are named according to their local origin (N'hangellite from Mozambique, Balkashite from Turkestan, etc.).

Productivity data are only available for laboratory flask culture under continuous illumination and for natural and unpredictable algal blooms, neither of which are much use for estimating the productivity of *B. braunii* in mass culture. An Australian report suggests that the following investigations merit early attention: laboratory and field studies of the conditions necessary for maximum growth and oil-production rates; a pilot-plant study to determine whether intensive cultivation can produce *B. braunii* growth rates comparable to those reported for other algae; development of appropriate cultivation, harvesting and processing techniques, and determination of suitability of the resulting 'solar crude' as an alternative source of fuels and lubricants.

The unicellular green alga, *Dunaliella bardawil*, is being tested on a pilot scale in Israel for its ability to utilize solar energy to produce glycerol, an important organic compound for industrial use. *Dunaliella* grows and multiplies on a broad range of salt concentrations from the low salinity of the ocean to the almost saturated salt solution of the Dead Sea, accumulating free glycerol to counterbalance the salt concentration of the growth medium. Under optimum conditions of cultivation and high salinity, 85% of its dry weight is glycerol. Investigations so far suggest that *Dunaliella* requires no more than sea water, carbon dioxide and sunlight, which makes it an extremely promising biosynthetic feedstock for the conversion of solar energy to chemical energy in the form of glycerol.

After processing, the alga may be used as animal feed as it lacks the indigestible cell wall found in most algae and, in addition, the pigment β-carotene can be extracted in worthwhile amounts. Thus the three products, glycerol, pigment and protein, seem likely to make this recent pilot-scale venture economically viable but as yet no definitive yield and cost data are available.

However, one need not go to algal or microbial culture for production of oils or hydrocarbons. A variety of higher plants also produce suitable precursors for direct production of hydrocarbon-type liquid fuels.

2.4.4   *Oil plants*

Vegetable oils can be extracted from an impressive range of plants. Besides sunflowers, palm, coconut, olives, and peanuts among the better known plants, there are more exotic species such as jojoba and guayule—both desert plants—castor bean, rapeseed, milkweeds, eucalyptus, squashes, copaiba, malmeleiro, babassu nut, etc. Oils from these plants, and almost certainly from others, are being investigated in many countries. But before we can turn confidently to these oils as fuels with sustained production we need to know how much they cost, as well as how much energy we have to invest in growing and processing the plants. There is no point in growing plants to produce vegetable oil for fuel if 2 l of premium fuel are consumed in the process of 'growing' one litre of oil. Here the key factor is the productivity—how much oil farmers can produce on a hectare of land—and the cost and ease of production. Trials in South Africa and the mid-west of the US with sunflowers and soya beans show that yields can be 1 t of oil per hectare with unsophisticated processes for extracting oil (something like 40–50% of weight of the harvested crop is oil), while yields as high as 2 t/ha are not uncommon, and claims of up to 5 t/ha have been reported. Oil produced in this way costs about \$2/gal (20p/l) and the ratio of energy produced in comparison with that put into growing and processing the crop varies between 3 and 10 to 1.

The main problem to further commercial development is still the present low bioproductivity of the plants now available. The recent success of a UK/Malaysian research and development programme in selecting high-yielding clones of oil palm (up to a record 14 t oil ha^{-1} a^{-1}—over 2–3 times current average yields) and their rapid propagation using tissue culture techniques is a striking example of what can be done to exploit oil-producing (and other) plants once such a programme is initiated and fully supprted.

There are some advantages in making fuel from palm oil rather than making ethanol from sugar and starch crops: it is easier to extract the fuel from palms; year-round production is possible by continuous harvesting; there are not so many polluting byproducts; and one of the 'leftovers' from the process is a protein-rich animal food. On top of that, growers can use the whole plant to make a number of products, the energy return is very good, and water is not needed during processing of the fuel.

Vegetable oil can be used directly or mixed with diesel oil in compression ignition engines. However, performance and engine life are improved if the oil is used to produce methyl or ethyl esters. On the other hand, many species of plants produce hydrocarbons which can be used as fuels and chemicals; probably the best known is natural rubber from the cultivated tree *Hevea brasiliensis*. Such hydrocarbons are chemically more reduced than carbohydrates, and can be used directly rather than, as is necessary with carbohydrates, after a microbiological or thermochemical conversion.

Currently, serious efforts are being made to select and establish trial plantations of plants which produce hydrocarbons of lower molecular weight than rubber with the aim of extracting liquids which will have properties very close to those of petroleum.

Systematic searches for species of plants with high hydrocarbon content have been made sporadically in the past. In 1921, a monograph was published on the 'rubber' content of N. American plants and, in the four years before his death in 1932, the American inventor Thomas Edison examined about 2000 plants; he found many hydrocarbon-containing species but only one or two in which the molecular weight of the hydrocarbon was large enough to be considered as a possible candidate for substitution for natural (*Hevea*) rubber. Such screening efforts have generally been directed towards latex-producing plants, since latex, a milky emulsion of about 30% hydrocarbons, is the raw material of cultivated natural rubber. However, many species of plants, particularly in the family Euphorbiaceae, produce in their latex hydrocarbons with much lower molecular weights than rubber (10–20 000, rather than $1–2 \times 10^6$) and it is these molecular weight products which are now being sought as 'oils'.

In the genus *Euphorbia*, almost every species is a latex-producing plant; the latex of about a dozen species of *Euphorbia* has now been analysed, and most samples contain hydrocarbons of a much lower molecular weight than those of rubber.

The reduced organic materials (those extractable by acetone and benzene) average $10(\pm 5)\%$. From the value of 10% oil content, a rough calculation has been made (Calvin, 1976): for biomass production, an annual yield per acre of 10 t (dry) was expected, giving an oil production of 1 t (7 barrels). Annual cultivation costs per acre were estimated to be $150, or about $20 per barrel. Processing costs were not included, but neither was the value of the residual biomass. The possible added value of the oil for uses other than fuel has yet to be determined. This preliminary cost estimate was, then, close enough to the current price of petroleum to warrant further investigation. However, recent field trials show that a yield of 1.5 t oil/ha is only possible under irrigation, but dry land cultivation gives very low yields.

The question of whether there is a net energy benefit from such an enterprise, i.e. an energy output:input ratio greater than one, has apparently not been resolved as yet. The by-products may have monetary values and energy contents which could change the picture significantly.

Natural rubber can be produced from the desert shrub guayule and large quantities have been so produced in the past; because much of the research needed to make guayule a practical proposition was done by the USDA during World War II, it is currently the most viable possibility as a multi-use oil and hydrocarbon crop. Guayule research,

Component	Rubber crop		Oil plus by-product rubber crop		Oil crop		Gutta crop	
	Yield kg/ha/a	Composition dry basis %	Yield kg/ha/a	Composition dry basis %	Yield kg/ha/a	Composition dry basis %	Yield kg/ha/a	Composition dry basis %
Total dry matter	13 500	100	17 900	100	22 500	100	11 500	100
Crude protein	1485	11	1610	9	1350	6	1150	10
Rubber	1350	10	360	2	—	—	—	—
Gutta	—	—	—	—	—	—	1380	12
Oil	810	6	2150	12	2250	10	920	8
Polyphenol	945	7	1250	7	4050	18	805	7
Extracted residue	10 395	77	14 140	79	16 200	72	8395	73

*Yields are based on harvesting and using the entire aerial plant.

Table 2.8. Oil and hydrocarbon crop models, yield and composition* to be taken as specifications for new crops (Buchanan et al., 1978)

however, has a more general applicability and it is especially important for its contribution to the long-term objective of producing hydrocarbons from new types of plants.

Buchanan *et al.* evaluated a large number of plants for possible introduction as oil- and hydrocarbon-producing crops. They anticipated that a 50% improvement in dry-matter yield and a two- to threefold increase in oil and hydrocarbon content should be possible during domestication. Such projected increases in yields seem justified in view of past accomplishments by plant breeders and agronomists working on conventional crops such as *Hevea*, but they must produce at least 2 t ha^{-1} a^{-1} of oil plus hydrocarbon to be comparable with *Hevea* and meet the specifications in Table 2.8.

A few of the large number of species examined were selected for more detailed studies as potential rubber crops (thirteen spp.), potential oil + rubber crops (eleven spp.), oil crops (nine spp.) and 'gutta'-rubber crops (three spp.). Table 2.8 shows theoretical values for yield and composition and indicates desired specifications for new crops.

Hydrocarbons and oils may be used after separation and chemical treatment, possibly hydrocracking and/or esterification. However, at present the total world production of vegetable oils is only around 40×10^6 t/a, and other systems are not yet developed. This leaves the fermentation of sugar to ethanol as the only major commercial route of large-scale production of liquid transport fuel from biomass raw materials. However, as already mentioned, a major problem with use of sugar or starch is again the limited production and competition for use as food. The lignocellulose remains as the only major potential fermentative feedstock. The conversion of such raw material into a suitable liquid feed for fermentation is a major challenge to biotechnology.

2.5   Conversion to fuels

2.5.1   *Feedstock preparation*

As far as fuel alcohol is concerned, almost all existing projects are based on the use of molasses, sugar-cane juice, maize starch, or to a limited extent, cassava. The fermentation step is but one stage in the overall process, which may involve agricultural production of raw materials, harvesting and transport to the factory, feedstock preparation, fermentation, distillation, dehydration, denaturation, blending and distribution. In addition, a major consideration may be the disposal or treatment of the aqueous waste (stillage or slops) from the distillery. Obviously, it is not possible to cover all aspects of this in detail, so attention will be focused on those aspects which are important in terms of technical feasibility, energy balance and economics but at the same time reflect the microbiological aspects of the system—in other words, those aspects of the process which could be improved by developments in biotechnology. These factors can be divided into two groups: the first relates to the nature of the feedstock, which should contain the required concentration of fermentable sugars produced at low cost with low energy inputs; the second group relates to the separation of the alcohol where both the economic and energy costs could be reduced if the product stream from the fermenter has a high alcohol concentration.

For the production of ethanol from sugar crops the standard techniques for juice extraction in sugar production are used. Simple sugars may be extracted by mechanical expression of the juice for sugar cane or by diffusion for beet. Starch crops require mechanical grinding to produce a slurry which is then heated to disrupt the starch grains. A variety of hydrolysis procedures may follow, based on combinations of acid and/or enzyme treatment. A typical process would use a thermophilic bacterial amylase at 90 °C for thinning, followed by saccharification (hydrolysis of the dextrins to glucose) by a glucoamylase at 50–60 °C. On a glucose conversion basis the weight yield is 51%. However, since about 5% of the sugar is used for cell growth and maintenance energy or for synthesis of other organic compounds such as glycerol, acetic acid, acetaldehyde and fuel oil (mainly higher alcohols), the maximum yield is about 48% of the feed sugar. The weight yield will also depend on the nature of the feed:

$$1 \text{ kg invert sugar} = 0.484 \text{ kg ethanol} = 0.61 \text{ l}$$
$$1 \text{ kg sucrose} \quad = 0.510 \text{ kg ethanol} = 0.65 \text{ l}$$
$$1 \text{ kg starch} \quad\;\; = 0.530 \text{ kg ethanol} = 0.68 \text{ l}$$

These differences arise as a result of the incorporation of water during the hydrolysis of oligosaccharides and higher polymers.

For small-scale manufacture, the extent of purification of the feed stream from feedstocks such as maize may be minimal; in large factories, on the other hand, conventional wet milling as used in the production of high quality glucose syrups has the advantage that it produces a number of valuable by-products such as corn oil and gluten. The sale of such products helps to offset the cost of alcohol production. Even in small-scale whole grain fermentation the residual solids plus yeast will have value as animal feed. The sugar level in the feed will vary depending on the raw materials used but where practicable the fermentable solids should be in the range of 16–25% in order to achieve a high alcohol content in the brew. This is important because the steam consumption required in a conventional still system for production of 96% ethanol from a 5% beer is about twice that required for production from a 10% beer.

At present, the major problems associated with power alcohol production lie in the fact that the raw materials are potential sources of human food or animal fodder. Hence, a competition for raw materials exists, resulting in a high cost of feedstock. Depending on the nature of the raw materials, between 60 and 85% of the final alcohol selling price in present systems can be attributed to feedstock costs. Furthermore, with the exception of the sugar-cane based power alcohol factory in which the fibrous part of the stem (bagasse) may be used as fuel, the overall process of alcohol production uses more energy than is recovered in the ethanol produced. These problems could be overcome if cellulose could be used as a feedstock and an alternative method derived for recovering the alcohol at lower energy input cost.

The trouble with cellulose is that, in its native state in the plant cell wall, it occurs as an insoluble complex together with hemi-celluloses and lignin. Furthermore, because of hydrogen bonding, the individual cellulose molecules become orientated in microfibrils which show varying degrees of crystallinity and prevent attack by hydrolytic agents. Even when the plant cell material is completely hydrolysed, the yeasts which are currently used in alcohol production are not capable of using the five-carbon sugars, uronic acids or phenols derived from the non-cellulosic constituents of the plant cell walls.

A variety of approaches are being taken to solve this problem. In some procedures, the objective is to produce a glucose syrup and, hence, the main consideration is hydrolysis of cellulose. In other systems, the aim is to ferment as great a proportion as possible of the lignocellulose material, either following chemical separation of the various constituents or by use of bacteria capable of direct metabolism of both cellulose and hemicellulose. Milling is the most effective physical pretreatment, but is expensive in terms of both money and energy, although it has been suggested that the use of cryo-milling may be more cost effective.

Chemical treatment with materials such as sodium hydroxide, peracetic acid, or sodium hypochlorite may be used for delignification. Swelling and separation of fibres can be accomplished by alkali treatment. Hydrolysis may be accomplished using either mineral acids or biological methods, using cellulytic enzymes from fungi such as *Trichoderma*, *Aspergillus* or *Sporotrichum*. Separation of cellulose and hemicellulose may be followed by yeast fermentation of the glucose stream, for example, to form ethanol, and fermentation using bacteria such as *Klebsiella* or *Aeromonas* to form butandiol from, say, the pentoses. Alternatively, lignocellulose may be used directly as substrate for bacteria such as *Clostridium thermocellum*.

For a typical glucose-from-cellulose process, mechanical pretreatment, steam explosions or chemical delignification may be used. All have problems of energy input needed; acid processes show problems of corrosion and wastes from neutralization, whereas solvent processes may be costly in terms of solvent recovery. The characteristics of the product which determine their accessibility to enzymes are degree of crystallinity, extent of lignification, and particle size.

Most work has been done on the cellulase complex of *Trichoderma viride*, or *T. reesei*. Kinetics of action have been studied in detail and mutants or varieties of much higher activity, such as the C-30 strain, have been isolated. The *T. reesei* cellulase complex has three components: an endo-enzyme (Cx fraction), which attacks the substrate at random, an exo-1,4 glucanase (C$_1$ fraction), which produces cellobiose, and a 1,4 glucosidase, which converts this to glucose. The level of glucosidase is low and may have to be supplemented by enzyme from other fungi such as *Aspergillus*. The enzyme of the brown rot fungi differs in that these fungi contain a system which generates H_2O_2 and results in a more rapid initial breakdown.

Problems arise in the use of such enzyme complexes from fungi since both enzyme activity and enzyme synthesis are inhibited by end-products. For instance, a 0.01% solution of cellobiose can cause over 75% inhibition of cellulase activity. Modification of enzyme activity, regulation and synthesis by mutation, selection and recombinant techniques will, it is hoped, produce more active enzyme preparations of higher stability and greater tolerance to end-product inhibition. In the meantime, a wide range of other organisms have been investigated in

Table 2.9. Microorganisms investigated in biological breakdown of lignocellulose

Cellulose and cellulases
  Trichoderma viride
  T. reesei
  T. koningii
  Coniphora cerebella
  Sporotrichum pulverulentum
  Polyporus adustus
  Myrothesium verrucaria
  Penicillium funiculosum
  Fusarium solani
  Aspergillus wentii
  Coniphora thermophila
  Thielavia terrestris
  Phanaerochaete chrysosporium
  Clostridium thermohydrosulfuricum
  Clostridium thermocellum
  Clostridium thermosaccharolyticum
  Thermomonospora spp.
Xylanose
  Chaetomum trilaterale
Glucosidase
  Aspergillus phoenicus
Lignase
  Erwinia spp.
  Trichospora fermentans

respect of their use for degradation of lignocellulose. Some of these are listed in Table 2.9. Lignin degradation may be brought about by strains of *Erwinia*, or by the yeast *Trichospora fermentans*, whereas a number of fungi, including the well known genera of *Mucor*, *Penicillium*, *Trichoderma* and *Aspergillus*, have xylanase activity.

2.5.2 *Fermentation*

The production of methane (biogas) by anaerobic digestion and the formation of ethanol (power alcohol) by yeast-based fermentation are the principal biological technologies currently available for improving the quality of fuels derived from plant materials.

If it is assumed that the initial substrate is glucose, the overall conversions may be summarized as follows:

Biogas production $C_6H_{12}O_6 \rightarrow 3(CH_4) + 3(CO_2)$
Ethanol formation $C_6H_{12}O_6 \rightarrow 2(CH_3CH_2OH) + 2(CO_2)$

Although details of the biochemical pathways involved in these conversions will not be considered here, it should be noted that a part of the small amount of energy released during these anaerobic transformations is trapped as reductant or energy source (ATP) and used to support growth or cell maintenance of the bacteria or yeast. Hence, although these microorganisms are metabolically inefficient in terms of substrate utilization for growth, compared with organisms which grow in the presence of oxygen and respire glucose stoichiometrically to carbon dioxide and water, the processes achieve the objective of forming products efficiently and with a higher energy density than that of the initial biomass substrate. Depending on the phase (liquid, solid or gas), temperature and pressure of the reactants and products, the heats of combustion (calorific values) are approximately 16 kJ/g for glucose, 30 kJ/g for ethanol and 56 kJ/g for methane.

Although both anaerobic digestion and ethanol fermentation depend on microbiological conversion to produce a high-grade fuel from biomass, this is about the only similarity.

2.5.3   *Ethanol*

The production of ethanol by yeast is well established technology (see also Chapter 3). In order to produce a fuel, the following sequence of processes must be carried out (as illustrated in Fig. 2.5): feedstock preparation; fermentation; stripping and rectifying; dehydration (if to be used in a petroleum blend); denaturation and storage; stillage treatment and disposal or recovery for animal feed, etc. At each stage there are specific points which should be considered as they relate to the feasibility, energy balances and economics of using fermentation as a route to liquid fuel.

The scale of operation of power alcohol plants may be very large, with inputs measured in thousands of tonnes per annum and outputs measured in millions of litres per annum. The greatest contribution of power alcohol to a country's energy budget occurs in Brazil, where over 5×10^9 l were produced in 1982 and the expanding Proalcool programme is expected to produce over 11×10^9 l in 1985. This programme is primarily based on factories fermenting cane juice directly, with some contribution from molasses. The second big national programme is that of gasohol production in the USA currently based on the fermentation of maize starch—current (1984) production is about 5×10^8 gal/a.

The major use of power alcohol is as a fuel in internal combustion engines as a blend or, where suitable engines are available, as a fuel in its own right. The difference between the two modes of use is that anhydrous alcohol is required for blending with petroleum in order to avoid problems of phase separation. On the other hand, a purpose-built engine can run on the 95% ethanol/water mixture produced from the primary distillation. Although ethanol can be used for cooking, heating, lighting or generation of steam and electricity, there is no

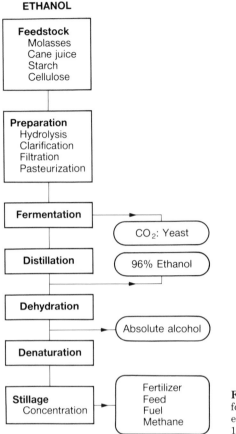

ETHANOL

Fig. 2.5. Flow diagram for the production of ethanol (from Coombs, 1980).

particular advantage in this as a considerable net energy loss occurs during the conversion of biomass to ethanol.

All stages of the processing require an input of energy, the major requirement being for the concentration and dehydration of the ethanol for distillation; this energy may come from crop wastes such as bagasse, straw, etc., from the burning wood, or combustion of fossil fuels, gas, oil or coal. In general, the energy input necessary to process the feed material is approximately the same as the energy output in the form of ethanol; thus, it requires a waste or very cheap fuel to power the overall process.

Most commercial-scale fermentation alcohol is currently produced using yeasts (*Saccharomyces*, generally *S. cerevisiae*, but also *S. uvarum* (carlsbergensis) and *S. diastaticus*). One objective is to match the yeast with the substrate. *S. cerevisiae* can use glucose, fructose, maltose and maltotriose, i.e. those sugars which appear in the traditional feed from sugar or starch crops. *S. diastaticus* can also use dextrins, whereas the species *Kluyveromyces fragilis* and *K. lactus* can use lactose.

The production of ethanol by yeasts is an anaerobic process, although growth of new cells requires oxygen, and traces of oxygen may be needed to support alcohol-producing cells. At the metabolic level the regulation of ethanol production from glucose is complex; the concentration of substrate, oxygen and product (ethanol) all affect yeast metabolism, cell viability, cell growth, division and alcohol production. Selection of suitable strains of yeast with a tolerance for higher concentrations of both substrate and alcohol has been of particular importance in increasing yields.

Where practicable the level of fermentable solids used is in the region of 16–25% (w/v), giving final concentrations of ethanol of 6–12%. With most yeasts the preferred temperature is between 25 and 33 °C, and operation at a pH of between 4 and 5 will reduce possibilities of contamination and preclude the development of so-called 'killer yeasts'. Additional nutritional requirements depend on the nature of the raw materials used to prepare the substrates; e.g. molasses may require only the addition of a source of nitrogen and phosphate, whereas a starch hydrolysate of high purity may require the addition of essential trace elements and vitamins.

Essentially, there are three methods of fermenting sugar feedstocks, namely batch, batch with cell recycle, and continuous. In batch fermentation the substrate is fermented out by growth of a freshly cultured inoculum initially under aerobic conditions. This results in an anaerobic environment in which the rest of the substrate is converted to alcohol. The yeast is discarded at the end of the fermentation and hence a new culture must be grown for the next batch of alcohol production. This continual propagation of cells is costly in terms of substrate.

With yeast grown in a batch system, about 5% of the sugar may be used for cell growth and maintenance energy or for the synthesis of other compounds such as glycerol, acetic acid, acetaldehyde and fusel oil (mainly higher alcohols). The maximum weight yield is thus about 48% of feedstock. With yeasts, the productivity on a dry cell weight basis may vary between one and two grams of ethanol per hour per gram of cells. Higher productivities (2.5–3.8) have been recorded for bacteria such as *Zymomonas mobilis*. However, such organisms have yet to be developed commercially. Productivity in the reactor will reflect the mode of operation, the strain of yeast, the cell density and the nature of the substrate, varying from around $1 \mathrm{~g~l^{-1}~h^{-1}}$ to over $10 \mathrm{~g~l^{-1}~h^{-1}}$. As alcohol concentrations are increased and the substrate concentration decreases, the sugar may not be completely utilized within a reasonable time.

The disadvantages of long fermentation times and poor substrate utilization can be partially overcome by use of a batch recycle process in which the yeast is separated from the fermented mash at the end of one cycle and retained for the next fermentation. However, the advantage of the batch process is that it is very resistant to process abuse and can be operated by less skilled personnel, but is characterized by low volumetric productivity; e.g. a 36 h fermentation yielding a 5% (w/v) ethanol solution gives an average of about 1.4 g ethanol $l^{-1} h^{-1}$. This may be increased to over 10 g $l^{-1} h^{-1}$ by batch recycle or continuous systems where the cells are removed from the brew, recycled and maintained at a high cell concentration.

Such systems will require higher operator skills, plus higher capital costs and more sophisticated control devices. With continuous operation the extent to which the substrate is converted to ethanol depends on the dilution rate. High productivity requires high dilution rates, but under such conditions the average substrate residence time is reduced. In other words, optimization of the two factors of volumetric productivity and completeness of sugar utilization oppose one another. The choice of operating parameters for the fermentation must thus reflect the balance between substrate costs, capital costs and operating costs, which in turn have to be considered in relation to the total production system.

At the end of the fermentation, the alcohol concentration lies between 6 and 12% depending on the yeast strain and starting sugar concentration. A high alcohol concentration is important, since the steam consumption for distillation increases rapidly, e.g. from around 2.25 kg steam per litre of 96% alcohol for a 10% beer to over 4 kg steam for a 5% beer. Additional steam is needed to produce anhydrous ethanol from the constant boiling water/ethanol azeotrope. The most common method involves the addition of benzene at ten times the volume of water present. The benzene/water/ethanol azeotrope distills first at 64–84 °C. When all the water is exhausted, a second azeotrope (ethanol/benzene) is formed distilling at 68.25 °C. When all the benzene is exhausted, absolute ethanol remains and is drawn off from the bottom, and the benzene is recycled.

The major by-products are CO_2, yeast, fusel oil and stillage. These have by-product value, although stillage may be an embarrassment in its dilute form. In most cases, since potable ethanol attracts a high rate of tax, it is necessary to denature the alcohol by the addition of bittering agents although blending with petrol will do the same thing.

2.5.4   *Energy balances*

The net energy production in both ethanol fermentation or anaerobic digestion may be small or even negative. The reason for this is that considerable amounts of process energy are required for production of raw materials, processing, separation, purification or compression of the product. This energy may come from crop wastes such as bagasse or straw, from the burning of wood, from waste heat produced by associated manufacturing processes, from combustion of fossil fuels, or from the electricity grid supply.

Where ethanol production alone is considered, the amount of energy consumed in the overall production of a litre of alcohol is equal to or exceeds the energy content of the ethanol. Hence, if there is to be a net gain, process energy must be derived from a renewable source or the by-products must have an energy content. At present a sugar-cane based system is the only one in which a positive net gain of energy is achieved, because the process energy comes from combustion of fibrous parts of the cane (bagasse). On the other hand, systems using starch or sugar beet feedstocks, for example, require an alternative fuel input. If oil is used for this purpose, there is little or no net liquid fuel energy gain. On the other hand, if the overall process is fuelled by coal, for example, it represents one method of converting coal to a liquid transport fuel, with a net saving of oil-based energy. In addition, for sugar or starch based systems the cost of raw materials may contribute between 60 and 85% of the final overall production costs and are thus amenable to price and productivity gains. Thus, for reasons of energy quality, value and economics, the process is receiving considerable attention.

Research objectives are to increase the efficiency of substrate utilization, increase productivity on a volumetric basis, increase concentration in the product stream and work at higher temperatures. These actions will save costly substrate by decreasing the need to produce cell material (recycle or immobilization) or destroy the side reactions which produce unwanted products, such as glycerol, acetate, etc., and at the same time save energy by reducing the need for fermenter cooling and decreasing energy input into distillation. This can be brought about by both microbiological and engineering changes.

From the microbiological point of view, changes are needed in the response to O_2 concentration, substrate concentration and specificity, ethanol concentration, temperature, ratio of fermentation to cell growth, amount of by-products, and flocculation characteristics. This last is important since flocculating organisms may be separated by simple settling rather than energy- and cash-expensive centrifuges.

Selection of tolerant yeast has been and will remain the major area of improvement. Intolerance is due in part to osmotic effects and in part to the effect of alcohol concentration on membrane permeability. One approach is to extend the range of organisms used (e.g. *Z. mobilis*, *Clostridium* spp.), the other to use genetic manipulation to move characteristics from various yeasts and bacteria into *S. cerevisiae*. This latter is attractive due to the wide commercial experience with this organism. Again, as far as yeasts are concerned, much is known about their life cycle and genetics and plasmid vectors are available. Many of the industrial strains of yeast are, however, polyploid, do not have a mating type and show low spore viability. Thus, genetic analysis and improvement by conventional crossing is difficult—hence the attraction of recombinant techniques. The transfer of genes for the synthesis of amylase and cellulase into yeasts is already under way.

From the fermentation viewpoint, much of the research work relates to immobilized cell systems which can be run as a continuous process without the need for cell recycle, or systems in which cells are recycled and the alcohol removal is effected under reduced pressure (vacuum fermentation). Problems with both types of systems lie in the rapid production of CO_2, which must be removed, in questions of control of cell growth and the need for clean, particle-free sterile feed streams. A further problem arises where the substrate is expensive, since high productivity requires high dilution rates, but high dilution gives poor substrate utilization; hence, a serial system may be needed. Again a back-up system will be required in case of infection. Consequently, as yet, no such systems would appear economically competitive. The solution of such problems remains a challenge to the biotechnologist's art, both at the microbial and at the engineering level.

2.5.5   *Methane by anaerobic digestion* (see also Chapter 6)

Anaerobic digestion produces a mixture of methane gas and carbon dioxide as the result of the breakdown of complex substrates by a mixed population of diverse microorganisms. As the required product is in the gas phase it just bubbles off, so separation is not a problem, although for more sophisticated uses or pipeline distribution, scrubbing of impurities and compression may be necessary. The possible raw materials that may be used in an anaerobic digester vary greatly, from agricultural wastes (spoilt crops or food materials), processing wastes, sweet waters from sugar factories or palm oil extraction, garbage, municipal solid waste and distillery slops, to purpose grown crops, including more

placed in aerobic, sterile medium. The open-circuit voltage was 0.3–0.5 V and a small current of 0.2 mA was observed. Although there were reports of a number of different types of biofuel cell using different organisms and fuels over the next fifty years, the space research programme greatly stimulated interest in this area in the late 1950s and early 1960s. The many different configurations investigated fall into three basic types, described below.

The product biofuel cell

Organisms or enzymes are used to convert non-electroactive fuels to electroactive species which are then oxidized in a conventional cell. For example:

$$\text{Glucose} \xrightarrow{\substack{\text{fermentative} \\ \text{microorganisms}}} H_2$$

$$H_2 \longrightarrow 2H^+ + 2e^- \ \text{(anode)}$$

$$2H^+ + \tfrac{1}{2}O_2 + 2e^- \longrightarrow H_2O \ \text{(cathode)}$$

The depolarizer biofuel cell

Here, organisms or enzymes are involved in catalysing the electrochemical reaction at one or both of the electrodes. For example, a hydrogen cell can be constructed using the enzyme hydrogenase to effect the anodic reaction, whilst the enzyme laccase catalyses the cathodic one:

$$H_2 \xrightarrow[\text{hydrogenase}]{} 2H^+ + 2e^- \ \text{(anode)}$$

$$2H^+ + \tfrac{1}{2}O_2 + 2e^- \xrightarrow[\text{laccase}]{} H_2O \ \text{(cathode)}$$

The regenerative biofuel cell

In this type of device, the biological component regenerates the electrochemically active compounds, which in turn interact with the electrode(s). An example of such a system is shown in Fig. 2.8, which depicts a recently described enzyme biofuel cell which uses methanol as the fuel.

Whilst a few biofuel cells have been commercialized and substantial electrode current densities (up to 40 mA cm^{-2}) and power outputs (about 1 kW) have been achieved, they have not yet been adequately perfected to allow any significant practical deployment.

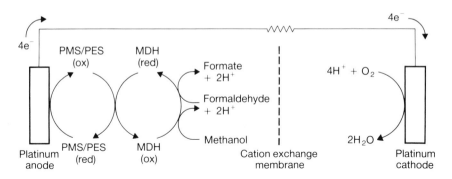

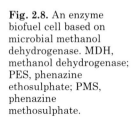

Fig. 2.8. An enzyme biofuel cell based on microbial methanol dehydrogenase. MDH, methanol dehydrogenase; PES, phenazine ethosulphate; PMS, phenazine methosulphate.

Recently, however, biofuel cell technology has advanced substantially both in the form of enzyme cells using alcohols and carbon monoxide as fuels and whole microorganism cells, both being of a regenerative type. In the latter case, redox mediators, which can reversibly penetrate the microbial wall and membrane, are used. These interact with components of the electron transport chain, substantially short-circuiting the biological electrochemical system and connecting it to the fuel cell anode (Fig. 2.9). Such devices use sugars or waste material as the fuel but theoretically could use any substance capable of supporting microbial growth. They can have substantial lifetimes (weeks to months) and the outputs of the most successful experimental cells are comparable with the best enzyme cells. Closely related photobioelectrochemical cells in which high potential electrons at the anode are derived from the electron transport chain of photosynthetic microorganisms have also been devised to convert light energy to electricity.

The current state of development of bioelectrochemical cells indicates real possibilites for the application of such devices where relatively lower power outputs are required, particularly for medical and military purposes and for electricity generation in remote areas. In

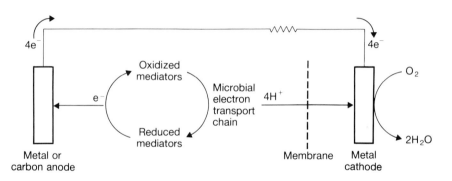

Fig. 2.9. Principle of a mediated, whole-organism microbial fuel cell.

the short term, biofuel cells are finding application as specific sensing devices since their electrical output is proportional to the amount of fuel added. The use of related bioelectrochemical cells for biochemical synthesis is discussed in Chapter 4.

2.8   Recommended reading

General

BARNARD G.W. (1983) Liquid Fuel Production from Biomass in the developing countries: An Agricultural and Economic Perspective. In *Bioconversion Systems*, (ed. Wise D.L.), pp. 112–268. CRC Press, Boca Raton, Florida.

BARNARD G.W. & HALL D.O. (1982) Energy from renewable resources: ethanol fermentation and anaerobic digestion. In *Biotechnology*, Vol. III, (Chapter 7), (ed. Dellweg H.). Verlag Chemie, Weinheim.

BARNETT A., PYLE D.L. & SUBRAMANIAN S.K. (eds.) (1978) *Biogas Technology in the Third World: a Multidisciplinary Review*. Intl Devl. Res. Council, 60 Queen St., Ottawa K1P 5Y7.

BioENERGY COUNCIL (1981, 1984) *International BioEnergy Directory*. Bio Energy Council, P.O. Box 12807, Arlington, Virginia 22209.

BOLAND D.J. & TURNBULL J.W. (1981) Selection of Australian trees other than *Eucalyptus* for trials as fuelwood species in developing countries. *Aust. For.* **44**, 235–246.

BUCHANAN R.A. *et al.* (1978) Hydrocarbon and rubber-producing crops. *Econ. Bot.* **32**, 131–153.

CANNELL M.G.R. (1982) *World Forest Biomass and Primary Production Data*. Academic Press, London.

CBNS (1982) *Economic Evaluation and Conceptual Design of Optimal Agricultural Systems for Production of Food and Energy*. Center Biol. Natural Systems report to US-Dept. Energy, Washington DC 20545.

CIQA (1978) *Guayule*. Centro Investi. Quimica Appl., Aldama Ote 351, Saltillo, Mexico.

COOMBS J. (1980) Renewable sources of energy (carbohydrates). *Outl. Agric.* **10**, 235–245.

COTE W.A. (ed.) (1983) *Biomass Utilization*. Plenum Press, New York.

EARL D.E. (1975) *Forest Energy & Economic Development*. Clarendon Press, Oxford.

GOLDMAN J.C. (1979) Outdoor mass cultures. *Water Resour., Wash.* **13**, 1–19.

GOOD N.E. (1981) Fuel from biomass. In *Beyond the Energy Crisis*, Vol II, (ed. Fazzolari R.A. & Smith C.B.), pp. 491–498. Pergamon Press, Oxford.

HALL D.O. (1980) Renewable resorces (hydrocarbons). *Outl. agric.* **10**, 246–255; (1981) *New Scientist*, **89**, 524–526.

HALL D.O. (1981) Solar energy through biology: fuel for the future. In *Advances in Food Producing Systems for Arid and Semi Arid Lands*, (ed. Mannasah J.T. & Briskey E.J.). Academic Press, New York.

HALL D.O. (1982) Biomass for energy: fuels now and in the future. *J. R. Soc. Arts* **130**, 457–471; (1982) *Experientia*, **38**, 3–10.

HALL D.O. (1984) Photosynthesis for energy. In *Advances in Photosynthetic Research*, Vol. II, (ed. Sybesma C.), pp. 727–740. Martinus Nijhoff, The Hague.

HALL D.O., BARNARD G.W. & Moss P.A. (1982) *Biomass for Energy in the Developing Countries*. Pergamon Press, Oxford.

HEDEN K. (1982) Swedish energy forestry. *Biomass*, **2**, 1–3.

HIGGINS I.J. & HILL H.A.O. (1979) Microbial generation and interconversion of energy sources. In *Microbial Technology*, (eds. Bull A.T., Ellwood D.C. & Ratledge C.). *Soc. gen. Microbiol. Symp.* **29**, 359–377.

HOBSON P.N., BOUSFIELD S. & SUMMERS R. (1981) *Methane Production for Agricultural and Domestic Wastes*. Appl. Sci. Publ., London.

HUGHES D.E. (ed.) (1982) *Anaerobic Digestion (1981)*. Elsevier, Amsterdam.

ICE (1978) *Alcohol Fuels*. Inst. Chem. Eng., NSW Branch, Sydney, Australia.

JOHNSON J. & HINMAN H.E. (1980) Oil and rubber from arid land plants. *Science*, **208**, 460–464.

KINGSOLVER B.E. (1982) *Euphorbia lathyrus* reconsidered: its potential as an energy crop for arid lands. *Biomass*, **2**, 287–295.

KLASS D.L. & EMERT G.H. (1981) *Fuels from Biomass & Wastes*. Ann Arbor Sci. Publ., Michigan.

LEACH G. (1976) *Energy & Food Production*. IPC Press, Guildford, UK.

LEWIS C. (1983) *Biological Fuels*. Edward Arnold, London.

LIPINSKY E.S. (1981) Chemicals from biomass: petrochemical substitution options. *Science*, **212**, 1465–1471.

LIPINSKY E.S. (1981) *Systems Study of Animal Fats and Vegetable Oils for Use as Substitute and Emergency Diesel Fuels*. Report to US Dept. Energy by Battelle, Columbus, Ohio 43201.

MCLAUGHLIN J. & HOFFMAN J.J. (1982) Survey of biocrude-producing plants from the southwest. *Econ. Bot.* **36**, 323–339.

MOSS P.A. & HALL D.O. (1982) Biomass for energy in Europe. *Int. J. Solar Energy*, **1**, 239–262.

NAIR P.K. (1980) *Agroforestry Species*. Intl Council Res. Agroforestry, PO Box 30677, Nairobi.

NAS (1977) *Methane Generation from Human, Animal and Agricultural Wastes*. Natl Acad. Sci. USA, Washington DC 20418.

NAS (1981) *Firewood Crops: Shrub and Tree Species for Energy Production*. Natl Acad. Sci. USA, Washington DC 20418.

NAS (1982) *Priorities in Biotechnology R & D for International Development*. Natl Acad. Sci. USA, Washington DC 20418.

NEPA (1981) Du petrole sur nos terres. *Motorisation et Technique Agricole Suppl. March 1981*. NEPA (S.A.) 10 rue Martel, Paris 75493.

OALS (1980) *A Technology Assessment of Guayule Rubber Commercialisation*. Office Arid Lands Studies, Univ. Ariz., Tuscon, Arizona.

OTA (1980) *Energy from Biological Processes*. Office of Technology Assessment, US Congress, Washington DC 20510.

PALZ W., CHARTIER P. & HALL D.O. (eds.) (1981) *Energy from Biomass*. Appl. Sci. Publ., London.

PIMENTEL D. & PIMENTEL M. (1979) *Food, Energy and Society*. Edward Arnold, London.

RABSON R. & ROGERS P. (1981) The role of fundamental biological research in developing future biomass technologies. *Biomass*, **1**, 17–38.

RICHMOND A. & PREISS K. (1979) The biotechnology of aquaculture. *Interdisc. Sci. Rev.* **5**, 60–68.

ROYAL SOCIETY LONDON (1982) Industrial & Diagnostic Enzymes. *Phil. Trans. Roy. Soc. Lond.*

SAE (1981) *Fuels from Crops*. Soc. Automative Engin., Natl Sci. Centre, Melbourne, Australia.

SCURLOCK J.M.O. & HALL D.O. (1984) Energy from Biomass in Europe. In *5th Canadian Bioenergy R & D Seminar*, (ed. Hosnain S.), pp. 56–65. Elsevier, London.

SHELEF G. & SOEDER C.J. (eds.) (1980) *Algae Biomass, Production and Use*. Elsevier. Amsterdam.

SILVERSIDES R. (1982) Energy from forest biomass—its effect on forest management practices in Canada. *Biomass*, **2**, 29–42.

SLESSER M. & LEWIS C. (1979) *Biological Energy Resources*. Spon, London.

Sɴ Iɴ. (1981) *Assessment of Plant Derived Hydrocarbons.* Report to US-Dept. Energy, Washington DC 20545.

Sᴛᴀꜰꜰᴏʀᴅ D.A., Wʜᴇᴀᴛʟᴇʏ B.I. & Hᴜɢʜᴇs D.E. (eds.) (1980) *Anaerobic digestion.* Appl. Sci. Publ., London.

Sᴛᴇɪɴʙᴇᴄᴋ K. (1981) Short-rotation forestry as a biomass source: an overview. In *Energy from Biomass* (eds. Palz W., Chartier P. & Hall D.O.), pp. 163–171. Appl. Sci. Publ., London.

Sᴛᴇᴡᴀʀᴛ G.A. *et al.* (1979) *The Potential for Liquid Fuels from Agriculture and Forestry in Australia.* CSIRO, Div. Chemical Technology, Canberra.

Sᴛᴇᴡᴀʀᴛ G.A. *et al.* (1982) *The Potential for Production of Hydrocarbon Fuels from Crops in Australia.* CSIRO Div. Chem. Technol, Canberra, Australia.

Sᴛʀᴜʙ A., Cʜᴀʀᴛɪᴇʀ P. & Sᴄʜʟᴇsᴇʀ G. (eds.) (1983) *Energy from Biomass*, Vol. 2. Appl. Sci. Publ., London.

Tᴇʀɪ (1982) *Biogas Handbook.* Tata Energy Res. Inst., Bombay 400023, India.

Tʀɪɴᴅᴀᴅᴇ J.C. (1981) Energy crops—the case of Brazil. In *Energy from Biomass*, (eds. Palz W., Chartier P. & Hall D.O.), pp. 59–74. Appl. Sci. Publ., London.

TRW (1980) *Energy Balances in the Production and End-use of Alcohols Derived from Biomass.* TRW Crop. US Dept. Energy, Washington DC 20545.

Zᴀʙᴏʀsᴋʏ O.R. (ed.) (1981) *Handbook of Biosolar Resources*, Vol. II, Resource Materials. CRC Press, Boca Raton, Florida.

Photosynthesis

Bᴇᴀᴅʟᴇ C.L., Lᴏɴɢ S.P., Hᴀʟʟ D.O., Iᴍʙᴀᴍʙᴀ S.K. & Oʟᴇᴍʙᴏ R.J. (1985) *Photosynthesis in Relation to Bioproductivity*, UN Environment Programme, PO Box 30552, Nairobi, (publ. Tycooly Intl, Dublin).

Cᴏᴏᴍʙs J. (1982) Improving biomass productivity. In *Energy from Biomass*, Vol. II, (eds. Strub A., Chartier P. & Schleser G.), pp. 105–110. D. Reidel Publishing Co., Dordrecht, Holland.

Cᴏᴏᴍʙs J., Hᴀʟʟ D.O. & Cʜᴀʀᴛɪᴇʀ P. (1983) *Plants as Solar Collectors.* D. Reidel Publishing Co., Dordrecht, Holland.

Cᴏᴏᴘᴇʀ J.P. (ed.) (1975) *Photosynthesis and Productivity in Different Environments.* Cambridge University Press.

Eʙᴇʟɪɴɢ W. *et al.* (1982) Improving agricultural productivity. *Econ. Impact*, **39**, 6–47.

Eᴅᴡᴀʀᴅs G. & Wᴀʟᴋᴇʀ D.A. (1983) *C-3, C-4: Mechanisms of Cellular and Environmental Regulation of Photosynthesis.* Blackwell Scientific Publications, Oxford.

Gᴏᴠɪɴᴅᴊᴇᴇ (ed.) (1982) Photosynthesis, Vol. 1 & 2, Academic Press, New York.

Hᴀʟʟ D.O. & DᴀSɪʟᴠᴀ E.J. (1983) Photosynthesis: a biosolar tool for development. *Nature Resour.* **19**(2), 2–10.

Hᴀʟʟ D.O. & Rᴀᴏ K.K. (1981) *Photosynthesis*, 3rd edn., Edward Arnold, London.

Hᴀʟʟ D.O. & Pᴀʟᴢ W. (eds.) (1982) *Photochemical, photoelectrochemical and photobiological processes.* D. Reidel Publishing Co., Dordrecht, Holland.

Hᴀʟʟɪᴡᴇʟʟ B. (1981) *Chloroplast Metabolism.* Clarendon Press, Oxford.

Osᴡᴀʟᴅ W.J. (1981) Algae as solar energy converters. In *Energy from Biomass* (eds. Palz W., Chartier P. & Hall D.O.), pp. 633–646. Appl. Sci. Publ., London.

Pɪʀsᴏɴ A. & Zɪᴍᴍᴇʀᴍᴀɴ M.H. (eds.) (1977) *Encyclopaedia of Plant Physiology: Photosynthesis*, Vol. 5. Springer Verlag, Berlin.

UK-ISES (1976) *Solar Energy—a UK Assessment.* UK Section Intl Solar Energy Soc., 19 Albemarle St., London W1.

Cell-free systems

Bᴏʟᴛᴏɴ J.R. & Hᴀʟʟ D.O. (1979) Photochemical conversion and storage of solar energy. *Ann. Rev. Energy*, **4**, 353–406.

BRAUN A.M. (ed.) (1983) *Photochemical Conversions*. UNESCO, Presses Polytechnic Romandes, Lausanne.

CALVIN M. (1982) Plants can be direct fuel source. *Biologist*, **29**, 145–148; (1980) *Energy*, **4**, 851–870.

CONNOLLY J.S. (ed.) (1981) *Photochemical Conversion and Storage of Solar Energy*. Academic Press, New York.

Ethanol

COOMBS J. (1981) Ethanol—the process and the technology for production of liquid transport fuel. In *Energy from Biomass* (eds. Palz, W., Chartier P. & Hall D.O.), pp. 279–291. Appl. Sci. Publ., London.

DA SILVA J.G. (1978) Energy balance for ethyl alcohol production from crops. *Science*, **201**, 903–906.

FERCHAK J.D. & PYE E.D. (1981) Utilization of biomass in the U.S. for the production of ethanol fuel as a gasoline replacement. I. & II. *Solar Energy*, **26**, 9–25.

KHAN A.S. & FOX R.W. (1982) Net energy analysis of ethanol production in NE Brazil. *Biomass*, **2**, 213–232.

KOVARIK B. (1982) *Fuel Alcohol*. IIED/Earthscan, 10 Percy Street, London W1.

LAMED R. & ZEIKUS J.G. (1980) Ethanol production by thermophilic bacteria. *J. Bacteriol.* **144**, 569–578.

LYONS T.P. (1981) *Gasohol, a step to energy independence*. Alltech Publ., Lexington, Kentucky 40503.

SERI (1981) *Alcohol Fuels Program Technical Review*. Solar Energy Res. Inst., Golden, Colorado 80401.

STONE J.E. & MARSHALL H.B. (1980) *Analysis of Ethanol Production from Cellulosic Feedstocks*. Dept. Energy, Mines and Resources, Ottawa.

Biofuel cells

ASTON W.J. & TURNER A.P.F. (1984) Biosensors and biofuel cells. *Biotechnol. genet. Eng. Revs.* **1**, 89–120.

BENNETTO H.P. (1984) Microbial biofuel cells. *Life Chem. Reps.* (in press).

HIGGINS I.J., HAMMOND R.C., PLOTKIN E.V., HILL H.A.O., UOSAKI K., EDDOWES M.J. & CASS A.E.G. (1980) Electroenzymology and biofuel cells. In *Hydrocarbons in Biotechnology*, (eds. Harrison D.E.F., Higgins I.J. & Watkinson R.J.), pp. 181–193. Heyden, London.

TURNER A.P.F., ASTON W.J., HIGGINS I.J., DAVIS G. & HILL H.A.O. (1982) Applied aspects of bioelectrochemistry: Fuel cells, sensors and bioorganic synthesis. *Biotechnol. Bioeng. Symp.* **12**, 401–412.

3   Food, Drink and Biotechnology

G. A. BEECH, M. A. MELVIN & J. TAGGART

3.1   Introduction

3.1.1   *Microorganisms and food*

Disregarding agricultural activities, which are covered elsewhere in this volume, the food and drink industry receives raw materials (mostly agricultural products) and converts them into food or drink products, or materials which will be used in the preparation of food and drink products. All organic materials used in food and drink manufacture are potential substrates for microbial activity. This single fact gives microbiology a key position in the food industry, with both positive and negative implications. The negative aspects are perhaps the more obvious, as precautions against microbial activity occupy such a crucially important place in food manufacturing and handling practice. Unwanted microbial activity in food may result in deterioration of the quality or appearance, or the production in the food of compounds which have toxic effects after ingestion. The overt spoilage of food and the consequent wastage and economic loss are obviously undesirable. However, toxin production is the most serious potential consequence of microbial growth in food, since certain microorganisms are capable of producing, under conditions favourable to them, toxins which cause severe illness, even death.

In so far as biotechnology involves the manipulation of micro-organisms such that they act to our advantage, food preservation offers less scope for its application than do the positive aspects of food production. Although some examples of the application of biotechnology to food preservation will be given, this chapter is concerned primarily with the positive role of microbial activity in food production, and the ways in which modern techniques and knowledge can enhance that role.

3.1.2   *The two scales of biotechnology*

If modern biotechnological practice is examined, a natural dichotomy becomes apparent which is related to the scale of the operation and to the value of the product. The differences are shown in Table 3.1. The small scale of biotechnology, with its expensive, specialized product, is not the kind of biotechnology found in the food and drink industry. The reason is one of economics. Food is produced from relatively cheap materials by relatively cheap processes. The value of the final product does not give justification or impetus to the sort of expensive research and development and expensive processing found in, for example,

Table 3.1. The two scales of biotechnology

	Small scale	Large scale
Vessel size	100–1000 l	10 000 l
Value of product	High	Low
Type of product	Medical, pharmaceutical, high specialization	Commodity transformation, low specialization
Main biotechnology research and development	Genetic manipulation	Fermentation technology, process engineering
Cost of research and development	Higher	Lower

antibiotic production. Even the production and use of microbial enzymes in the food industry, though sophisticated technology, are more correctly described as large-scale, low product value biotechnology. Anyone acquainted with practical biochemistry knows that it is possible to produce, often with considerable time and effort, enzymes of high purity and specificity. Such expensive preparations, however, do not find their way into food processing, where the enzymes employed are crude, and purified only to the extent necessary to enable them to be handled in relatively small volumes. An exception is the case where an enzyme is attached to a solid support, over which the substrate is passed. In this continuous reactor system the enzyme is effectively reused many times, and this justifies the use of highly purified enzyme preparations. That is not to say that there is not scope for biotechnological development in the food industry, but simply, and importantly, to point out that there are limits to the acceptable cost of a food process which restrict its likely complexity and demand a high yield: that is why larger-scale commodity transformation tends to dominate biotechnology in food production.

3.1.3 *The scope of biotechnology in the food and drink industry*

The range of food products to which microorganisms make a contribution is wide, from the most ancient of traditional fermented foods, such as bread, cheese, yoghurt, wine and beer, to the newest of food products, MycoProtein. The role played by microorganisms may include the use of microbial enzymes or other metabolites, the many food fermentation processes, and the cultivation of specific microorganisms as food products. The microorganisms employed in food production may be added to the process as specially prepared pure cultures, or may occur in sufficient numbers in the raw material to begin activity under the correct conditions. The latter case applies particularly to traditional food fermentations, developed in the complete absence of microbial knowledge. In the industrialized world, such processes have been mostly brought under much greater degrees of control, especially with regard to the strains and purity of microorganisms used.

The contribution of biotechnology to food manufacture has been to provide improved practices and greater sophistication in the use of microorganisms. The opportunity for the future is for further improvements in microbial strains for particular jobs, by genetic manipulation, and for further developments in fermentation technology, so that we may improve the yield and quality of existing food products and also provide some new ones.

3.2   Dairy products

3.2.1   *General points*

A major food fermentation is that of milk products. Milk fermentation, usually caused by streptococci and lactobacilli bacteria, generally causes the breakdown of lactose to lactic acid. Other reactions which may occur, either during the main fermentation or post-fermentation reactions, produce distinctive fermented milk products. These include butter-milk, sour cream, yoghurt and cheese. The final product depends on the character and intensity of the fermentation reactions. Those reactions which occur in addition to the main lactic acid fermentation often produce the individual nature of the product. For example, the secondary fermentations which occur in ripening cheeses are responsible for their individual characteristics. Some of the reactions occur with peptides, amino acids and fatty acids present in the product.

There are six major fermentation reactions which may occur in milk: lactic acid, propionic acid, citric acid, alcoholic, butyric acid and coliform-gassy. Of these, the lactic acid is the most important and occurs in all milk fermentations. Lactose present in the milk is hydrolysed to galactose and glucose. The galactose is generally converted to a glucose derivative before fermentation. The bacteria present in the milk ferment the glucose to lactic acid. The formation of a curd by casein at its isoelectric point (pH 4.6) by lactic acid is important in cheese production (see section 3.2.2).

The propionic acid fermentation is important in Swiss cheese production as the propionic acid and carbon dioxide formed lead to the typical cheese flavour and eye formation. The particular flavour associated with butter-milk, sour cream and cream cheese is due to a citric acid fermentation. The flavour results from a balance of diacetyl, propionic and acetic acids and other related compounds.

Milk products in which alcohol fermentation has occurred are not widely known in Europe and America. This fermentation occurs in milks used in Russia but it is considered a defect in other products and growth of the organism responsible (*Torula* yeast) is usually prevented. Both butyric and coliform-gassy fermentations are detrimental.

Milk fermentations are now carried out under controlled conditions. However, they have been occurring for many thousands of years, being produced by the indigenous bacteria present in milk. Nowadays, dairy starters are used to produce predictable qualities and characteristics in the various milk products. The cultures of active bacteria used may be of one strain of one species of microorganism (a single-strain culture) or a number of strains and/or species may be used (multi-strain or mixed-strain culture). Commercial starter cultures will consist of lactic acid and aroma bacteria. Table 3.2 shows some of the bacterial cultures used

Table 3.2. The functions of some bacteria used in milk fermentations

Culture	Function	Use
Propionibacterium		
P. shermanii	Flavour +	Swiss cheese
P. petersonii	eye formation	
Lactobacillus		
L. casei	Acid	Ripening
L. helveticus	Acid	Starter for Swiss cheese
L. bulgaricus	Acid	Swiss cheeses
L. lactis	Acid	
Leuconostoc		
L. dextranicum	Flavour substance	Sour cream,
L. citrovorum	from citric acid	butter-milk, butter
	(diacetyl most important)	and starter culture
Streptococcus		
S. thermophilus	Acid	Yoghurt + Swiss cheese
S. lactis	Acid	Cheese starter cultures
S. cremoris	Acid	

in fermented milk products, their major function and product use. The selection and combination of these bacterial strains and species will depend on product criteria or requirements, e.g. the rate of acid formation depends on the product in question.

The following sections will deal with the various fermented milk products, current practice and future development.

3.2.2   *Cheese*

Cheese production is one of the oldest fermentations known; it is a method of preserving the nutritional value of milk, and has been described by both Greek and Roman writers. Few varieties have been developed deliberately; most have arisen by chance. Cheeses vary from the very soft to the hard grating, the difference being that all soft natural cheeses contain large amounts of water (55–60%), whereas hard cheeses contain less (13–34%). Although cheese properties vary quite considerably, a number of production steps are common to all. The first is the preparation and inoculation of the milk with lactic acid bacteria. This is followed by curdling of the milk, which usually requires the enzyme rennin. After separation from the watery liquid (whey), the resulting curd is cooked and then pressed into forms, after which it is salted and allowed to ripen.

Inoculation

In the past, cheese makers depended on the bacteria naturally present in milk. However, the presence of desirable and undesirable organisms led to variations between batches.

The modern cheese maker prefers to use milk with the lowest practicable microbial content and uses starter cultures containing the desired balance of bacteria to give consistent quality. The types of organisms used will be determined by the temperature of the cooking stage (see Table 3.3).

Lactic acid formation in cheese production serves the following important functions:

(1) promotion of curd formation by rennin;

(2) promotion of whey drainage due to curd shrinkage;

(3) prevention of undesirable growth of microorganisms during manufacture and ripening;

(4) modification of curd elasticity and promotion of the fusion of the curd into a solid mass;

Starter	Curd cooking temperature (°C)
S. lactis or S. cremoris	> 38
S. thermophilus + S. lactis	42–46
S. thermophilus + L. bulgaricus or L. helveticus or L. lactis	49–54

Table 3.3. Starters used as a function of curd heat treatment in cheese production

(5) modification of the nature and extent of enzymatic changes during ripening, helping to determine the cheese characteristics.

Acid production at a specified rate for particular cheeses is dependent upon the use of a reliable starter. Recent advances in starter production include the development in concentrated form of deep-frozen or freeze-dried starters for direct inoculation into the cheese milk. The problem with dried cultures is to avoid damaging the cell. Dairy starter production has progressed from milk-based, phosphate-buffered media used in the 1960s to whey-based media. Following this, high-buffered media were used, which prevent the pH falling below 5.2. Another development is controlled bulk starter media, which maintain pH at 6.2.

Coagulation

After inoculation, rennet is usually added to produce the curds and whey. The function of the rennet is to convert liquid milk to a gel (the curd). Cheeses set without rennet extract form a calcium paracasein curd. The difference in the two methods of curd formation is that in the absence of rennet, curd formation may take up to 16 h, whereas with rennet curd formation occurs in only 15–30 min. The time interval between the addition of the starter culture and rennet addition is known as the milk ripening period and may last up to 60 min, depending on the particular cheese. The ripening period allows the bacteria to produce sufficient lactic acid to activate the rennin.

Rennet is obtained from the stomach of calves. During the last twenty years, shortages developed and this stimulated interest in rennet substitutes. The function of the rennin can be emulated by most proteinases. However, the enzyme rennin, in addition to coagulating milk, contributes to proteolysis in the cheese during ripening. Any rennet replacer must, in addition to coagulating the milk, produce the required proteolytic reactions. Most rennet substitutes are more proteolytic than rennet relative to their clotting activity; this excessive proteolytic activity can reduce cheese yield and fat retention and, in addition, may produce undesirable effects on flavour of the finished cheese due to the products formed by proteolysis. It is known that bitter flavours may arise from the presence of certain peptides that will be produced where cheeses undergo long ripening times. Short ripening and mould cheese do not suffer to the same extent as it is probable that insufficient proteolysis occurs during the ripening period. Many proteinases are capable of clotting milk and a large number of proteinases from plant, animal, bacterial and fungal sources have been examined as potential rennet substitutes. Most plant coagulants have proved to have too high a proteolytic activity. The most successful substitutes have been produced from microbial sources and are generally available commercially in America and Europe.

Two rennet substitutes which are currently available commercially are produced by a complex process from the mould *Mucor miehei*. Substitutes derived from *M. miehei* are currently considered more suitable for most cheeses than those from *M. pusillus*. In the USA, 60% of cheese is made with microbial coagulants. Future developments in this area include the use of genetic engineering to induce bacteria to make rennet. A review on this area is available (Green, 1977).

Cooking and pressing

After coagulation has occurred, the curds are cut, which leads to effective whey expulsion. Then the curds and whey are heated for a fixed time. The whey is drained after cooking, following which the next step permits the accumulating lactic acid to change the curd chemically. This operation is known as the 'knitting and transformation' step and this and the subsequent manipulations are responsible for the characteristic texture of the cheese. Salting and pressing occur after this, as well as any special applications which are necessary.

Special applications are given to individual cheese types and these include the following:

(1) inoculation of blue mould spores for Roquefort or blue cheeses;

(2) spraying of white mould spores on the surface of Camembert and Brie cheese;

(3) bacterial surface ripened cheeses, e.g. Port Salut, St. Paulin.

The characteristic organism in surface ripening is *Brevibacterium linens*.

Ripening

If required, the cheese is then ripened or matured. Cheddar and Swiss cheeses are in this category, whereas cream cheese is not. Ripening occurs by placing the cheese in a temperature controlled room for up to four years. Temperature can vary between 2 and 16 °C, depending on cheese type. Ripening allows the microorganisms and enzymes in the curd to hydrolyse fat, protein and other compounds present. The breakdown of these materials produces the characteristic flavours of the cheeses. Because of the time involved in ripening, major savings can be made by shortening this period. Most methods for accelerating cheese ripening are based on an increased rate of enzymatic breakdown of cheese protein or cheese fat. Cheeses with strong flavours, such as Danish Blue, are easier to work on than those with more subtle flavours; accelerated ripening of cheddar is more difficult. To ensure the natural balance of flavour compounds in the product, careful selection of enzymes is necessary. The production of peptides by the starter bacteria has been referred to earlier. These peptides may produce unpleasant bitter tastes which are detrimental to the cheese flavour. The introduction of extra proteinase will only be beneficial as long as the activity is not exceeded, leading to the unpleasant flavours. The development of suitable microbial peptidases may overcome this problem in the future. One proteinase which gives acceptable results is that derived from *Bacillus subtilis*—neutrase. Maturation time has been reduced from four to two months for a medium-flavoured cheese and from twelve to six months for a mature-flavoured cheese. Provided enzyme addition is at an appropriate level, no bitterness develops. An added advantage of neutrase is that when stored at 6 °C the ripening rate reverts to that of an untreated cheese, which has obvious advantages if cheese demand were to drop suddenly. The effects of lipases on ripening are uncertain, although they are said to accelerate cheddar ripening. A current problem with the use of proteinases added to the milk is that on separating the curds from the whey, the enzyme is lost into the whey. In addition to being uneconomic, the presence of enzymes in the whey creates problems in its disposal. Microencapsulation of the enzyme is one possible solution and this is one area of current investigation.

Shorter manufacturing and ripening times are also associated with the use of lactose-hydrolysed milk (LHM). The conversion of lactose to glucose and galactose is by β-galactosidase (lactase). Lactose hydrolysis can be achieved by *S. lactis* addition to raw milk and allowing up to 70% lactose breakdown. Faster manufacturing and ripening rates are the result of the stimulative effect of glucose on the growth and acid production of cheese starter bacteria. Research workers do not agree on the effect of LHM. Some suggest that although glucose will stimulate starter growth it does not promote accelerated ripening. Although the mechanism of action is uncertain, cheeses made from LHM do ripen faster than normal. Once the mechanism is established it may be exploited rationally.

Another method for acceleration of ripening is to increase the number of starter bacteria. The presence of proteinases and peptidases in the starter will accelerate ripening due to the increased concentration of these enzymes. However, it is important that the rate of acid production is unchanged as it can be critical for particular cheese varieties. Current methods to prevent increased acid production are by genetic modification of bacterial starters by heat shock, sensitization and solvent treatment. Certain bacterial mutants produce more protease or lipase without affecting acid production. Such mutants used in starters will cause more rapid ripening whilst leaving the early production method unchanged. With an increasing knowledge of the genetics of starter bacteria it may be possible to tailor organisms for specific requirements. A major advantage of using modified starters is that, unlike enzyme addition, they are natural constituents. Table 3.4

Table 3.4. Examples of the use of added enzymes to accelerate cheese ripening

| Types of cheese | Enzymes added | | Stage of addition |
	Type	Source	
Cheddar	Acid and neutral proteinases, peptidases, lipases, decarboxylases	Various commercial and animal enzymes	Curd
Cheddar Romano, Parmesan	Lipase, proteinase	Lamb gastric tissues	Milk and curd
Gouda	Proteinase	*Aspergillus oryzae*	Milk or curd
Edam Cheddar	Proteinase + peptidases	*Pseudomonas fluorescens*	Milk
Rossiiskii	Proteinase	Pancreatin	Milk
Mozzarella	Lipases, esterases (microencapsulated)	Calf pregastric secretions	Milk
Blue	Lipase	*Aspergillus* spp.	Curd

Reproduced from Law (1980).

gives some examples of the use of added enzymes to accelerate cheese ripening and further information can be found in the literature.

Whey

The production of whey as a by-product of cheese manufacture has been mentioned previously. Whey has, in the past, been considered an industrial waste material. Recent developments have occurred in whey utilization, and the production of sweeteners by the use of lactase is becoming important. The glucose and galactose produced are intended for use in the UK in baking, confectionery and ice-cream making. The enzyme may be isolated from *Aspergillus niger*. Production of these sugars from whey is occurring in both the USA (Corning/Kroger) and the UK (Corning/Dairy Crest).

3.2.3 *Yoghurt*

This is one of the older fermented products. After heat treatment, the milk is inoculated with 2–3% yoghurt culture. The essential bacteria are *Streptococcus thermophilus* and *Lactobacillus bulgaricus*. For the most desirable consistency, flavour and odour, the two organisms should be present in about equal numbers in the culture. Initial acid production is largely due to *S. thermophilus*. Blended starters require frequent renewal because repeated transfers affect the desirable ratio between the bacterial species and strains. Without renewal, *L. bulgaricus* becomes dominant.

3.2.4 *Butter*

Butter is one of the simplest of dairy products. The cream from milk is concentrated from 30–40% to 30–32%, depending on the desired composition of the final butter. The churning of this cream changes an oil-in-water emulsion into a water-in-oil type.

Selected butter cultures may be used in butter production to enhance its flavour and keeping qualities. Flavour improvements have resulted by the development of special cultures of bacterial species selected for their ability to form desirable flavour compounds. The first cultures used were *Streptococcus lactis* or related types. Later, mixed cultures were introduced which contained *Streptococcus lactis*, *Leuconostoc citrovorum* and *L. dextranicum*.

In addition to improving flavour, certain off-flavours can be masked. The addition of lipases to butter is a potential modification which would enable butter to be spread directly on removal from the fridge.

3.2.5 *Cultured butter-milk*

Cultured butter-milk is prepared by souring true butter-milk or, more commonly, skimmed milk with a butter starter culture. The starter used consists of a mixture of lactic streptococci (*S. lactis* or *S. cremoris*) with aroma bacteria (*L. citrovorum* or *L. dextranicum*). Both types are essential to produce the characteristic flavour and aroma of butter-milk, but the streptococci greatly outnumber the others. The function of the lactic streptococci in the starter is to produce lactic acid, which is necessary to give the desired sour taste, to curdle the milk and to lower the pH to the point where the aroma bacteria produce maximum amounts of volatile acids.

3.2.6   *Cultured sour cream*

This is made in a similar manner to cultured butter-milk. The milk is inoculated with 0.5–1% butter starter and incubated until the acidity reaches 0.6%.

3.2.7   *Novel products*

A modified milk suitable for people suffering from lactose intolerance can be produced by the use of β-galactosidase which reduces the lactose content. An inexpensive industrial process is needed if this is to become a viable process and increase milk consumption. Current sources of the enzyme include yeast, moulds and bacteria.

3.3   Cereal products

3.3.1   *Bread and baked goods*

In the UK, most products are made by the Chorleywood Bread Process, although a wide variety of breadmaking technologies are in use around the world. The dominant yeast strain used in breadmaking remains *Saccharomyces cerevisiae*, usually grown on beet or cane molasses in batch fermentation.

In simple terms, a dough of flour, water, yeast and salt is prepared and mixed at room temperature. During this process, the constant folding of the dough creates nuclei for gas cell production and growth. The mixed dough is allowed to rest, and is then divided into pieces of the desired weight, moulded (a process of rolling and folding to enhance the texture) and allowed to prove in a humid atmosphere. During the proving stage, and the first part of baking which follows, the gas cell nuclei induced during mixing and moulding fill with the carbon dioxide produced by the anaerobic fermentation of flour glucose and maltose. The expanded structure is baked, a thermal process which gelatinizes the starch, inactivates the yeast and drives water from the dough: the result of these actions is a stable rigid product with a light cellular texture and a flavour-enriched crust.

In addition to the production of carbon dioxide, the anaerobic fermentation produces a variety of organic acids, alcohols and esters which contribute significantly to flavour development in the food.

In Germany and the USA, sour breads are produced from rye flour; the basic technology of production remains as for wheat flour, but the product incorporates a pre-fermented mixture of rye flour and water, inoculated with a mixed *Lactobacillus* culture. The acid products of the pre-ferment or 'sour' donate a unique taste to the bread.

3.3.2 *Starch hydrolysates*

The production of the three major starches of commerce is effected by conventional technology. Manufacture of starches from maize, wheat and potato has been comprehensively reviewed (see 'Recommended reading'). The hydrolysis of these starches on an industrial scale is performed by a variety of techniques employing all-acid, acid-enzyme and all-enzyme processing. The development of relatively sophisticated industrial technology from the crude mineral acid processes of the first quarter of the century represents a well-documented example of biotechnological development in response to consumer and industrial demand.

By the mid-1960s, all-enzyme processing of starch, using sequentially α-amylase from *B. subtilis* and amyloglucosidase from *A. niger* or *A. oryzae*, was being introduced to replace the all-acid and acid-enzyme processes. The major benefits were rapid processing, with little contamination by reversion products, and the ability to produce materials of high dextrose equivalent (D.E.). The D.E. of an hydrolysate (a scale in which glucose (dextrose) represents 100) is a measure of the extent of hydrolysis from starch, which is assumed to have a D.E. of 0.

The next major development was the introduction of thermostable α-amylase preparations, notably from *B. licheniformis* and *B. amyloliquefaciens*, capable of operating at temperatures above 100 °C and able to give materials with D.E. values approaching 100, after treatment with amyloglucosidase.

High-temperature liquefaction of starch is now used almost universally in industry, but the second stage, saccharification, is still conducted using 'traditional' *A. niger* glucamylase batch digestion. Proposals for the use of immobilized enzymes have been made, but they have not been introduced to any great extent.

In addition to glucose manufacture, the most notable development in this area of industry has been the introduction of glucose–fructose mixtures, called variously isoglucose or high-fructose corn syrup (HFCS). The industrial potential of this technology has been most effectively expressed in the US, where there is an abundance of starch-bearing crops and little indigenous sugar; HFCS is able to replace sucrose in most food applications. Isoglucose production in Europe has been retarded by fiscal barriers.

The starting material for HFCS is deionized starch hydrolysate, produced by the method described above, and with a D.E. as near to 100 as possible. Isomerization is effected by enzymes from a variety of organisms, each having attractions in its ease of use, freedom from and reliance on cofactors, and stability; the organisms most frequently used are *Streptomyces* spp., *B. coagulans*, *Actinoplanes missouriensis* and *Arthrobacter* spp. After isomerization, further processing can give materials containing up to 90% fructose, although the initial product is the reaction equilibrium mixture of glucose (51%), fructose (42%) and oligosaccharides (7%).

By the use of various amylases, transglucosylases and debranching

Fig. 3.1. Stages in starch hydrolysis.

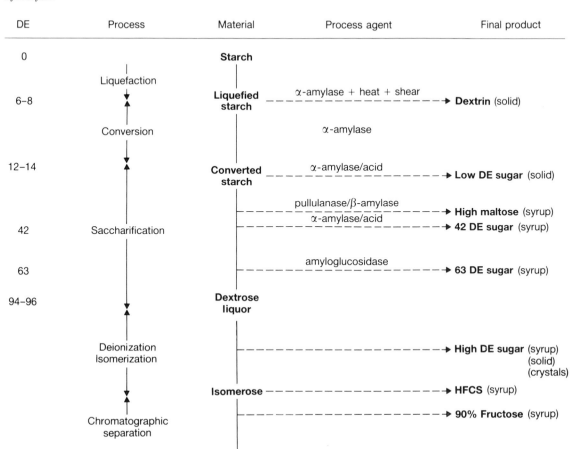

DE	Process	Material	Process agent	Final product
0		**Starch**		
	Liquefaction			
6–8		**Liquefied starch**	α-amylase + heat + shear	**Dextrin** (solid)
	Conversion		α-amylase	
12–14		**Converted starch**	α-amylase/acid	**Low DE sugar** (solid)
			pullulanase/β-amylase	**High maltose** (syrup)
42	Saccharification		α-amylase/acid	**42 DE sugar** (syrup)
63			amyloglucosidase	**63 DE sugar** (syrup)
94–96		**Dextrose liquor**		
	Deionization Isomerization			**High DE sugar** (syrup) (solid) (crystals)
		Isomerose		**HFCS** (syrup)
	Chromatographic separation			**90% Fructose** (syrup)

enzymes, it is possible to manufacture a wide variety of starch hydrolysates having a range of industrial applications. The interrelations of raw material, intermediate product and finished product are shown in Fig. 3.1.

3.4   Brewing

3.4.1   *Alcoholic beverage production*

The production of beverages by alcoholic fermentation is one of the oldest fermentations known. Beer and wine were probably the two earliest alcoholic drinks produced. Little was known of the actual processes and mechanism of production until Pasteur's work in the late nineteenth century. He showed that living yeast cells caused fermentation in the absence of air, converting sugar into ethanol and carbon dioxide. Work later in the nineteenth century showed that the fermentation resulted from the action of substances contained within the yeast cells. One of the major discoveries of fermentation microbiology was made by Hansen at the Carlsberg centre in Copenhagen while working on 'wild' yeast. These wild yeasts were known to give problems during beer fermentation and, by isolating pure cultures of yeast, which he then used in the brewing process, Hansen initiated the use of pure cultures in beer production.

Alcoholic beverages are produced by the alcoholic fermentation of a sugar-containing material to ethanol and carbon dioxide. The fermentation is carried out by species of the yeast *Saccharomyces*. In some cases sugar is present naturally, such as in grapes used in winemaking; in others, sugars are produced from starches in cereals, as in beer production. The presence of free sugar is essential for alcoholic fermentation by *Saccharomyces* as the species is not capable of hydrolysing polysaccharide material. The production of ethanol from glucose occurs via the Embden–Meyerhof–Parnas pathway (Fig. 3.2).

The yeasts used in the manufacture of alcoholic beverages are strains of *Saccharomyces cerevisiae* or *Saccharomyces carlsbergensis*. The definitive difference between the yeasts is that *S. carlsbergensis* ferments raffinose completely, whereas *S. cerevisiae* does not.

3.4.2   *Beer*

The production of sugars from brewery raw material is the first requirement in the alcoholic fermentation. Barley has been the traditional source of polysaccharides but other materials known as 'adjuncts' (also carbohydrate-containing materials) are also used.

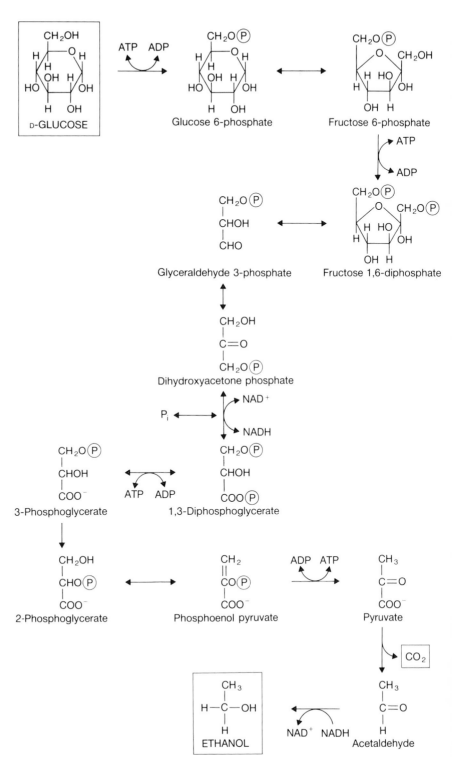

Fig. 3.2. The Embden–Meyerhof–Parnas Pathway (after Rose, 1977).

Malted barley is still the major component of beer. Malted barley and other adjuncts are crushed and then mashed with water at temperatures up to 67 °C. The mashing process allows the natural indigenous enzymes in the malted barley to degrade the carbohydrates present in the grain. The separation of the soluble from the insoluble spent grain completes this process. The soluble material is known as 'wort'. This material is boiled in a copper with hops. To produce a beer of a known alcohol content, the wort (after boiling) is adjusted to an appropriate specific gravity. The specific gravity of the wort is dependent upon the sugars extracted and available for fermentation. At an appropriate time when the fermentation is complete, the yeast is separated from the beer, which is then allowed to mature for a suitable period. After filtration and other treatments the beer is now ready. Fig. 3.3 shows diagrammatically the operations performed.

The use of selected yeast strains for beer production, initiated by Hansen, is now common. The two cultures used are *S. cerevisiae* and *S. carlsbergensis*. The latter is a bottom fermenting yeast and is used for lager production, whereas the former is both a top and bottom fermenting yeast. Top fermenting species are used for ale production.

Although yeast genetics has been studied for many years, it is only recently that the possibility of breeding yeasts for brewing purposes has been exploited. As more becomes known about the properties of yeasts and the characteristics that they give to the final products, it is possible to carry out strain selection on existing brewery yeasts and choose one that will yield an ideal product. The requirement for ideal yeasts will vary according to the fermentation system used and the beer characteristics required. Some of the important characteristics are flocculation properties, haze level and ability to ferment maltotriose, etc. Taste panel evaluation results are also important so that appropriate flavour components in beer are produced. In the past, strain selection from existing brewery yeasts has been the main source of strains yielding particular characteristics. Strain selection has been more profitable than hybridization, partly due to the poor sporulating ability of brewer's yeast and the low viability of the ascospores which do form. Of the one to four ascospores formed in each ascus, not all are released when the asci reach maturity. *Saccharomyces* yeasts reproduce vegetatively, producing spherical, ellipsoidal or, more rarely, cylindrical cells by multi-lateral budding. As development in brewery technology may require yeasts with different properties from traditional yeast, hybridization may be necessary. The major impact of biotechnology in the brewing industry will be on the production of yeast strains that will give desired beer characteristics.

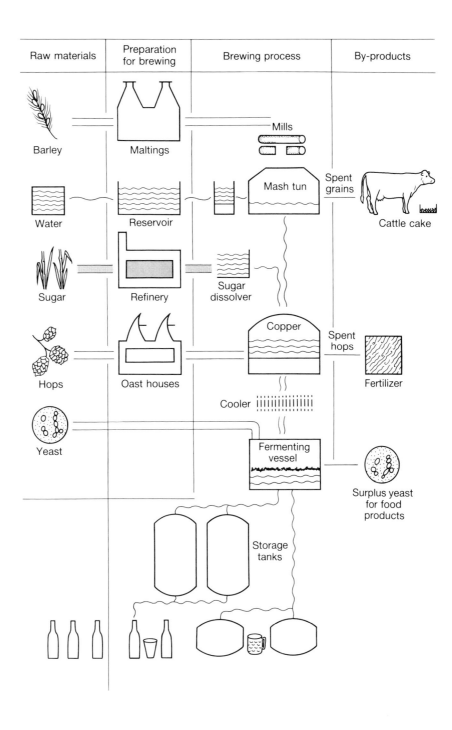

Fig. 3.3. Schematic outline of operations carried out in beer production. Reproduced by courtesy of Dr Anna M. MacLeod.

3.4.3 *Wine*

All alcohol fermentations require a sugar substrate and for wine production this sugar is obtained from grape juice. Nearly all of the world's wine is made from one species of grape, *Vitis vinifera*. Wine grape juice is a perfect medium for wine production because of its high nutrient concentration, a natural acidity which inhibits undesirable microbial growth, high sugar concentration and the production of pleasant aromas and flavours.

Wine production, unlike beer production, has relied until recently upon endogenous yeasts in the environment for the fermentation. The only pretreatment was the addition of sulphur dioxide to the grapes before crushing to inhibit browning. Additionally, sulphur dioxide tends to inhibit the action of non-wine yeasts, allowing the wine yeasts, which are more tolerant to sulphur dioxide, to carry out the necessary fermentation. The stems, seed and skin remain in contact with the fermenting 'must' for red wine production until the end of the alcohol fermentation whereas for white wine production only the juice is used. Sulphur dioxide is usually added before crushing, but can sometimes be added later.

Table 3.5 shows some of the more commonly found yeasts on grape skin. In the past it was the naturally occurring yeasts that carried out the alcohol fermentation. However, in newer wine-growing areas the use of yeast starter culture is quite widespread. One of the reasons is that the desirable microflora may not be present and hence inoculation with a standard yeast culture will ensure production of a wine with required characteristics. In addition, legislation restricting the level of use of sulphur dioxide will encourage the use of yeast starter cultures. The winemaker will be reluctant to depend on the natural wine yeasts if the competition from non-wine yeasts would no longer be inhibited. There are a number of advantages using starter cultures compared with spontaneous fermentation: the lag period of the yeast growth cycle is shortened and the fermentation yields a product of known characteristics. There is also a decreased possibility of off-flavour development since fermentation by the wild non-wine yeasts does not occur. The use of specific strains will become increasingly common in order to ensure the required flavour characteristics. Mixed starter cultures can give flavourful end-products, which pure strains cannot do.

The grape variety is responsible for some flavour compounds. For example, the flavours of the Muscat varieties are attributed to the terpene alcohols, linalool and geraniol. Other flavours are formed during the ageing process, the interaction with wood and air during storage in wooden casks being the most well known.

After the alcohol fermentation, the new wine must be stored under suitable conditions to prevent spoilage. Sulphur dioxide is added to prevent oxidative browning if the wines are not expected to undergo a malo-lactic fermentation. Before addition of sulphur dioxide, the wine is separated from the yeast to stop further fermentation.

Premium wines undergo various ageing treatments, depending on the wine type, whereas the less expensive wines are bottled early, usually within the same year of production. A particular problem with

Table 3.5. Some yeasts found on grapes and in musts and wines (after Kunkee & Goswell, see Rose, 1977)

Genus and species	Where reported
Candida pulcherrima (now *Metschnikowia pulcherrima*)	Common in musts and wines; on grapes
C. scottii (now *Leucosporidium scottii*)	In Czechoslovakia; on green grapes from New Zealand
Debaryomyces phaffii	On grapes in Cognac region
Endomycopsis lindneri	On grapes and in must in Brazil
Hansenula saturnus	In wine made from late-harvested grapes
Kloeckera africana	Fairly common on grapes and in musts and wine
K. apiculata	Very common
Pichia fermentans	On grapes, and in musts and wine
Saccharomyces acidifaciens (now a variety of *S. bailii*)	In spoiled wines and on grapes; resistant to sulphur dioxide (a fructophile)
S. bayanus (now includes *S. oviformis*)	On grapes and in musts and wines
S. bisporus	On grapes; low alcohol yield
S. carlsbergensis (now *S. uvarum*)	On grapes and less often in wine (used for lager beer fermentation)
S. cerevisiae	The classic wine yeast, and possibly the most widely distributed
S. chodatai (now *S. italicus*)	On Georgian (USSR) grapes
S. coreanus	Rare, from grapes
S. elegans (now *S. bailii* var. *bailii*)	On grapes and in wine (a fructophile)
S. elegans var. *intermedia* (now *S. baillii* var. *baillii*)	On grapes in Brazil
S. globosus	On grapes and grape juice and wine, rare
S. heterogenicus	On grapes, in musts, and in grape juice, but rarely
S. italicus	On grapes and in grape juice from warm climates
S. oviformis (now *S. bayanus*)	On grapes, in musts, grape juice and wines; common
S. pastorianus (now *S. bayanus*)	In grapes, in musts and frequently in Loire fermentations
S. rosei	Common on grapes and in wines
S. rouxii	From overripe grapes
S. steineri (now *S. italicus*)	In wine and on grapes
Torulopsis bacillaris (now *T. stellata*)	On grapes and in wine (a fructophile)
T. burgeffiana (now *M. pulcherrima*)	From grapes and in musts

the cheaper wines is their susceptibility to malo-lactic fermentation, which tends to occur at bottling time. If wine is susceptible to this fermentation it is encouraged before bottling and if unlikely the fermentation is inhibited. In the premium red wines such a fermentation is desirable, expected and usually occurs during storage. The bacteria involved in this fermentation include strains of the lactic acid bacteria, *Leuconostoc*, *Lactobacillus* and *Pediococcus*. Induction of the malo-lactic fermentation does not occur under low pH conditions and this technique may be used to inhibit the fermentation. It occurs less frequently in white wines due to their lower pH value. Further developments on the induction of this fermentation include the use of solid support enzymes rather than bacteria.

A reaction which has led to specialized wine types such as Sauternes is that caused by the fungus *Botrytis cinerea*. Infection by this fungus leads to an increase in the sugar concentration caused by dehydration of the grapes; this is responsible for the sweetness of the wine. Infection other than at harvest time is not desirable. Another process of interest is the carbonic maceration process. Red wines that are to be ready by 15 November in the year of production do not undergo a standard process. The grapes are not crushed but placed whole in a fermentation tank and covered with carbon dioxide. The fermentation either occurs inside the grape as an anaerobic fermentation or grape juice seepage is caused by the disruption of grape skin cells by carbon dioxide. No microbiological studies have been made on this process.

Fortified wines are those in which some of the alcohol is derived from the yeast fermentation of grapes and some from the addition of distilled spirits. They include port, sherry and madeira (see section 3.4.4).

Future developments for the wine industry will include the production of more sophisticated wine yeast strains and commercial yeast starter cultures and, therefore, of wines with particular characteristics.

3.4.4   *Spirits*

Although steeped in history, the production of distilled spirits is newer than that of non-distilled alcoholic beverages. Distillation is necessary to produce a beverage with an alcohol content of 40% by volume. Modification of the still designed by Coffey in 1830 is the basic design for stills today and is known after him. The difference between spirits tends to depend on the raw material used and whether or not the final product is matured. Strains of *Saccharomyces* suitable for distillery fermentation are used and large distilleries maintain their own selected yeasts on culture medium. The selection of distiller's yeast strains depends on an optimum performance under the specific conditions of distillery worts, which should have a good fermentation vigour and be capable of a yield of alcohol approaching the theoretical maximum. Although the choice of raw material can be varied, certain spirits are usually produced from particular materials. Brandy, a distilled wine product, is manufactured from grapes and malt whisky from barley. Products such as grain whisky, gin and vodka, although usually produced from grain such as maize, may be manufactured from other suitable materials. Rum is usually made from either cane or beet molasses. When cereals such as wheat and maize are used, the starch must be hydrolysed into sugars before fermentation can occur, and for this reason whisky is considered to be the distillation product of an unhopped beer. The early stages of whisky production are identical to those for the preparation of worts in beer making. However, when maize and other similar cereals are used, the mashing is preceded by a cooking process, when the starch in these grains is exposed to the malt enzymes in the mash tun.

The use of molasses as a raw material for spirit production does not require such preparation, as the sugars are present in the correct form for fermentation. Certain pretreatments are necessary, however, and these include clarification and heating and finally dilution with water to a suitable sugar concentration for fermentation. After preparation of the raw materials, a suitable yeast strain can be added and fermentation occurs. At the end of the fermentation period, when the ethanol content is between 9 and 11% (v/v), the alcohol is separated from the yeast, usually by allowing the fermented mixture to stand so that the yeast cells can settle out. The remaining material may be separated by centrifugation before distillation. Two distillation procedures are used, the pot still method and the continuous distillation system (the Coffey patent). After distillation, the product is either stored to mature (e.g. whisky) or finished and bottled directly (gin and vodka). Spirits are usually diluted to be sold at the standard alcohol content of 40% (v/v). Some of the more important individual differences in spirit production are outlined below.

The selection of a suitable yeast strain is important. In rum production, *Schizosaccharomyces* strains are usually used for the heavy aroma rums and the quick-fermenting *Saccharomyces* are better for the lighter rums. In addition, the bacterium *Clostridium saccharobutyricum* accelerates the formation of alcohol during fermentation by yeast and the best rum yield is obtained when the ratio of bacteria to yeast is 1:5.

The bacteria are added when the ethanol concentration is between 3.5 and 4.5% and the sugar content is 6% (w/v).

Yeast strains used for spirit production must have the ability to withstand ethanol concentrations of 12–15% (v/v). In addition, when the raw material is a grain source, the yeast must have the capacity to hydrolyse lower oligosaccharides to glucose in order to maximize starch conversion to ethanol and carbon dioxide.

The distillation step is an expensive part of the process due to the high energy requirement. Yeasts tolerant of higher alcohol concentrations would reduce the energy consumed during this stage, as a result of the removal of lower volumes of water to achieve the same final alcohol concentration. Further developments of yeast strains tolerant of high ethanol concentrations are expected.

Fermentation will not commence until the carbohydrate is converted to a suitable substrate for the yeast. The addition of starch-degrading enzymes will hasten this process and amylase, produced in bacterial culture by *Bacillus subtilis* strains, and amyloglucosidase, derived from fungal culture of strains of *Aspergillus niger* or its variants, are often used. Enzyme treatment can occur in both beer and spirit production (see section 3.3.2 for further information on this subject).

3.4.5   *Cider*

The fermented juice of the apple is commonly known as cider. Cider and wine production show certain similarities.

In cider making the apples are initially milled to a fine pulp from which the juice is then extracted. Machines used are either of a pressure or non-pressure type. Juice preparation before fermentation varies considerably, from no treatment to the elimination of the natural microflora and substitution by suitable yeast strains. The most common treatment, sulphur dioxide addition, inhibits *Kloeckera apiculata*, which is detrimental to the final flavour. Fermentation can now occur either using the natural yeast flora or by the addition of yeast starter cultures. The qualities required in yeast for cider making are similar to those in other fermentations and include complete fermentation, flocculation ability and rapid fermentation rate. As the natural growth of *Saccharomyces* is slow, the large cider manufacturers tend to add pure yeast cultures of their own choosing to the sulphated apple juice. Different yeast strains produce aroma compounds and in the same way that brewers may choose yeasts to yield beers with particular flavours, the cider producer may do the same. For the production of a particular cider, the yeast used must predominate over the naturally occurring yeasts in the fermentation so that it will outgrow these and impose its own characteristics. The establishment of a dominant yeast is probably the most widely used method of inoculation among larger cider makers. A particular characteristic necessary in a yeast is the production of polygalacturonase, to effect the degradation of de-esterified pectin to galacturonic acid. If this does not occur, the cider will not clarify at the end of fermentation. The pectin-destroying enzymes, including poly-galacturonase, which may be added are usually of fungal origin (see section 3.7.3).

Sulphite addition in cider making is similar to that for wine production as it is possible to prevent development of all but fermenting yeasts. If sulphiting and strict cleaning procedures are not maintained, 'wild' yeast cells may develop and overgrow the added culture. Other requirements in choosing particular yeast strains include the need to produce the desired flavour compounds in the product, which may be termed as high or low 'fusel' cider. Particular yeast strains can be chosen to give the required product specification.

After fermentation, the cider is removed from the yeast and clarified. The presence of malic acid makes the product susceptible to the malo-lactic fermentation, as with wine. The fermentation will not occur if the cider is extremely acid or kept cold.

As in the case of beer production, advances in this area are likely to involve the development of yeast strains that can give the product desirable characteristics as far as the consumer is concerned.

3.4.6 *Vinegar*

Vinegar, although not an alcoholic beverage, has been included in this section because one of its two production steps includes an alcoholic fermentation.

Vinegar is defined as a product containing not less than 4% (w/v) acetic acid. Two fermentations are necessary; the first is an alcoholic fermentation which converts sugar-containing material (the substrate can be any material that can be fermented to yield alcohol) to alcohol using *S. cerevisiae*. After the alcohol fermentation is complete, the yeast is allowed to settle and the supernatant removed. The alcohol content is adjusted to 10–13% and, if sulphur dioxide has been added to control bacterial growth during the alcohol fermentation, it must be removed before the next stage—the conversion of ethanol to acetic acid via acetaldehyde. All vinegar processes require mixed *Acetobacter* micro-flora and some processes require inoculation. The organisms usually associated with this stage are *A. schuetzenbachii*, *A. curvim*, *A. orleanense* and related organisms. This step is an aerobic fermentation which demands a high oxygen requirement and generates considerable heat. In the past, the second stage was slow but improvements have been made to increase the conversion of ethanol to acetic acid. The submerged fermentation technique could not be adapted to vinegar production until after World War II when antibiotic studies on aeration had been undertaken. Nowadays, this technique is employed extensively.

3.5   Protein products

3.5.1   *Traditional fermented protein foods*

The use of microorganisms in the preparation of protein foods has a history which pre-dates the origins of microbiology. The major products concerned are cheese, in its various forms, and fermented soya bean products. In both of these groups of food, protein is the main nutritional component, and their preparation is described in sections 3.2 and 3.7 of this chapter. They are examples in which microbial activity is used to alter substantially the nature of the proteinaceous raw material to give a food product which may be stored for a longer period (cheese) or which is in a more acceptable form (soya bean curd). In some preserved meat products, microbial activity plays a part. For example, some types of sausage, such as bologna and salami, undergo an acid fermentation during their curing. This generally involves mixed lactic acid bacteria and the resultant acid product enhances keeping quality and contributes to the characteristic flavour. Similarly, the brining of meat (another preservation-based practice) involves some activity by acid-producing bacteria (cf. section 3.7.1 on preserved vegetables). Also, there are various kinds of oriental fermented fish products, involving the growth of moulds or yeasts.

These are all examples of protein modification by microorganisms. Apart from cheese, discussed in section 3.2.2, the opportunities for the application of modern biotechnology to the processes seem limited. The more sophisticated application of biotechnology is in the growth and harvesting of microbial mass which then becomes the protein food product.

3.5.2   *Single cell protein*

The nutritional value of microbial material can be quite high with respect to many important factors; not the least is protein, which represents a large proportion of the cell dry weight of most species. The use of microbial protein to contribute to the world's protein supply has been a subject for discussion and experimentation for several decades. The route for its use may be either indirect, as a protein component of animal feedstuffs, reducing the requirement for such materials as soya bean meal and fishmeal, or as a material for direct inclusion in the human diet. Proteins for animal feeds are discussed in Chapter 9, and we will concern ourselves here chiefly with microbial mass as a directly consumed foodstuff.

To distinguish it from proteins originating from higher, multicellular plants and animals, microbial protein has been termed 'single cell protein' or SCP. Its production depends upon the large-scale growth of suitable microorganisms, and their subsequent harvesting and processing to a food product. The central operation is fermentation technology, a specialization born of brewing and antibiotic production, and demanding particular interdisciplinary skills when the object is the optimum conversion of substrate to microbial mass. Growing microorganisms for food has two main attractions. First, the growth rate of microorganisms is very much faster than those of animals or plants, the doubling time being measured in hours, thus potentially shortening the time needed to produce a given mass of food. Second, a range of materials may be considered as suitable substrates, depending on the microorganisms chosen. The two chief strategies with regard to substrate are to consider low grade waste materials or to use readily available carbohydrate to produce microbial material containing high quality protein. In all cases, fermentation technology is of major importance, including the specification of medium formulation and growth conditions, the design and operation of a suitable fermentation vessel and associated control systems, and the separation of the cell mass from the fermentation broth. On a commercial scale, SCP is produced invariably in submerged liquid culture, and batch or continuous culture techniques may be used. Continuous culture, though much more demanding than the batch method, offers considerable advantages in terms of overall productivity of the fermentation system, and has been the chosen method of production in commercial SCP systems.

Continuous culture

Continuous culture is a technique which depends upon the maintenance of a dynamic steady state. In a mixed, submerged culture of constant volume, this means a constant growth rate of the microorganisms, maintained by a constant rate of volumetric dilution of the culture by the addition of growth medium. Media used in continuous culture are invariably designed such that one substrate (frequently the carbon source) is growth limiting, and will be found in minimal concentration in the culture supernatant. The technique has found much application as an experimental tool in the field of microbial physiology. Even in the microbiology laboratory, where the handling of pure cultures by aseptic techniques is everyday routine, the practice of continuous culture requires particular care and attention to apparatus design and operating technique if microbial contamination is to be avoided. The importance of asepsis in continuous culture becomes apparent when it is realized that the technique is of value either for experimental or commercial purposes, only if the desired steady-state conditions can be maintained for days, weeks or even months, without interruption. During this period, the culture must be stirred and supplied continuously with sterile growth medium and air, and culture will be removed continuously. The temperature must be held constant and it is common also to control the pH value. There will also be regular sampling to check the constancy of conditions and purity of the culture. On a commercial scale, this requires highly specialized plant, bearing in mind also that it must be able to withstand initial sterilization procedures before each production run. This later demand applies also to any sensor or probe fitted into the vessel for the purpose of monitoring or controlling parameters of the culture.

Safety considerations with SCP

In traditional food fermentations, the microorganisms involved often become incorporated as part of the final food product (the proportion of the total being fairly small). By virtue of long established practice and custom, the safety of these products is assumed. The differences with SCP, however, are firstly that the food product consists virtually entirely of microbial mass and secondly that the microorganisms chosen are likely to have no history of usage or accepted occurrence in food. Understandably, therefore, government regulatory bodies concerned with food demand that the introduction of a SCP food be backed by a great deal of data based on a wide range of testing of the new product for safety. This exercise is inevitably very expensive, and constitutes a definite limitation on the degree of development which has occurred with SCP products, particularly for direct human consumption. As a consequence, the emphasis in SCP development has been more towards products for animal feed use rather than towards proteins for direct food use. As well as the fact that the safety and testing requirements are less stringent for materials intended for animal feeds, aesthetic factors are less important, allowing the consideration of a wider range of potential substrates, including organic waste materials. Examples of SCP products which have reached commercial production as animal feed ingredients are ICI's Pruteen (a bacterial product grown on methanol), BP's Toprina (a yeast grown on n-alkanes) and a fungal product of the Finnish Pekilo process. This last process uses as a substrate sulphite liquor, a waste product of paper manufacture. All of these SCPs are produced as lightly coloured, powdery materials.

SCP products for human use are few in number. An established microbial product is yeast extract (hydrolysed brewers' yeast), which is not regarded as a protein food, but is used in small amounts for its flavour and vitamin content. Also, it is a by-product of brewing rather than the primary product of a SCP process. During the Second World War, *Candida* yeast was grown in Germany as a protein food, but the practice has not subsequently become established. In more recent times, Hoechst have developed a 90% protein isolate prepared from methanol-grown bacteria. This protein material is prepared as one of a number of products of fractionation of the harvested bacterial cells. It is said to possess desirable functional properties and is intended for use as a food ingredient. The Hoechst process results in a group of high added value products and thus falls more into the small-scale category of 'Biotechnology' in Table 3.1. However, the only novel microbial protein food to receive government approval for human use is MycoProtein, developed in the UK by Ranks Hovis McDougall.

3.5.3   *MycoProtein*

MycoProtein is a food product which consists basically of fungal mycelium. The organism used is a strain of *Fusarium graminearum*, which was isolated originally from a sample of soil, and the process and product are the results of an extensive programme of experimentation, development and testing. MycoProtein is produced at pilot plant scale by continuous fermentation, using glucose as substrate, with other nutrients, and ammonia and an ammonium salt as the nitrogen source. A flow diagram of the production line is given in Fig. 3.4. Following the fermentation stage the culture is subjected to heat treatment to reduce the ribonucleic acid content, and the mycelium is then separated by vacuum filtration.

In comparison to animal proteins, the production of MycoProtein shows several advantageous features. In addition to the growth rate

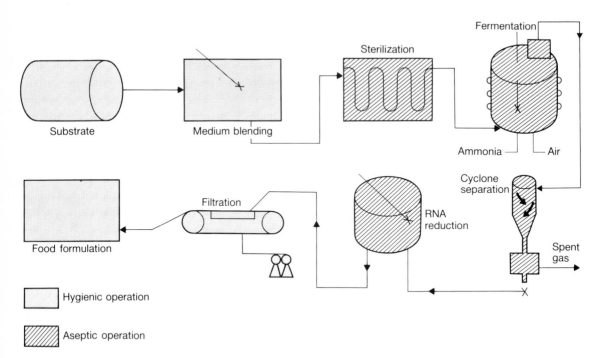

Fig. 3.4. Flow diagram of the MycoProtein production process.

advantage common to all SCP products, the conversion of substrate to protein shows a much greater efficiency than conversion of feed by farm animals. This is shown by the figures in Table 3.6 and it should be remembered that animal feeds need to contain a proportion of protein, possibly as much as 15–20%, depending on the species and the method of husbandry. The filamentous morphology of the culture (Fig. 3.5) also means that the mycelial mass has a natural texture which can be used to advantage. It is possible to give a meat-like texture to the product, which may be flavoured and coloured accordingly. The coarseness of the texture depends on the hyphal length in the culture, which in turn may be controlled via the growth rate. The texture of the final, formulated product may be seen in the example shown in Fig. 3.6. The normal

Table 3.6. MycoProtein and animals. Conversion rates in protein formation

	Starting material	Product	
		Protein	Total
Cow	1 kg feed	14 g	68 g beef
Pig	1 kg feed	41 g	200 g pork
Chicken	1 kg feed	49 g	240 g meat
Fusarium graminearum	1 kg carbohydrate + inorganic nitrogen	136 g	1080 g wet cell mass

Fig. 3.5. Scanning electron micrograph showing hyphal structure of *Fusarium graminearum.*

Fig. 3.6. Example of a MycoProtein product.

method of preservation is by freezing, but a spray-dried powder is an alternative product.

The testing of MycoProtein for nutritional value and safety was extensive and has resulted in permission from the Ministry of Agriculture, Fisheries and Food to sell MycoProtein in the UK. The main nutritional characteristics of MycoProtein are shown in Table 3.7, which also gives a comparison with beef.

| Component | % Content (dry weight basis) | |
	MycoProtein	Raw lean beefsteak
Protein	47	68
Fat	14	30
Dietary fibre	25	Trace
Carbohydrate	10	0
Ash	3	2
RNA (ribonucleic acid)	1	Trace

Table 3.7. Typical composition of MycoProtein compared to beef

3.6   Food additives and ingredients

3.6.1   *Acidulants*

Acidulants are used mainly as flavouring agents to impart a 'sharp' taste to food, but their use probably derives from the widespread adoption of organic acids as food preservatives.

The most widely used acidulant in modern food processing is probably citric acid. This important compound was produced by pressing Italian lemons, and over three-quarters of the world's needs were met by this route until the early 1920s. Today it is produced by the fermentation of molasses and glucose hydrolysate by *A. niger*. However, the fermentation process must be carefully controlled because, unlike many other secondary metabolites, citric acid is an important metabolic regulator. Malic acid is used widely as an acidulant in tomato canning and is produced from *A. flavus*. Other acid products used widely in the food industry are itaconic acid, produced by *A. terreus*, gluconic acid, which is used as gluconolactone, produced by *A. niger*, and fumaric acid from *Rhizopus* spp.

3.6.2   *Amino acids* (see Chapter 4)

World production of amino acids, principally for the supplementation of feeds and food, now exceeds 200 000 t/a. A majority of this tonnage is supplied by fermentation, but chemical synthesis and enzymic transformation are increasingly important techniques. The major fermentation products are glutamic acid from *Corynebacterium glutamicum* and lysine from *B. flavum*. Glycine and methionine, the other two amino acids made in large tonnage, are still produced by chemical synthesis.

3.6.3   *Vitamins and pigments*

Although the bulk of industrial requirements are met from natural products or chemical synthesis, two products have traditionally been produced using biotechnology: β-carotene and riboflavin. The latter is produced in submerged culture from *Eremothecium ashbyii* or *Ashbya gossypii*. Considerable potential exists for the production of toxicologically safe food-grade colours from microorganisms; at present no β-carotene is produced in any quantity, and the important group of red colours is produced from grape extract.

3.6.4   *Flavourings*

The 5' nucleotides, notably 5' inosinate and 5' guanylate, are produced industrially by the action of nuclease from *Penicillium citrinum* on nucleic acid. These materials are used as flavour enhancers; products which amplify the flavour notes present naturally in foodstuffs. A major flavour enhancer is monosodium glutamate (MSG), which may be produced from *Micrococcus glutamicus*. The use of flavour enhancers originated in Japan but the principle has been used in many food formulations on a worldwide basis.

3.6.5   *Oil and fats*

The potential of microbial oils and fats, particularly those containing unusual polyunsaturated fatty acids and which might command a premium price, has been explored and reviewed. It appears that, under present economic conditions, these materials will not displace lipids from plant and animal sources.

3.6.6   *Gums and thickeners*

The natural gums from plant exudates and seaweeds are among the oldest known food additives, and were certainly in widespread use in China in the first century BC; it has been suggested that the Old Testament contains reference to their use. Gums are included in the additive class known as hydrocolloids; these compounds are used to enhance the physical properties of food principally through their ability to thicken and to form gels. These functions stabilize food structure, aid palatability and improve the appearance of foods.

Of significance in this area has been the industrial scale development of bacterial polysaccharides from *Pseudomonas* spp. to replace xanthum gum, one of the most widely used thickening agents. This product has been described in detail. Dextran from *Leuconostoc mesenteroides* is used extensively as a stabilizer in confectionery and in ice-cream (see Chapter 5).

3.7   Fruit and vegetables

3.7.1   *Preserved vegetables*

As is the case with many raw food materials, the need to preserve vegetables so that they are available out of season has led to the development of some distinctive new products. Before the arrival of canning and freezing, salt was commonly used as a preservative for vegetables. At low concentrations of salt (2–2.5%), acid fermentation by bacteria can occur. In low protein vegetables containing sugars this can yield attractive results, but in high protein vegetables (e.g. peas and beans) putrefaction of protein occurs. When these vegetables are preserved with salt, enough is added to inhibit all fermentation. Where low salt levels are used, lactic acid bacteria predominate and lactic production from sugar reduces competition from coliforms, proteolytic bacteria, anaerobes and spore formers.

Sauerkraut is made from fresh, shredded cabbage. After salting, *Leuconostoc mesenteroides* predominates early, anaerobically converting sugars to lactic acid, acetic acid, ethanol, mannitol, esters and CO_2. Then *Lactobacillus plantarum* continues lactic acid production from the sugars and also from the mannitol (removal of mannitol is important as it imparts a bitter flavour). Although there is some control maintained over the fermentation conditions, the addition of starter cultures is not necessary, nor even advantageous in sauerkraut production.

Pickles are prepared from brined cucumber. The final product may be unfermented, partly fermented or fully fermented. Where fermentation occurs, lactic acid bacteria are again important, fermenting the sugars present. Lower concentrations of salt may be used and the brine acidified with acetic acid at the start. When dill and spices are added the product is dill pickles.

The preparation of olives involves brining and lye treatment. With green olives there is a six to ten month lactic fermentation stage involving *Leuconostoc mesenteroides*, succeeded by *Lactobacillus plantarum*. Ripe olives are given little or no fermentation. In both cases the treatment with lye is important for the removal of oleuropein, a bitter-tasting glucoside.

Microorganisms are active at certain stages in the preparation of beverages, especially coffee and cocoa, where lactic acid bacteria and yeasts grow during the soaking of berries. This activity probably assists in the removal of the pulp around the seeds, but the contribution by microorganisms to the nature of the final product is tenuous.

These examples of the use of microorganisms in vegetable processing indicate simple, low-technology operations. Whilst, in the case of pickling, microbial activity is a major determinant of the nature of the final product, the means of influence is largely acid production, and very similar products can be prepared in shorter periods of time by adding appropriate organic acids to fresh vegetables and eliminating the fermentation stage. It is worth noting also that despite the predominance of canning and freezing as methods of preserving vegetables, a demand for pickled products remains because of their special characteristics.

3.7.2   *Soya products*

The soya bean has long been a staple food crop in Asian countries, especially China and Japan. Long before its use in the West as a source of protein and oil, it was known as a major contributor of these nutritional factors to the oriental diet. There are many traditional Eastern foods based on soya beans, commonly depending on microbial activity for their special characteristics. In this context, fungi predominate, in particular species of *Aspergillus*.

Soy sauce is based on a mash of soaked and cooked soya beans, with roasted wheat. This is inoculated with a starter of mixed organisms, *Aspergillus oryzae* predominating. After incubation at 25–30 °C for 3–5 days, there is a heavy surface growth of *A. oryzae*. The mash is then made up to a 20% salt brine and left for a period of six months to two years. With newer methods, in which pure cultures of *A. oryzae* are used, this period is reduced to between one and three months. Finally, the liquor which is drained or pressed from the final mash is the soy sauce product. Apart from quickening the process by the use of pure cultures, methods have been developed for producing extracted soy hydrolysates by purely chemical means. This gives a non-fermented soy sauce product.

Tamari is a similar Japanese product, based on a mash of soy beans sometimes including rice. The fermentation time is shorter than for Chinese soy sauce, *Aspergillis tamarii* being the predominant organism. Miso is a Japanese preparation in which cooked soy beans are inoculated with *A. oryzae*, mixed with salt and water and allowed to ripen for several days before being ground to a paste. To make natto, boiled soya beans are wrapped in rice straw and fermented for 1–2 d. *Bacillus subtilis* is the dominant organism here, producing a slimy coating.

Soya bean cheese or curd is common in the Eastern diet. The soya beans are first soaked and ground to a paste before being filtered through linen. The solids are curdled by the addition of a calcium magnesium salt, and cut into blocks. After one month at about 14 °C, white moulds are active. Final ripening is carried out in either brine or a specially prepared wine.

The potential impact of modern biotechnology on these traditional processes appears to be limited. Where technology has touched them, it has generally been in the direction of reducing lengthy fermentation stages, sometimes eliminating the microbial role altogether. There is undoubtedly further scope for progress by strain improvement and better control of fermentation. The soya bean may also be a material from which, with the oriental traditions as a base, quite new fermented food products may be developed for the Western market, as well as the Eastern. Although that concept refers to whole bean products, there are examples of biotechnology-based new soya protein products. These are produced by controlled hydrolysis of soya bean protein using microbial enzymes, and can be tailored to a limited degree in terms of solubility, functionality and taste. The most promising example is known as isoelectric soluble soya protein hydrolysate (ISSPH), which has applications as a more sophisticated meat extender than soya bean meal and for fortification of soft drinks for societies with protein deficient food resources.

3.7.3   *Fruit juices—application of enzymes*

By both recent history and immediate future prospect, the application of enzymes of microbial origin is one of the main ways in which biotechnology can bring innovation to the food processing industry. An area which has benefited greatly is the preparation of fruit juices, where the range of enzymes which can be used includes pectinases, cellulases, hemicellulases, amylases and proteinases. The use of these enzymes, in addition to assisting long established processes, has extended the range of fruit products and brought greater yields of product from the basic fruit materials. Enzymes are employed in the following main stages of fruit processing.

(1) Mash treatment: to digest the flesh for production of pulps and nectars; to increase juice yield and aid in the extraction of colour and flavour.

(2) Juice treatment: to reduce viscosity and aid in the preparation of concentrates; to improve clarification, ease of filtration and stabilization.

Three main principles underlie the choice of enzymes and the ways in which they are used to give desirable results in fruit juices:

(1) enzyme activity—the concentration of enzyme, temperature and reaction time must all be considered for their effect on the product;

(2) pectin degradation—the type of pectin present and the particular enzymes used can greatly affect the resultant product;

(3) clarification mechanism—there are three distinct, successive phases, the recognition of which is vital to success with fruit juices: destabilization, coagulation, and sedimentation.

Dealing with the pectic substances in fruits is a very important part of juice production. In excessive concentrations, pectic substances can give unacceptably high levels of viscosity and turbidity. Their control enables considerable improvement in yield and quality. Pectic substances or pectins, are heteropolysaccharides with a galacturonan backbone. They form the substrates for a range of enzymes including pectin lyases, pectate lyases, polygalacturonases and pectinesterases. These enzymes are derived mostly from bacteria and fungi and their action is to depolymerize by splitting glycosidic linkages and, in the case of pectinesterases, to break esterified carboxyl groups.

3.8   Overview

3.8.1   *Short-term prospects*

It has been estimated that 15% of overall sales in the food industry may be attributed to biotechnological processes, despite which the influence of this technology in the industry remains no greater than it was twenty-five years ago (Tonge & Jarman, 1981). The reasons for this are threefold.

(1) Food processing remains a labour-intensive, craft-based low-technology industry. Many of the manufacturing processes are scaled-up culinary operations and the underlying science is ill-defined and only partially understood.

(2) In the scale of biotechnology, described at the beginning of this chapter, food processing is in the large tonnage, low added value sector. In effect, the low added value of food products precludes investment in the research and development and safety testing necessary to render biotechnological products and processes competitive.

(3) The main thrust of work in relation to food processing will be to

suppress or exclude undesirable changes attributable to biologically active artefacts; this is now called 'negative biotechnology', which is a euphemism for food preservation. Whatever its name, it remains outside the mainstream of biotechnology. In the short term, therefore, the major benefits of the new technology will be seen in high added value areas such as food additives, flavour principles, colours and functional agents. The development of these products is akin to innovation in pharmaceuticals; safety testing is the norm and there is demand for secure sources of supply at high but acceptable cost.

3.8.2   *Long-term prospects*

The scale of investment in large tonnage biotechnology processes is such that only companies capable of implementing long-term strategic financial programmes are able to take part in the exploitation of the technology. The main inherent advantage is that the productivity of a fermenter or bioreactor is so much greater than that of the plant or animal that provision of commodites via this route should ultimately be the most cost-effective route.

Ironically, it is the implication of this productivity advantage that proves to be the most important factor in the rate of implementation of the technology. The potential displacement of established commodities has resulted in fiscal changes which distort the economics; this distortion is relatively easy to effect in a low added value system. In recent years, this intervention has been applied to isoglucose where the technology was retarded in the EEC by a quota/levy arrangement, and to power alcohol (Economic Development Administration, 1981) where the technology was encouraged in the US by capital grants and levy exemption at national and regional levels. These artificial distortions of economic viability can be expected to continue in view of the perceived importance of agricultural production within national economies.

However, in the longer term, the inherent productivity advantage will manifest itself as demand outstrips the ability of traditional systems to supply. The fact that citric acid is now produced by fungal culture and that the surface of Europe is not covered by lemon trees is one small demonstration that biotechnology can succeed in the food and drink industry.

3.9   **Recommended reading**

ARIMA K. (1977) Recent developments and future directions of fermentations in Japan. *Devs ind. Microbiol.* **18,** 78–117.

AUNSTRUP K. (1979) Production of extracellular enzymes. In *Applied Biochemistry and Bioengineering*, Vol. 2, (eds. Wingard L.B., Katchalski-Katzir E. & Goldstein L.), pp. 27–69. Academic Press, London.

BIRCH G.G., BLAKEBROUGH N. & PARKER K.J. (eds.) (1981) *Enzymes and Food Processing*. Appl. Sci. Publ., London.

BEECH F.W. (1972) Cider making and cider research: a review. *J. Inst. Brew.* **78**, 477–490.

BEAUCHAT L.R. (ed.) (1978). *Food and Beverage Mycology*. AVI Publishing, Westport, Connecticut.

COLLINS T.H. (1968) *A Summary of Breadmaking Processes*. Flour Milling and Baking Research Associaton Report Number 13.

ECONOMIC DEVELOPMENT ADMINISTRATION (1981) *Alternative Utilization of Wheat Starch*. US Department of Commerce, Washington DC.

GLICKSMAN M. (1969) *Gum Technology in the Food Industry*. Academic Press, London.

GREEN M.L. (1977) Review of the progress of dairy science: milk coagulants. *J. Dairy Res.* **44**, 159–188.

GREENSHIELDS R.N. (1978) Acetic acid: vinegar. In *Primary Products of Metabolism* (ed. Rose A.H.), pp. 121–186. Academic Press, London.

HERBERT D., ELSWORTH R. & TELLING R.C. (1956) The continuous culture of bacteria; a theoretical experimental study. *J. gen. Microbiol.* **14**, 601–622.

HERSOM A.C. & HULLAND E.D. (1980) *Canned Foods. Thermal Processing and Microbiology*, p. 43. Churchill Livingstone, Edinburgh.

HIROSE Y., SANO K. & SHIBAI H. (1978) Amino acids. In *Annual Reports on Fermentation Processes*, Vol. 2, (ed. Perlman D.), pp. 155–189. Academic Press, London.

HOWLING D. (1979) The general science and technology of glucose syrups. In *Sugar: Science and Technology*, (eds. Birch G.G. & Parker K.J.), pp. 259–285. Appl. Sci. Publ., London.

JARMAN T.R. (1979) Bacterial alginate synthesis. In *Microbial Polysaccharides and Polysaccharases* (eds. Berkeley R.C.W., Gooday G.W. & Ellwood D.C.), pp. 35–50. Academic Press, London.

JARVIS B. & HOLMES A.W. (1982) Biotechnology in relation to the food industry. *J. Chem. Tech. Biotechnol.* **32**, 224–232.

JARVIS B. & PAULUS K. (1982) Food preservation: an example of the application of negative biotechnology. *J. Chem. Tech. Biotechnol.* **32**, 233–250.

KOOI E.R. & ARMBRUSTER F.C. (1967) Production and use of dextrose. In *Starch: Chemistry and Technology*, Vol. II, (eds. Whistler R.L. & Paschall E.F.), pp. 553–568. Academic Press, London.

KUNKEE R.E. (1967) Malo-lactic fermentation. In *Advances in Applied Microbiology*, Vol. 9, (ed. Umbreit W.W.), pp. 235–279. Academic Press, London.

LANCRENON X. (1978) Recent trends in the manufacturing of natural red colours. *Process Biochem.* **13**, no. 10, 16.

LAW B.A. (1980) Accelerated ripening of cheese. *Dairy Inds. Intl*, **May 1980,** 17.

LAW B.A. (1981) Short cuts to faster flavour. *Food*, **March 1981,** 17–19.

LORENZ K. (1981) Sourdough processes—methodology and biochemistry. *Bakers' Dig.* **55**, 32–36.

MACLEOD G. & SEYYEDAIN-ARDEBILI M. (1981) Natural and simulated meat flavors. *Crit. Rev. Food Sci. Nut.* **14**, 309–437.

MAGA J.A. (1974) Bread flavor. *Crit. Rev. Food Technol.* **5**, 55–142.

RADLEY J.A. (ed) (1976) *Starch Production Technology*. Appl. Sci. Publ., London.

RANKINE B.C. (1972) Influence of yeast strain and malo-lactic fermentation on composition and quality of table wine. *Am. J. Enol. Vitic.* **23**, 152–158.

RATLEDGE C. (1979) Resources conservation by novel biological processes. 1. Grow fats from wastes. *Chem. Soc. Rev.* **8**, 283–295.

REED G. (ed.) (1975) *Enzymes in Food Processing*. Academic Press, London.

REED G. (ed.) (1982) *Prescott and Dunn's Industrial Microbiology*, 4th edn. Macmillan, London.

ROBINSON R.K. (1981) *Dairy Microbiology*, Vol. 2. Appl. Sci. Publ., London,

ROSE A.H. (ed.) (1977) *Alcoholic Beverages*. Academic Press, London.

ROSE A.H. (ed.) (1982) *Fermented Foods*. Academic Press, London.

SCHIERHOLT J. (1977) Fermentation processes for the production of citric acid. *Process Biochem.* **12**, no. 9, 20–21.

SCHLINGMANN M., FAUST U. & SCHARF U. (1981) *Bioproteins for human nutrition*. FIE Conference, 1981, London.

SHARMA S.C. (1981) Gums and hydrocolloids in oil–water emulsions. *Food Technol.* **35**, no. 1, 59–67.

SITTIG W. (1982) The present state of fermentation reactors. *J. Chem. Tech. Biotechnol.* **32**, 47–58.

SODECK G., MODL J., KOMINEK J. & SALZBRUNN W. (1981) Production of citric acid according to the submerged fermentation process. *Process Biochem.* **16**, no. 6, 9–11.

SPINKS A. (Chairman) (1980) *Biotechnology: Report of a Joint Working Party*, p. 35. HMSO, London.

STEVENS T.J. (1966) A method of selecting pure yeast strains for ale fermentations. *J. Inst. Brew.* **72**, 369–373.

TONGE G.M. & JARMAN T.R. (1981) *Opportunities for biotechnology in the food industry*. FIE Conference, 1981, London.

WHITWORTH D.A. & RATLEDGE C. (1974) Microorganisms as a potential source of oils and fats. *Process Biochem.* **9**, no. 9, 14–22.

WIERZBICKI L.E. & KOSIKOWSKI F.V. (1972) Food syrups from acid whey treated with beta-galactosidase of *Aspergillus niger*. *J. Dairy Sci.* **56**, 1182–1184.

WILLIAMS A. (1975) The history of the Chorleywood bread process. In *Breadmaking: The Modern Revolution*, (ed. Williams A.), pp. 25–39. Hutchinson Benham, London.

YAMADA K., KINOSHITA J., TSUNODA T. & AIDA K. (1972) *The Microbial Production of Amino Acds by Fermentation*. John Wiley & Sons, New York.

4   Chemistry and Biotechnology

D. J. BEST

4.1   Introduction

All the chapters in this book are concerned in some way with the exploitation of the remarkable catalytic and recognition properties of biological systems. Perhaps the greatest potential for such exploitation in modern society is in the chemical industry. Of course, it is self-evident that biomass itself is a complex chemical system and that most of the activities and products of biotechnology are of a biochemical nature, whether they be fuel chemicals (Chapter 2), fermentation products in the form of foodstuffs or drinks (Chapter 3), biopolymers (Chapter 5), organisms involved in recycling chemicals on earth (Chapter 6), complex biochemical products used in medicine (Chapter 8) or agriculture (Chapter 9). This chapter is primarily concerned with the underlying concepts, prospects and technology of biotechnological chemical production and, in general, concentrates upon chemical products and processes not discussed elsewhere in the book.

Modern applied microbiology and, hence, biotechnology have their origins in the chemical industry in the early years of this century with the advent of industrial processes for producing chemical feedstocks (e.g. acetone, ethanol, butanediol, butanol and isopropanol) from plant carbohydrates. This important development is also discussed in Chapter 1. This substantial industry was dramatically displaced by the ensuing rapid development of the oil-based petrochemical industry. However, there is now increasing competition for fossil reserves for use in the production of chemicals, energy and even food. This is associated with increases in the price of oil and coal and makes the adoption of new processes for deriving chemicals from renewable biomass increasingly likely, including reassessment of improved processes essentially similar to those used at the beginning of the century.

New ground is already being broken, e.g. with the algal production of glycerol, and there is much current activity aimed at exploiting the lignin component of wood. This is a technically difficult problem but, when solved, it will offer a new route to many of the important aromatic intermediates so central to the modern chemical industry.

In addition to offering new routes to chemical feedstocks from biomass, biotechnology can offer more effective and efficient catalysis for chemical interconversions. This can take two forms: in the first, the specificity of one or more enzymes *in vivo* or *in vitro* is used to catalyse a simple but technically difficult chemical interconversion, as in the hydroxylation of steroid molecules for drug synthesis; the second type of process involves the complete biosynthesis from simple building blocks of complex and valuable fine chemical products. An existing example is the production of antibiotics by fungal secondary metabolism. Table 4.1 summarizes the different categories of biocatalytic activity. A promising area for future development is the generation of valuable plant products such as terpenes (for aromas) and alkaloids (for drugs) using large-scale plant cell culture.

Table 4.1. Biotechnology and the chemical industry

Growth-related biocatalysis	
No constraints imposed	Single cell protein
Environmental manipulation to induce metabolic imbalance	Organic acids, amino acids
Generation of mutants for the overaccumulation of metabolic intermediates	Amino acids
Induction of particular enzyme pathways	
Inhibited fermentations	Antibiotics
Directed synthesis from precursors to bypass metabolic control	Amino acids, antibiotics
Post-growth biocatalysis	
Single-step conversions, avoiding enzyme purification or preserving enzyme stability	Glucose isomerisation, gluconate production, steroid transformation
Multi-step enzymic conversions	Steroid transformation
In vitro *biocatalysis*	
Isolated enzyme for single-step conversion with natural substrate	Glucose isomerase, amino acids
Single-step generation of chemical intermediates from unnatural substrates using isolated enzymes of broad specificity	Hydroquinone formation using glucose oxidase
Multi-step semisynthetic metabolic pathways	
Biologically derived chemical catalysis	
Chemical mimicry of enzyme catalysis	
Fusion of biological specificity with chemical catalysts to produce 'bioorganic complexes'	

4.2   The development of current chemical biotechnology

4.2.1   *The solvent fermentations*

The development of fermentative production of glycerol is discussed in Chapter 1, and ethanol—a solvent, food component, chemical intermediate and fuel—is discussed in detail in Chapter 2. The remaining important solvent fermentation is that of acetone and butanol, which was first developed for industrial use in Manchester by Weizmann during the First World War. Acetone was important in the manufacture of cordite and as a propellant for heavy artillery shells and, until the outbreak of hostilities, had been imported from Germany. Low grade acetone could be obtained from the destructive distillation of wood but high quality solvent was required.

The fermentation was based upon the utilization of starch, at concentrations of up to 3.8% (w/v), by the anaerobic, spore-forming bacterium, *Clostridium acetobutylicum*, and resulted in a 30% conversion to a mixture of solvents (60% butanol, 30% acetone, 5–10% ethanol/isopropanol/mesityl oxide). The rest of the substrate was converted to hydrogen and carbon dioxide and the process is outlined in Fig. 4.1. Due to the enormous volumes of gas generated, the large-scale fermentations required no agitation, although control of foaming became a major problem. The acetone:alcohol ratio showed some strain-dependent variation and many starch degraders capable of solvent production could also ferment molasses (up to 6% (w/v) sugar in the medium). The factor controlling the amount of substrate used was the tolerance of the process organism to n-butanol (*c.* 1.2%, v/v), with an acetone concentration of 0.4% (v/v). The fermentation was not usually subject to contamination with aerobic bacteria but bacteriophage

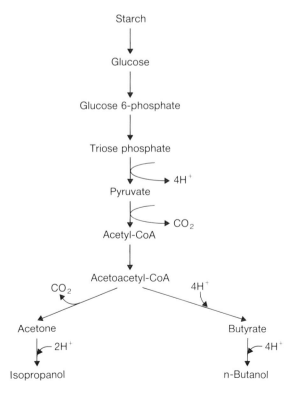

Fig. 4.1. The acetone–butanol fermentation.

infection proved to be a major problem. It was subsequently found that the process organism could be 'immunized' by several transfers of the organism in the presence of the bacteriophage. Phage infection did appear to be strain specific. Solvent recovery was achieved by distillation. At the end of the First World War, butanol assumed prime importance in the process, as it was used in the production of a wide variety of substances, including urea–formaldehyde resins, plasticizers and brake fluids. The by-product, hydrogen, was used in the manufacture of synthetic methanol and hydrogenation of edible oils, and carbon dioxide was either compressed or converted to dry ice. The residual solids contained large amounts of riboflavin (vitamin B_2) and could be used as a protein-rich animal feed supplement.

However, the fermentative production of these solvents declined after the Second World War due to the decrease in the relative cost of petrochemicals, compared to sugar polymers. Only in South Africa has production of n-butanol by fermentation continued. Economic prospects for reversion to the use of this fermentation for butanol production are improving and it may well be that the application of genetic engineering to the microorganisms involved will tip the balance in favour of this route. The main inadequacy of the existing strains is the lack of tolerance to the product and rather poor yields.

The production of butylene-2,3-glycol was extensively investigated during the Second World War and renewed interest in this microbiological conversion is being generated. Utilization of whey is a problem in the dairy industry and it can be used as a carbon source for the production of this glycol using *Klebsiella pneumoniae* or *Enterobacter aerogenes*. Butylene glycol is converted into a feed chemical for the production of synthetic rubber.

4.2.2   *Organic acid production*

The most important organic acid is acetic, with a current US annual market (excluding vinegar) of about 1.4×10^6 t (value about M\$500). In the past, microbiological ethanol oxidation has been the main source of acetic acid but, with the exception of vinegar manufacture, this process is not currently economic. It may become so again, however, as a result of current research into thermophilic bacteria that can convert cellulose to acetic acid, and strains of *Acetobacterium* and *Clostridium*, which can convert hydrogen and carbon dioxide to acetic acid. Industrial acetic acid is used in the manufacture of many chemical products, including rubber, plastics, fibres and insecticides. The conventional microbiological conversion of ethanol to acetic acid by strains of *Acetobacter* and *Gluconobacter* is an aerobic process and, therefore, not strictly speaking a fermentation process. Vinegar is, in its own right, a major microbiological product and is discussed in Chapter 3.

The commercial production of lactic acid, using *Lactobacillaceae* such as *Lactobacillus delbrueckii*, *L. leichmannii* and *L. bulgaricus*, started at the end of the nineteenth century and was one of the first processes to incorporate medium pre-sterilization by heating. This micro-aerophilic process is conducted at high operating temperatures (45–50 °C), using starchy materials that have been degraded either enzymatically or by acid hydrolysis (see section 4.2.6, 'Amylases and amyloglucosidases'). *L. bulgaricus* efficiently ferments lactose and can, therefore, use whey as a starting material. In other cases, sucrose at concentrations of 12–18% (w/v) is metabolized in 3–4 days, generating large quantities of carbon dioxide, which help to create the optimum semi-aerobic conditions. The conversion of 1,2-propanediol to lactic acid by *Arthrobacter oxydans*, *Alcaligenes faecalis* or *Fusarium solani* has also been described. L(+)-Lactic acid is the predominant isomer produced normally but adapted strains of *L. leichmanii* will produce the D(−) isomer. The production of lactic acid has been studied in continuous culture, where *L. delbrueckii* demonstrated a productivity of 89 g/l/d in a one-stage process. Immobilized preparations of lactic acid bacteria, using alginate as the support matrix, have achieved a conversion efficiency of 97%, with a 90% yield of the L(+) isomer, and have a half-life of 100 d. In these processes, lactic acid is formed as the calcium salt and the process filtrate is treated with sulphuric acid to recover the product. Lactic acid is used as an additive in soft drinks, essences, fruit juices, jams and syrups, in the decalcification of hides in the tanning industry, in the plastics industry, where the L(+) form is polymerized to polylactide for use in plastic foils, and its salts have therapeutic uses.

The production of citric acid by fungal fermentation (Fig. 4.2) was another early biotechnological process, first identified in 1893, whose commercial development has gone hand in hand with the evolution of many of the fundamental aspects of microbiology. Although early problems were encountered with contamination, it was found that the process could be operated at very low pH without significantly affecting the production of the acid by the moulds used. Under such conditions it was easier to achieve and maintain sterile conditions. Yields of 60% from high sugar concentrations were obtained in one to two weeks of fermentation, the highest yields being achieved when mycelial development was restricted in some way. The original production process

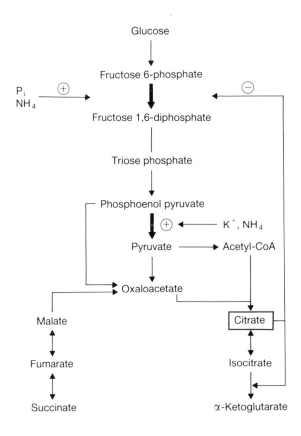

Fig. 4.2. Production of citric acid.

involved a surface fermentation, but the introduction of submerged fermentation in the 1950s represented a significant improvement. It was found that stable submerged fermentation was possible if the process was operated in two stages, one involving mycelial development and the other citric acid production in a phosphorus-free medium. Conditions were quickly elaborated for the utilization of cheap carbohydrate sources such as cane molasses, starch and glucose syrup. The presence of metal ions in the raw complex materials caused drastically reduced yields and required removal, either by precipitation using hexacyano-ferrate or by treatment with ion exchange resins or quaternary ammonium salts. Methanol, or other lower alcohols, are also widely used to remove these metals. The basis for the action of the alcohol is not known but it may have some effect on the cytoplasmic membrane. During the 1960s, new processes for citric acid production emerged, utilizing n-paraffins (C_{9-30}) and strains of *Corynebacterium*, *Arthrobacter* and *Brevibacterium*, but these were not commercialized. Citric acid production by *Candida* yeasts has also been examined. These yeasts produce a mixture of citric and isocitric acids, the proportion being related to genetic as well as process parameters. In this respect, the activity of aconitate hydratase was found to be crucial and mutants with low levels of this enzyme produced greater amounts of citric acid. Hydrocarbon-utilizing yeasts will also produce citric acid from glucose. *Trichoderma viride* produces large amounts of citrate and opens up the possibility of production from cellulose. There also exists a fermentative process for the production of L_5-*allo*-isocitric acid, a diastereomer of isocitric acid, using a *Penicillium* species.

Aspergillus niger is still the dominant fungus used in the commercial production of citric acid, although *A. wentii* also finds application. The fermentation process is complex, as citric acid is a product of the organism's primary metabolism and any appreciable excretion of this metabolic intermediate represents a severe metabolic irregularity, arising from some metabolic imbalance or genetic deficiency. Growth is usually restricted by some medium limitation (P, Mn, Fe, Zn). The growth substrate used must be available for rapid uptake and non-hydrolysed polymers are not normally used as extracellular hydrolysis would become rate limiting for the process. Citric acid overflow occurs as a result of phosphate deficiency, but under rigorous metal limitation phosphate does not need to be limiting. The role of metal ions in this regard is not fully understood. The optimum pH of the process is 1.7–2.0, more alkaline conditions causing the formation of substantial amounts of oxalic and gluconic acids. Exact control over the environmental conditions is, therefore, a prerequisite to overcoming metabolic regulation and promoting optimal citric acid formation. The prevailing conditions possibly stimulate glycolysis to provide an unlimited flow of carbon to intermediary metabolism. The extent of citrate accumulation will then be dependent upon the supply of oxaloacetate.

Manganese deficiency reduces the activities of enzymes of the tricarboxylic acid cycle with the concomitant reduction in anabolism. This disturbance in metabolism results in elevation of the intracellular levels of ammonium ions, which may serve to alleviate the inhibition of phosphofructokinase by citrate. Manganese may also be involved in the biochemistry of the cell surface and hyphal morphology. The high oxygen requirement of the process allows reoxidation of cytoplasmic

NADH without the generation of ATP and involves an alternative respiratory branch distinct from the standard respiratory chain. Metabolic flux through glycolysis without any significant metabolic control is stimulated. In combination with a constitutive pyruvate carboxylase, peculiar TCA cycle enzymes and unusual kinetics of the oxaloacetate metabolizing enzymes, this causes a high intracellular concentration of citrate, which serves to further its own accumulation by inhibition of isocitrate dehydrogenase.

Several different process configurations are used for the industrial production of citric acid. The Koji process is a traditional solid-state process and is similar in operation to the surface process. Submerged fermentation is more difficult technically than the surface procedure, but may be batch, fed batch or continuous in operation. Batch fermentations are used for glucose substrates, whilst fed batch is more applicable if molasses are used. Continuous culture with high productivity is possible but the industrial application of this process configuration is unlikely in the foreseeable future. Two rate maxima exist with regard to growth and product formation. In the first phase, there occurs significant product formation, which is dependent upon growth rate. In the second phase, no growth occurs and maximized product formation is dependent upon biomass concentration. At the end of the fermentation, the mycelial mass is removed by filtration and washed. Oxalate is then precipitated as calcium oxalate at a pH less than 3.0. The protein-rich mycelial mass can be used as an animal feed. Citrate is precipitated from the fermentation liquid as tricalcium citrate tetrahydrate. This is filtered off, washed, the acid freed by treatment with calcium sulphate and the free acid further purified using activated carbon and ion exchange resins. Alternatively, the acid may be recovered by solvent extraction.

Citric acid has a pleasant acid taste and is very soluble in water. It finds extensive application in the food industry, in pharmaceuticals and cosmetics. The esters of the acid are used in the plastics industry and, as a metal chelator, citric acid finds application in metal purification and as a readily biodegradable ingredient of detergents, replacing phosphates in this role.

4.2.3   *Other organic acids*

There are several other examples of organic acids for which processes based upon microbial fermentation have been described. These include gluconic acid and its derivatives, malic acid, tartaric acid, salicylic acid, succinic acid, pyruvic acid and kojic acid. Although some have

achieved commercialization, the majority are not economically viable
in the present climate.

D-Gluconic acid and its δ-lactone are simple dehydrogenation
products of glucose (Fig. 4.3) and the commercial production of the acid
from glucose by *Aspergillus niger* was implemented in the early 1920s,
high yields being obtained when the acid was neutralized. In submerged
cultures, a 90% conversion rate was achieved in 48 h. Pilot plant studies
demonstrated the production of the acid to 95% of the theoretical yield
from glucose solutions of 150–200 g/l within 24 h if the fermentation was
conducted under elevated pressures. The process could be run in a
semicontinuous mode using mycelial recycle up to a total of nine times.
In addition, glucose concentrations could be increased to 350 g/l if
boron compounds were used to complex out the gluconate as calcium
borogluconate. However, this process required the use of specially
resistant strains and was discontinued as the salt was found to be
harmful to animal blood vessels. Control of pH is maintained by the
addition of either calcium carbonate or sodium hydroxide. The production of gluconate was found to correlate directly with the levels of
glucose oxidase in the medium and the fermentation process is used as a
commercial source of this enzyme.

Sodium gluconate can be used as a metal sequestering agent and the
control of fermentation pH by sodium hydroxide addition leads directly
to this product. Continuous processes have been developed which
convert 35% (w/v) glucose solutions to the sodium salt with a 95% yield.
There have also been some attempts to operate the process with
immobilized systems, using both whole cells and immobilized glucose
oxidase. Sodium gluconate acts as a sequestrant for calcium in the
presence of sodium hydroxide and is an ingredient of alkaline bottle
washing preparations. It also sequesters iron over a wide pH range and,
in preventing iron deposition, is used in alkaline derusting preparations. Calcium and iron gluconate salts are used in oral and
intravenous glucose therapy, whereas the free acid is used in cleaning
applications in the dairy industry. Gluconolactone is used as a slow
acting acidulant in baking powders, in meat processing and in other
applications in the food industry.

Bacterial gluconic acid fermentations are only of importance in
some Eastern countries; e.g. the 'tea fungus', an association of yeast,
acetic acid bacteria and gluconic acid bacteria, converts sweetened
black tea to a mixture of these acids. The limited success of bacterial
glucose dehydrogenation reactions is due to secondary reactions which
occur to produce 2-oxogluconic acid, 5-oxogluconic acid and dioxoglu-
conic acid. Processes have been designed for the production of these

120

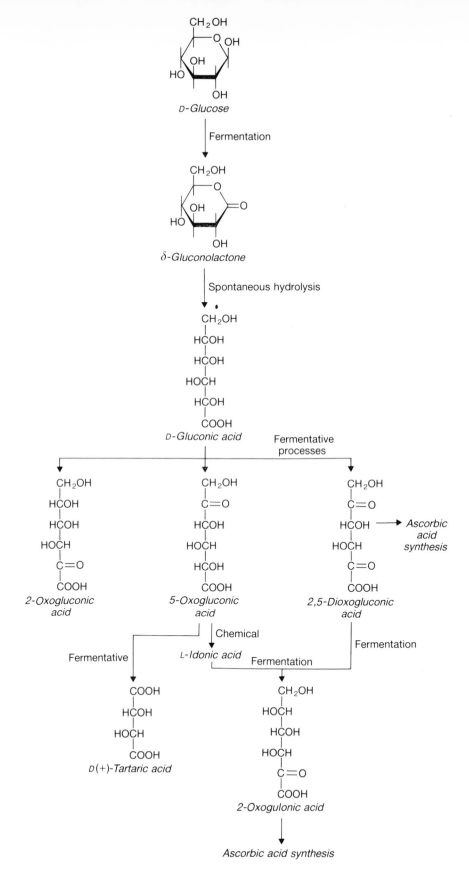

Fig. 4.3. Organic acids derived from glucose.

acids, which may themselves serve as substrates for further biological or chemical conversion. For instance, 5-oxogluconic acid can be chemically hydrogenated to produce L-idonic acid, which itself serves as a fermentation substrate for the production of 2-oxogulonic acid (Fig. 4.3). The same conversion can be achieved using totally biological means, in which glucose is transformed to 2,5-dioxogluconic acid by *Acetobacter* species, and this acid is converted directly to 2-oxogulonic acid using *Corynebacterium* or *Brevibacterium* in a two-step or single-step mixed culture process; the yields are poor, however. 2-Oxogulonic acid is important in that its methyl ester is easily transformed to ascorbic acid under alkaline conditions.

Tartaric acid is found extensively in nature as a by-product in the production of wine. It is also possible to effect the microbiological transformation of 5-oxogluconic acid, and strains that convert glucose to 5-oxogluconate via gluconate are used for the subsequent fermentation to produce tartrate. Mutants of *Acetobacter* and *Gluconobacter* are used for this purpose. Tartaric acid may also be produced from *trans*- or *cis*-epoxysuccinic acid. Despite the extensive use of tartrate in the food industry, it is not usually produced biotechnologically.

Malic acid, used as an acidifying agent in the food industry, can be produced from fumaric acid, either by fermentation using *Paracolobactrum* species or with immobilized fumarase. Processes involving production from n-paraffins by yeasts or from ethanol using *Schizophylum commune* have also been described.

Itaconic acid, used in the plastics and paint industry, is produced during the fermentation of glucose by *Aspergillus* spp. in high yield. 2-Oxoglutaric acid was formerly produced by biotechnological means from carbohydrate substrates and is also formed during the fermentation of C_{12-14} paraffins by *Candida hydrocarbofumarica* but has now been superceded by the catalytic oxidation of benzene.

The vast majority of the organic acids produced microbially have been derived from carbohydrate feedstocks, with the exception of those that can be produced from n-paraffins. The use of other hydrocarbon feedstocks as potential sources of more valuable organic intermediates has been recognized for many years, although few have been developed to commercial production stage. A case in point is the microbiological conversion of naphthalene to salicylic acid and other oxygenated intermediates (Fig. 4.4), which has been described extensively in the literature for over twenty years (Cain, 1980; Tangnu & Ghose, 1980; 1981).

Bacteria capable of the accumulation of salicylic acid (Fig. 4.4) during growth on naphthalene include many species of *Pseudomonas*,

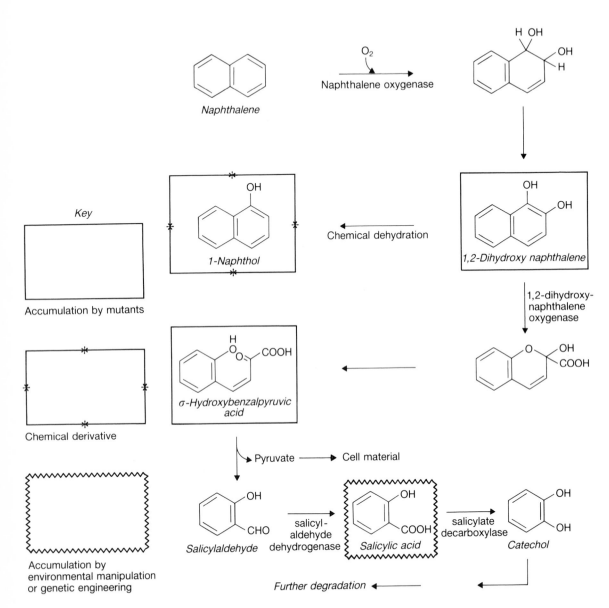

Fig. 4.4. Microbiological conversion of naphthalene to potentially useful organic intermediates.

Achromobacter and *Corynebacterium*. In addition, the accumulation of o-hydroxybenzalpyruvic acid and 1,2-dihydro-1,2-dihydroxynaphthalene by a *Nocardia* species has been patented. Most strains of wild-type naphthalene degrading bacteria in a well-aerated simple salts medium rarely produce salicylate in yields of more than 1%, but enhanced yields may be obtained by various environmental manipulations after suitable strain selection.

Substrate presentation is one critical factor affecting yield, and the accumulation of salicylate is dependent upon the continued presence of the naphthalene, to prevent further oxidative metabolism. The selection of successive degradative pathways depends upon the relative concentrations of naphthalene and salicylic acid. However, the polyaromatic naphthalene is extremely insoluble in water and is usually presented in the fermentation medium as a finely pulverized powder. Addition of emulsifiers such as Span 80, Span 20, lecithin, cephalin and other polyvinyl alcohols to the fermentation liquor significantly enhances salicylate accumulation by increasing the availability of the substrate. Pure naphthalene is commonly used as the substrate but it is possible to produce salicylate from crude naphtha fractions without interference from impurities such as alkylnaphthalenes, thiophene, benzothiophene and cresols.

As the fermentation progresses, the pH drops rapidly, requiring the use of heavily buffered media, employing high concentrations of phosphates, or the addition of urea or calcium carbonate. Various metal ions are required for maximum salicylate accumulation and specific additions such as organic and inorganic derivatives of aluminium or boron, pantothenic acid and various nutrients have been reported to further increase yields. The fermentation is, however, governed by product accumulation rather than substrate limitation and removal of salicylate from the medium results in the release of growth inhibition and the further generation of salicylic acid. Removal of product can be achieved in one of two ways. The addition of an anion exchange resin, such as Amberlite IRA-400, either directly to the medium or spatially confined within a dialysis bag, causes the absorption of salicylic acid. An alternative arrangement involves the circulation of the fermenter liquor to an ion exchange column in series with the production fermenter. In this way product concentration within the fermenter is kept low, with up to a sixfold increase in yield. This method also enhances product recovery. An alternative method of product removal is by dialysis fermentation and a small-scale pilot study of this process found an increase in salicylate yield from 10 g/l to the equivalent of 206 g/l. This method is advantageous in that it avoids the possible toxic effects of ion exchange resins but does require substantially more liquid medium, with resultant lower product concentrations. Other more traditional methods of product recovery, such as solvent extraction, have also been used. Salicylate-producing fermentations also become unstable due to phage infections; the selection of phage-resistant mutants then becomes necessary.

The processes of naphthalene and salicylate degradation have been found to be plasmid encoded in several genera. The NAH plasmid specifies the conversion of naphthalene to salicylate, coding for enzymes responsible for this conversion, such as naphthalene oxygenase, 1,2-dihydroxynaphthalene oxygenase and salicylaldehyde dehydrogenase. Thus in many naphthalene utilizers the information allowing the use of this substrate is plasmid encoded, although this is not exclusively the case. Utilization of plasmid encoded degradative information may allow the insertion of such into host strains, converting them to salicylate producers. The ability to utilize or oxidize naphthalene is usually induced by growth upon naphthalene, salicylate or a salicylate analogue such as benzoate or aminobenzoate. However, some interesting co-oxidative processes have been described in which mutant strains of *Pseudomonas putida*, grown on glucose as sole source of carbon and energy, have been used to effect the oxidation of naphthalene to dihydroxy-1,2-dihydronaphthalene and 1,2-dihydroxynaphthalene, following post-growth enzyme induction by naphthalene or other inducers. The former compound undergoes acid dehydration to yield α-naphthol, an important monooxygenated derivative of naphthalene (Fig. 4.4).

4.2.4 *Amino acids* (see also Chapter 3)

Amino acid production from bacteria and their mutants

All proteinaceous amino acids are L-α-amino (or imino) acids and are important as nutrients, seasonings, flavourings and precursors for pharmaceuticals, cosmetics and other chemicals. They may be produced either by isolation from natural materials (originally from the hydrolysis of plant proteins) or by chemical, microbial or enzymatic synthesis. Whereas chemical synthesis produces a racemic product, which may require additional resolution (see section 4.2.6, 'Other enzymes of commercial importance'), the latter two procedures give rise to optically pure amino acids.

The majority of microbial production procedures outlined in this chapter so far have relied on the alteration of the environmental conditions under which the parent organism is grown to achieve overproduction of the target product. The desired metabolic imbalance has usually been delineated by empirical manipulation of factors such as substrate concentration, pH, product concentration or critical levels of other nutrients (metal ions, organic supplements). The commercialization of biological processes for the production of amino acids has led to new approaches to the generation of the correct metabolic imbalance within the producer organism, in order to produce large quantities of

Amides are less susceptible to spontaneous hydrolysis than esters and this method resolves amino acids to a higher optical purity.

(4) *N*-Acyl-L-amino acids are stereospecifically hydrolysed by aminoacylases or by carboxypeptidase to L-amino acids. Acylases can be obtained from a variety of microbial sources and fungal aminoacylase has been immobilized and used in a continuous enzymatic resolution procedure.

The resolution of racemic mixtures of terpene esters by hydrolases was commercialized in the late '60s, following the discovery that enzymatic hydrolysis of these esters only proceeded if the ester bond was equatorial or could easily attain this conformation. Optical resolution also depended upon the position of the diastereomeric centre and was found to be good with the epimers of secondary alcohols. Polyurethane entrapped yeasts are used as biocatalysts and with DL-menthol succinate as substrate in water-saturated n-heptane, a 72.6% conversion ratio was achieved to yield L-menthol of 100% optical purity. This resolution procedure has been patented on an 800 kg scale. Resolution of racemic terpenes can also be achieved by reversed hydrolysis, in which a lipase-mediated esterification is followed by chemical hydrolysis.

Future applications of immobilized enzyme technology

At present, there are only four significant industrial applications of immobilized enzyme technology: glucose isomerase, aminoacylase, penicillin acylase and lactase. The latter has been immobilized on silica particles and is used to convert lactose in whey to glucose and galactose. Applications of immobilized proteins in the foreseeable future include the following.

(1) Cholinesterase, for the detection of pesticides: inhibition of this enzyme is monitored either electrochemically or colorimetrically.

(2) Other enzymes may be used in a similar way for the detection of toxic chemicals. Carbonic anhydrase is very sensitive to low concentrations of chlorinated hydrocarbons, hexokinase to chlordane, lindane and toxaphene.

(3) Immobilized diisopropylphosphofluoridase, from squid nerve cells, could be used to detoxify organophosphate nerve gases, such as Soman and Sarin.

(4) Immobilized heparinase may be used in the prevention of blood clotting in extracorporeal blood circulators.

(5) Immobilized bilirubin oxidase will remove bilirubin from the blood of jaundiced infants.

(6) A novel use of immobilized haemoglobin has been proposed. The protein, entrapped in a polyurethane matrix, forms a haemosponge, which extracts oxygen directly from water with an efficiency of 80%. Oxygen can later be released from the polymer by the passage of a weak electrical charge or exposure to a vacuum. This system has been mooted as a potential oxygen supply for divers or for underwater combustion.

(7) It will soon be feasible to use immobilized systems composed of more than one enzyme. For example, three enzymes—urease, glutamate dehydrogenase and glucose dehydrogenase—enclosed within microcapsules could be used in the removal of urea from the blood of patients with kidney failure.

(8) In the dairy industry further application of immobilized enzymes may be forthcoming, with the immobilization of the milk clotting proteins pepsin and rennin for cheese production and the hydrolysis of butterfat by immobilized lipases and esterases.

(9) A wide range of immobilized enzymes will eventually be used in biosensors for specific, rapid, real-time analysis. Currently, only a few enzymes are exploited in this way but the number will increase as stabilization procedures are improved. Enzymes from extreme thermophiles may prove particularly useful because of their remarkable stability.

4.3   The generation of chemicals from biomass

4.3.1   *Plant cell biotechnology*

The plant kingdom has been a rich source of chemicals for various sectors of the chemical industry for many years, not only in providing raw materials such as sugars, but also in the generation of many complex secondary metabolites, including rubber, cocaine, dyes, flavours and fragrances. Alternative methods of chemical synthesis of these products are often not feasible if the natural structure is complicated and, with the advances in biotechnological methods in recent years, attention is being focused once again upon the plant kingdom not only for improved processes for the generation of established products, such as ajmalicine and codeine, but also for novel products and biotransformations. The expression of plant genes in bacteria is still some years away, primarily due to the lack of knowledge concerning gene expression in plants and also to the fact that these secondary metabolites are the products of multistep processes, the regulation of which is often poorly understood. Therefore, plant tissue culture may present a viable alternative for the generation of high value chemicals, especially pharmaceuticals, and as an aid to plant strain improvement. The use of plant tissue culture will allow control over chemical production and freedom from the vagaries of pest infection and climatic change that dramatically affect the traditional product.

Plant tissue cultures can be initiated from any known plant species, using a wide range of media. Elucidation of plant culture physiology and biochemistry has allowed great advances in terms of culture yield and productivity. However, many cultures are not capable of true autotrophic maintenance, requiring an exogenous carbon source, such as glucose or sucrose, together with a nitrogen source and a wide range of minerals and growth promotants. High biomass yields (25–30 g/l) can now be obtained but enhanced levels of product yield are usually only achieved at the expense of biomass yield and growth. Tenfold enhancement over whole plant yields can be attained but appropriate screening and selection strategies require development, especially if the product sought is produced in only very low amounts.

The first plant tissue culture derived product to achieve commercialization illustrates these aspects of plant cell biotechnology. Shikonin and its derivatives are found in the roots of the plant *Lithospermum erythrorhizon*. The plant itself has been traditionally used in Japan as a medicine, due to the antibacterial and anti-inflammatory effects of shikonin, which also finds use as a dye, as it is a bright red naphthoquinone compound. The plants are grown for five to seven years before the concentration of shikonin in the roots rises to 1–2%. As the plant cannot be grown in commercial quantities in Japan, it is imported from China and Korea and the pure natural product is valued at $4500 per kilogram and was, therefore, considered an attractive commercial target for production by plant tissue culture. Plant cells do not excrete their secondary products, sequestering them in vacuoles and organelles and making product recovery problematical. However, the accumulation of the dye shikonin in the cells simplified the identification of over-producing cell lines, due to their bright red appearance, and a cell line was established that accumulated up to 15% dry weight as the product. Subsequent medium optimization resulted in 13-fold increased productivity and a two-stage fermentation process was developed, the first allowing optimum generation of biomass whilst the second maximized secondary product formation. The plant cells are grown in suspension culture and air-lift fermenters are most suited to this type of culture, striking a balance between effective mixing, cell separation and minimizing shear damage of the relatively fragile cells. The production process for shikonin uses a 200 l vessel for growth, which is used to inoculate a 750 l production vessel. The product is then extracted by conventional chemical means in yields of 5 kg per run; this process has caused the price of the product to plummet.

Other approaches to product recovery may involve changes in the pH of the production medium or the addition of chemicals such as dimethyl sulphoxide to permeate the cells. This may be particularly important after plant cell immobilization, where selective permeabilization will allow continuous removal of the secondary metabolite from the intracellular vacuoles without complete disruption of primary metabolism. Indeed, immobilization itself, usually only by the most mild techniques, may produce conditions favouring extracellular accumulation of products by creating a microenvironment that mimics the productive parts of the plant. Growth may be slowed, diverting nutrients into secondary product formation. *Catharanthus* cells immobilized in alginate/acrylamide cloth liberate substantial amounts of ajmalicine and serpentine. The periwinkle *Catharanthus roseus* is a possible candidate for the production of a range of indole alkaloids, such as ajmalicine, vinblastine and vincristine, the latter two finding approved application in cancer therapy. However, the concentration of such alkaloids is often very low and they have to be resolved from the large number of other alkaloids also produced. In addition, tissue cultures do not produce vinblastine or vincristine but an alternative route does exist for their synthesis. These alkaloids are asymmetrical dimers of catharanthine and vindiline and the former can be produced in culture in enhanced quantities amongst a relatively limited mixture of other alkaloids. The dimers can then be synthesized chemically.

Plant tissue culture can also be used as a biotransformation system, as exemplified by the 12-β-hydroxylation of digitoxin to produce the heart stimulant digoxin, using either suspended cell cultures or immobilized systems of *Digitalis lanata* (foxglove). *De novo* synthesis of novel cardenolides has also been achieved with this system and with immobilized carrot cells.

Plant tissue culture techniques have also been applied to the improvement of plant crops, in the enhancement of disease resistance, sucrose and starch content, stress tolerance and crop yield. These techniques are also used in the improvement of woody plants, the source of lignocellulose, and of plants producing elastomers (guayule, milkweed) and lipids (oil palm, soybean).

The future potential of plant cell biotechnology is thus extremely exciting, encompassing the production of novel pharmaceuticals, sweeteners, agrochemicals, perfumes and aromas, and the advanced technology of hybrid formation by protoplast fusion enhances the scope of this fast developing field.

4.3.2   *Chemicals from biomass*

Since the Second World War, natural gas and petroleum have provided the chemical industry with high purity raw chemicals in large amounts at relatively stable prices. However, these criteria no longer apply and, coupled with the advances in biotechnology over the last fifteen years, the reassessment of the use of biomass for the generation of chemicals has been stimulated. In the years before the advent of petrochemicals, biomass did, of course, supply the raw materials for the chemical industry, in the form of fats and oils for soaps and detergents, methanol from the destructive distillation of wood, and starches and sugars for solvent fermentations. However, biomass as a source was subject to wide fluctuations in price and availability. As has been seen in this chapter and in Chapter 5, biomass in the form of fermentable sugars plays an increasing role in chemical and polymer production for utilization in many industries but this section is primarily concerned with the utilization of the greatest resource of biomass, lignocellulosic material. This involves the conversion of biomass to fermentable substrates, which may be used as starting materials for many microbially produced chemicals, fuels and foodstuffs. Each component of the biomass—cellulose, hemicellulose and lignin—has to be efficiently utilized. Lignin protects the polyglucan from enzymatic attack and for effective utilization of the recalcitrant cellulose polymer hemicelluloses need to be removed, followed by the disruption of the lignocellulosic complex and the destruction of the crystalline cellulose structure. There are a variety of chemical, physical and microbiological approaches to these problems.

Hemicelluloses are readily removed by solvation in dilute acid, often used in biomass pretreatment. Dilute acid hydrolysis, involving high temperatures and pressure, is relatively efficient in the production of glucose from wood and has been used commercially in several countries. Hydrolysis by concentrated acids is far more efficient but requires high capital expenditure for process containment. Hydrofluoric acid has been effectively used and the anhydrous acid can be reclaimed. Lignocelluloses have also been processed by grinding and by exposure to high intensity radiation, but the type of lignin liberated by these chemical processes varies. Reactive lignins are used in glues, resins, adhesives, printing inks, as dye dispersants, absorbents, sequestrents, an oil recovery aid, in asphalt emulsions and polymer applications. The recovery of lignin by solvent extraction yields a crystalline cellulose product that is used for the paper pulp industry and the economic viability of the process depends upon the efficiency of solvent recovery.

The phenol fractionation process is a single mild digestion step in which lignin and hemicellulose are dissolved in phenol at 100 °C and the residual cellulose filtered off and used as paper pulp. As the phenol extract cools, the hemicellulose preferentially partitions to the aqueous phase, leaving a fuel oil and a lignin that can be used to produce further phenol by hydrocracking. The process is thus self-sufficient with regard to energy and phenol requirements. Steam explosion processes convert lignocellulosics to forms in which they are more susceptible to enzymatic hydrolysis, whether in the rumen or not, whilst maximizing the production of reactive lignins for recovery in dilute alkali. Freeze explosion methods utilize liquid ammonia. Autohydrolysis by such steam cracking methods produces residues suitable for subsequent microbial degradation from temperate hardwoods, whereas conifers and tropical hardwoods, which are richer in lignin, often respond poorly to this treatment.

The enzymatic conversion of cellulose to glucose requires the costly production of cellulases and may require the selection of hypercellulolytic mutants. The rate of cellulose hydrolysis achieved is still slow and the hydrolysis can be combined with the yeast fermentation so that glucose is produced and utilized in the same fermenter vessel. This process has now achieved pilot stage evaluation. Considerable research effort has also been directed towards the cloning of the cellulase genes to suitable host organisms to confer the ability to ferment cellulosic substrates. This approach facilitates the fermentative production of chemicals from biomass on both a broader and more efficient basis. An alternative approach to this direct microbial conversion is the development of thermophilic, anaerobic processes, using *Clostridium thermocellum*, which combine hydrolysis and fermentation.

Hemicelluloses can be enzymatically degraded to pentose sugars using xylanases and pentose fermenting yeasts have been found which have a low alcohol tolerance. Genetic manipulation may be used to improve this tolerance or to confer on *Saccharomyces* the ability to utilize xylose by cloning the xylose isomerase gene. This will give the organism the ability to convert xylose, a substrate on which it cannot grow, to xylulose, a substrate which it is able to ferment.

The biodegradation of lignin is an oxidative process, primarily carried out by fungi, but which apparently does not provide sufficient energy to support microbial growth. The microbiology and biochemistry of lignin degradation is receiving increasing research effort. Extracellular enzymes from the fungus *Phanerochaete chrysosporium* have been shown to degrade lignin model substrates, an activity that requires hydrogen peroxide and oxidatively cleaves the propyl side chains from lignin.

The production of pure cellulose from sugars and starch has been described in *Acetobacter xylinum* and opens the way for the production of pure polymer, used for rayon, cellophane and other cellulosic polymers, from lignocellulose, following the removal of lignin.

It is arguable whether or not chitin, a long-chain polymer of *N*-acetylglucosamine, and chitosan, polyglucosamine with only a few *N*-acylated groups, are more abundant sources of biomass than cellulose, although they certainly are less available resources. Chitin can be obtained from Antarctic krill, from the waste of the shellfish industry or from the fungal mycelial waste of the fermentation industry. Both could be considered for a variety of applications, including adhesives, coagulants, drug carriers and paper and textile additives.

4.4 Future prospects

The future impact of biotechnology upon the chemical industry will rely upon the successful integration of basic microbiology and biochemistry with chemical technology. The prime motivation in using biological catalysts in chemistry resides in the ability of enzymes to catalyse enantiomerically pure, stereochemically predictable syntheses. At present, the economics of many biological processes are such that they are unable to compete with the chemical alternatives. It is commonly felt that an added value of approximately £500–1000/t of product over the cost of the starting material is necessary for a large-scale bioprocess for chemical production to be viable. There are, however, several areas in which a potentially exciting future may be identified.

(1) Reaction type: direct oxidation/oxygenation reactions; use of common enzymes for unexpected chemistry; protein engineering to alter catalyst properties; use of biological catalysts in non-aqueous environments.

(2) Reactor configuration: optimization of biocatalysts by genetic engineering, selection of thermostable systems or use of immobilized configurations; development of cheap, stable, efficient and generally applicable methods of cofactor recycling; chemical engineering in relation to large-scale biocatalytic systems (see Chapter 10).

4.4.1 *Reaction type*

Direct oxidation/oxygenation reactions

The use of oxidoreductases in organic synthesis will involve enzymes with strict structural, regio- and stereo-selectivity, for the transformation of complex molecules, and those with a more catholic substrate specificity, which may find application as general reaction catalysts. Nucleotide-dependent dehydrogenases could be used for alcohol/carbonyl conversions. For instance, horse-liver alcohol dehydrogenase is well characterized in terms of its substrate specificity and stereochemistry and will accept acyclic, mono-, bi- and tetracyclic structures. An active site model exists to allow prediction of reactivity with new substrates (see section 4.4.1, 'Use of common enzymes for unexpected chemistry'). Acyclic secondary alcohols are, however, poor substrates and other dehydrogenases (possibly thermophilic) could be sought for this synthetic application. The direct incorporation of oxygen into organic substrates at high efficiency and usually with great selectivity is a characteristic of oxygenases. Such direct oxyfunctionalization of unactivated organic substrates remains largely an unresolved challenge to the synthetic chemist. The possible range of such reactions is wide, including the conversion of alkanes to alcohols, olefins to epoxides, sulphides to sulphoxides, aromatic ring hydroxylation and cleavage and oxidative demethylation. It is possible that a comprehensive understanding of the mechanisms of enzymic oxygen activation and insertion may provide the basis for the construction of future chemical catalytic systems. Much recent interest has focused upon the use of various microbial oxygenase systems in the stereoselective epoxidation of olefins, for the generation of polymeric building blocks. Alkane and alkene-oxidizing microorganisms possess monooxygenases capable of catalysing this transformation, but it is unlikely that any of these processes will be economically viable given the current state of the technology. Due to the instability of the oxygenase enzymes, these processes were considered as whole organism biocatalysis. An interesting enzymatic alternative has been proposed involving a three-step process and oxidases (Fig. 4.14). A pyranose-2-oxidase, isolated from a basidiomycete fungus, promotes the formation of hydrogen peroxide using glucose as a substrate and energy source. A haloperoxidase subsequently mediates the reaction of the peroxide with the alkene and a halogen ion to form the alkene halohydrin, from which the hydrogen and the chlorine are stripped off to produce the alkene oxide. Fructose and gluconic acid are generated as by-products. The stereoselectivity of oxygenation catalysis may be harnessed for the production of stereoregular polymers, by the synthesis of diepoxide monomers. Alternatively, monomers for the synthesis of polymers may be generated by the action of dioxygenases on benzene (see Chapter 5) or the double monooxygenation of biphenyl to produce 4,4-dihydroxy-biphenyl.

Use of common enzymes for unexpected chemistry

Enzymes are, in fact, used relatively little in organic syntheses, primarily because most of the substrates of interest are not the natural substrates of enzymes. It therefore becomes necessary to explore the substrate specificity and reaction conditions of the more commercially available enzymes to see if they can be coerced into performing catalyses of more interest to the synthetic chemist. For instance, glucose oxidase is a commercially available enzyme which is very specific with regard to its electron donor, D-glucose. However, it possesses a wide specificity for the electron acceptor and the natural acceptor, oxygen, may be replaced by artificial dyes. Utilizing this knowledge the enzyme has been used to reduce benzoquinone to hydroquinone (Fig. 4.15), in a reaction that occurs three times faster than with oxygen, and the enzyme will also reduce a number of other quinones, some of which have biomedical applications, and aromatic nitro compounds. Galactose oxidase has been found to catalyse the stereospecific conversion of non-sugar aliphatic alcohols to aldehydes, glycerol being converted to the S-($-$)-glyceraldehyde. Xanthine oxidase could be used for the resolution of β-arylacroleins, as the *trans* isomer is oxidized one hundred times faster than the *cis* isomer.

Altering the reaction conditions under which an enzyme is used can also radically affect the chemistry of the reaction catalysed. Horseradish peroxidase is able to hydroxylate aromatic compounds in low yield and with low specificity. Lowering the reaction temperature from 25 to 0 °C eliminates the side reactions, causing *para*-substituted aromatics to be exclusively hydroxylated at the *meta*-position and vice versa. Hog-liver carboxyl esterase can catalyse transesterification as well as ester hydrolysis, but the latter usually predominates, as the poorly soluble alcohols cannot compete effectively with water molecules. The enzyme can be induced to catalyse transesterification using a biphasic system, in which the enzyme is suspended in a minimal amount of water and the bulk organic phase is composed of the alcohol and the ester.

Protein engineering

Protein engineering can involve chemical modification of an existing protein or genetic engineering to create a modified variant of a natural molecule.

The construction of an appropriate biological catalyst could involve the combination of the specificity of a protein with the catalytic activity

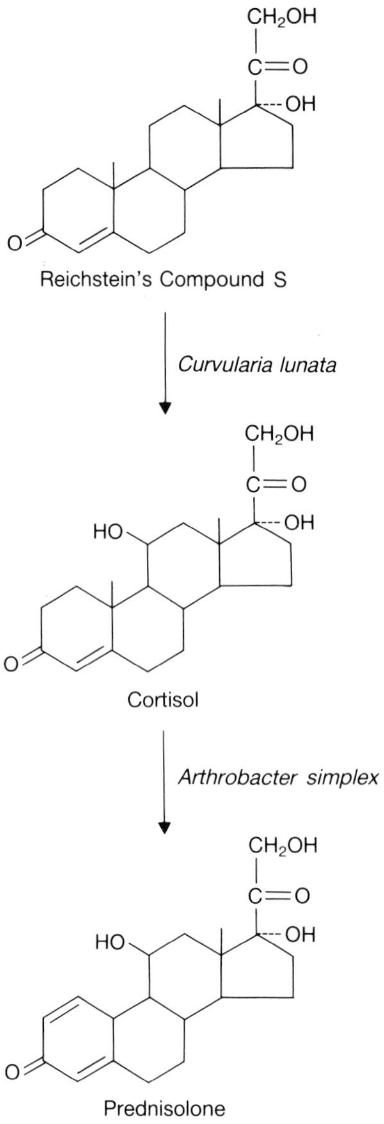

CH₂OH
C=O
---OH

Reichstein's Compound S

Curvularia lunata

CH₂OH
C=O
---OH
HO

Cortisol

Arthrobacter simplex

CH₂OH
C=O
---OH
HO

Fig. 4.13. Conversion of
Reichstein's Compound S
to prednisolone.

Prednisolone

of an organometallic complex. There are several examples of this type of
modification to produce a 'semisynthetic bioorganic complex'. For
example, sperm whale myoglobin can bind oxygen but possesses no
biocatalytic activity of its own. Combining this biological molecule
with three ruthenium electron transfer complexes ($Ru(NH_3)_5^{3+}$, bound
through surface histidine residues, produced a complex that was
capable of reducing oxygen whilst oxidizing various organic substrates,
such as ascorbate, at rates almost equivalent to natural ascorbate

oxidase. Proteins can potentially be modified in other ways. Papain is a well characterized proteolytic enzyme, whose three-dimensional structure has been determined. It possesses an extended groove in the vicinity of the cysteine-25 residue, which is known to participate in the proteolytic reaction and which can be alkylated with a flavin compound without destroying the accessibility of the binding region to potential substrates. The modified flavopapains have been used to oxidize N-alkyl-1,4-dihydronicotinamides and certain of these modified proteins have been shown to have greatly enhanced rates of catalysis over the natural flavoprotein NADH dehydrogenase, demonstrating the construction of a very efficient semisynthetic enzyme. Use of flavins with very strong electron withdrawing substituents in particular positions may possibly allow the generation of efficient catalysts for nicotinamide reduction.

Recent advances in the chemical synthesis of DNA offer the exciting prospect of developing protein engineering to the stage where novel proteins not found in nature could be created. The technology should also be developed to the stage where gene modification could be engineered to alter the protein in a predictable manner to improve some targeted functional property, such as turnover number, K_m for a particular substrate, thermostability, temperature optimum, stability and activity in non-aqueous solvents, substrate and reaction specificity, cofactor requirement, pH optimum, protease resistance, allosteric regulation, molecular weight and subunit structure. Conventionally, this has been achieved by mutation and selection programmes and, more recently, by chemical modification and immobilization. In order to effectively engineer a particular protein molecule, it is necessary to ascertain a series of ground rules concerning the structural features of proteins that confer some specific desirable property. Knowing the accurate crystal structure of the target protein it may then be possible to identify those areas where a predictable modification may be effectively made to improve the catalytic ability of that protein molecule. This will then be accomplished by alteration of the amino acid sequence of the protein.

The first controlled protein modification was performed in the mid-'60s when Koshland & Bender used a chemical modification to change a hydroxyl group to a sulphydryl group in the active site of the protease subtilisin. This thiol-subtilisin did not, however, retain its protease activity. Chemical modifications are, in general, not only harsh and non-specific but are also incapable of effecting many desired changes, particularly if the amino acid residue is buried deep within the protein tertiary structure. Protein engineering by gene modification is, therefore, necessary and, at present, this may be accomplished by two well established genetic engineering techniques (see Chapter 7). Site-directed mutagenesis involves the cloning of the gene for the protein of interest and its incorporation into a suitable genetic carrier. An oligodeoxynucleotide primer, incorporating the desired mutation, is then synthesized and this sequence of ten to fifteen nucleotides retains sufficient homology to the naturally occurring gene to hybridize to the appropriate part of that gene. Polymerases then use this synthetic primer to begin the synthesis of a complementary copy of the vector, which must then be separated from the original and used in the controlled production of the mutant protein. An alternative to this procedure is the cleavage and excision of the site identified for mutagenesis and its replacement with a synthetic segment containing the desired sequences.

Tyrosyl tRNA synthetase catalyses the aminoacylation of tyrosine tRNA by the activation of tyrosine with ATP to form tyrosyladenylate.

(i)

$$CH_2{=}CH{-}CH_3 \ + \ O_2 \longrightarrow CH_2{-}CH{-}CH_3 \ + \ H_2O$$

Propene   NADH₂   NAD   1,2-Epoxypropane

e.g. methane monooxygenases
alkane monooxygenases
alkene monooxygenases

(ii)

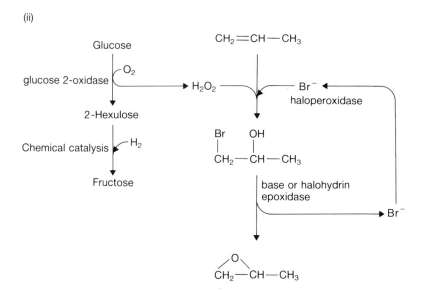

Fig. 4.14. Microbiological production of epoxide monomers. (i) Direct oxy-functionalization using broad specificity monooxygenases. (ii) Multi-step process.

The gene for this enzyme has been cloned from *Bacillus stearothermophilus* and inserted into bacteriophage M13. Specific site engineering has allowed the alteration of the catalytic properties of this enzyme, especially with regard to substrate binding. For example, threonine-51 was replaced by alanine and resulted in a twofold increase in substrate binding due to the removal of possible hydrogen bonding between the residue and tyrosyladenylate. If the residue is replaced by proline, the backbone of the enzyme molecule is disrupted but substrate binding is improved 100-fold because of the increased interaction between the substrate and histidine-48. Similar site-specific changes have been made in the enzyme β-lactamase, usually resulting in the inactivation of the mutant protein. The replacement of serine-70 with cysteine produces a β-thiol lactamase, which possesses a similar binding constant to the wild-type enzyme but has only 1–2% of its activity towards penicillin. The mutant protein is, however, at least as active or more active against certain activated cephalosporins and is also more resistant to proteases.

Mutations brought about by such site directed methods at present serve to confirm structural studies and, in some cases, have established that protein structural stability and catalyst activity can be uncoupled. By extending our knowledge of the interrelationship between protein structure and function it will be possible to finely tune biological catalysts and construct completely synthetic mimics. The cloning of the first synthetic gene for an enzyme, coding for an active ribonuclease fragment, has recently been reported.

Use of biological catalysts in non-aqueous environments

Biological catalysts exist and function primarily in an essentially aqueous environment but the scope of the use of these catalysts in biotechnology cannot be limited by the prerequisite for such a medium. Many desired transformations are of lipophilic or water-insoluble compounds, such as steroids and hydrocarbons. The use of organic solvents may not only increase the catalytic activity of a given biological catalyst, by improved substrate availability, but may also serve to alter the chemical equilibrium of a reaction.

The parameter of prime importance is the stability of the biological catalyst, be it whole organism or isolated enzyme, in the organic solvent. Catalytic activity in many water-immiscible organic phases often dramatically diminishes. Various *Nocardia* spp. have been used in benzene–heptane environments to catalyse the dehydrogenation of steroids. Immobilization has served to increase the stability of these catalysts, but it is then necessary to consider the hydrophobicity of the immobilization matrix. Organisms entrapped within hydrophobic gels showed higher catalytic activity than those entrapped in hydrophilic environments, due to the partition of the substrate between the gel and the external solvent. It is now possible to prepare matrices with controllable degrees of hydrophobicity by using photocrosslinkable resin prepolymers and urethane prepolymers. Enzyme immobilization on or in a rigid support affords increased resistance to distortion in organic solvents, reducing the loss of protein tertiary structure in such environments. Many of the enzymes catalysing the transformation of lipophilic substrates *in vivo* are membrane bound and immobilization serves to mimic this stabilized, membrane-bound state. Chymotrypsin has been used in chloroform to synthesize the ethyl ester of N-acetyl-L-tryptophan from ethanol and N-acetyl-L-tryptophan. Lipase in n-hexane and immobilized in a hydrophobic polymer catalyses an interesterification reaction for the synthesis of a cocoa-butter-like fat from olive oil and a saturated higher fatty acid.

Combination of one or more of the approaches discussed in this section could well lead to a dramatic extension of the range and applicability of biological catalysts in the production and analysis of chemicals for industry.

4.4.2   *Reactor configuration*

Biocatalyst optimization

The type of bioreactor configuration employed in any biotechnological process will be determined by the biochemical and biophysical characteristics of the chosen biological catalyst. The nature of the biocatalyst will also predetermine downstream processing of the desired product. In this regard the improvement of the catalyst by genetic engineering with regard to the physical parameters under which it is capable of operating, the specificity and productivity of the catalyst and the location of the eventual product (e.g. extracellular production of plant cell products) will assume great importance in conceiving the overall process. The importance of genetic manipulation in biotechnology is discussed comprehensively in Chapter 7.

Manipulation of the process characteristics will also be achieved by prudent choice of catalyst and much effort is now being devoted to the isolation, characterization and use of thermophilic organisms and the enzymes derived therefrom in many of the already commercially feasible biocatalytic reactions. The advantages of so doing reside in inherent increased stability of the proteins and their superior catalytic rates. Process improvement may also be sought by use of organisms derived from other extreme environments such as low pH or high salt habitats. Acidophilic halophiles may well receive consideration for organic acid production and halophilic species are already being considered for the production of polyols, as seen in the production of glycerol from *Duniella*.

The use of immobilized biocatalysts, both whole organism and purified enzyme, will have an increasing effect on the design and operation of biotechnological processes as such configurations will enhance the ease of handling, biocatalyst recycling and product isolation. The immobilization process itself may also enhance the stability or performance characteristics of the catalyst, allowing critical reassessment of the economic viability of a particular bioconversion.

Cofactor recycling

Thirty per cent of the known enzymes require one of the five known cofactors (NAD, NADP, ATP, FAD, CoA) for catalytic activity and the use of such enzymes for biocatalysis will depend either on the efficient recycling of these expensive reactants or on ways of circumventing the requirement. Three approaches to this problem exist: enzymatic, chemical and electrochemical.

The enzymatic regeneration of cofactors such as NADH can be achieved by a number of commercially available enzymes that use cheap substrates as the reductant source. Considerable effort has been devoted for this purpose to the use of formate dehydrogenase, an enzyme catalysing the reduction of NADH during the conversion of formate to carbon dioxide. As the product of the reaction is gaseous it may easily be lost from the reaction system and does not complicate product recovery from the reaction mixture. However, loss of the cofactor itself from the reaction system is a major problem, and attempts to recycle the regenerated cofactor within the reactor system have involved the attachment of the cofactor to larger molecules, whilst still retaining its catalytic activity. This immobilized complex may then be retained within the reactor configuration by an ultrafiltration membrane or hydrophobic liquid membrane. NAD has been attached to polyethylene glycol (Mr 10 000–40 000) and utilized in a membrane reactor. Unfortunately, the catalytic activity of immobilized cofactors is often reduced to only several per cent of that of the free form. Coenzymes have also been entrapped within gels with the appropriate regeneration enzymes.

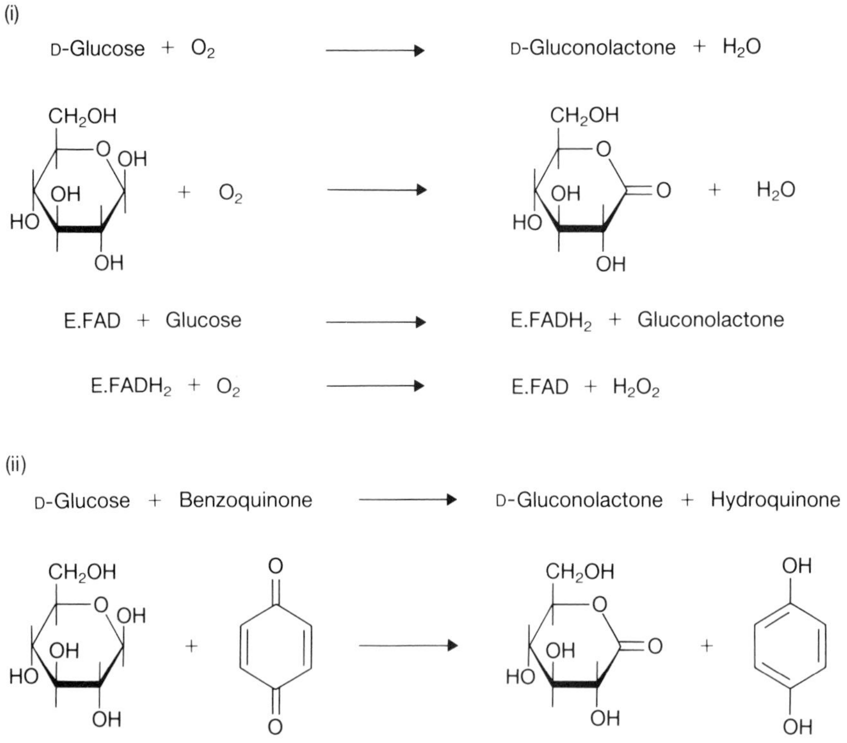

(i)

D-Glucose + O$_2$ $\longrightarrow$ D-Gluconolactone + H$_2$O

E.FAD + Glucose $\longrightarrow$ E.FADH$_2$ + Gluconolactone

E.FADH$_2$ + O$_2$ $\longrightarrow$ E.FAD + H$_2$O$_2$

(ii)

D-Glucose + Benzoquinone $\longrightarrow$ D-Gluconolactone + Hydroquinone

Fig. 4.15. Synthetic organic chemistry using glucose oxidase and artificial electron acceptors. (i) Normal reaction. (ii) Reaction with artificial electron acceptor.

The chemical reoxidation of NADH can be achieved chemically by linking it to oxygen via electron acceptors such as PMS/DCPIP or FMN.

Electrochemical oxidation of NAD(P)H and reduction of NAD(P)$^+$ shows considerable promise. A variety of electrodes have been developed for the oxidation of the reduced coenzymes. Stereospecific electrochemical reduction has, however, proved more difficult. Processes have recently been developed in which enzyme electrodes can be used to effect this stereospecific reduction via the electrochemical reduction of enzyme redox centres. This has been facilitated by the discovery of methods for transferring electrons between electrodes and these redox centres. In some cases, this permits catalysis to proceed in the absence of cofactor. Such cofactor replacement by modified electrodes is proving easier to develop for reductive rather than oxidative reactions.

4.4.3   *Future impact of biotechnology on the chemical industry*

The source of raw materials for the chemical industries will remain closely linked to petroleum for the foreseeable future and chemicals manufactured at low cost from these sources are not appealing targets for an alternative technology. Factors that will be important in establishing the place of biotechnology in this area will include shortages of raw material supply, rising energy costs and the continual need for efficient waste treatment. Decreasing availability of fuel sources will lead to a shift towards the use of biomass resources and fermentation and enzyme technology will continue to compliment the overall chemical technologies. However, there would appear to be only a limited prospect for the large-scale application of biotechnology to high volume commodity chemicals or polymers. Low volume, high added value fine chemicals still remain attractive economic targets for the very special advantages of biocatalytic processes.

4.5   Recommended reading

Current microbiological technology

ATKINSON B. & MAVITUNA F. (1983) *Biochemical and Bioengineering Handbook*. Nature Press, UK.

CAIN R.B. (1980) Transformation of aromatic hydrocarbons. In *Hydrocarbons in Biotechnology*, (eds. Harrison D.E.F., Higgins I.J. & Watkinson R.), pp. 99–133. Heyden, London.

DELLWEG H. (ed.) (1983) *Biotechnology*, Vol. 3: Biomass, Microorganisms for Special Applications, Microbial Products 1, Energy from Renewable Sources. Verlag Chemie, Weinheim.

EVELEIGH D.E. (1981) The microbiological production of industrial chemicals. *Scient. Am.* **245**, 154–178.

GODFREY T. & REICHELT J. (1983) *Industrial Enzymology: The Application of Enzymes in Industry*. The Nature Press, Macmillan, UK.

KIESLICH K. & SEBEK O.K. (1979) Microbial transformation of steroids. In *Annual Reports on Fermentation Processes*, Vol. 3, (ed. Perlman D.), pp. 267–297. Academic Press, New York.

PEPPLER H.J. & PERLMAN D. (eds.) (1979) *Microbial Technology*, Vols. 1 and 2: Microbial Processes. Academic Press, London.

ROSE A.H. (ed.) (1978–1980) Economic Microbiology Series. Vol. 2: *Primary Products of Metabolism*. Vol. 3: *Secondary Products of Metabolism*. Vol. 5: *Microbial Enzymes and Bioconversions*. Academic Press, London.

SCHINDLER J. & SCHMID R.D. (1982) Fragrance or aroma chemicals: microbial synthesis and enzymatic transformation—a review. *Process Biochem.* **17**(5), 2–8.

SCOTT C.D. (ed.) (1983) *Fifth Symposium on Biotechnology for Fuels and Chemicals. Biotechnology and Bioengineering Symposium Number 13*. John Wiley & Sons, New York.

SEBEK O.K. & KIESLICH K. (1977) Microbial transformation of organic compounds: alkanes, alicyclics, terpenes and alkaloids. In *Annual Reports on Fermentation Processes*, Vol. 1, (ed. Perlman D.), pp. 267–297. Academic Press, New York.

SMITH J.E., BERRY D.R. & KRISTIANSEN B. (eds.) (1983) *The Filamentous Fungi: Volume IV Fungal Technology*. Edward Arnold, London.

The generation of chemicals from biomass

CURTIN M.E. (1983) Harvesting profitable products from plant tissue culture. *Biotechnology*, **1**(8), 649–657

EVELEIGH D.E. (1983) Biological routes for cellulose utilisation. In *Biotech '83, Proceedings of the International Conference on the Commercial Applications and Implications of Biotechnology*, pp. 539–548. Online Conferences Ltd., Middlesex, UK.

FLINN J.E. & LIPINSKY E.S. (1983) Production of chemicals via advanced biotechnological processes. In *Biotech '83, Proceedings of the International Conference on the Commercial Applications and Implications of Biotechnology*, pp. 523–538. Online Conferences Ltd., Middlesex, UK.

FOWLER M.W. (1981) Plant cell biotechnology to produce desirable substances. *Chem. Ind.* **7**, 229–233.

JONES L.H. (1983) Plant cell cloning and culture products. In *Biotechnology*, (eds. Phelps C.F. & Clarke P.H.), pp. 221–232. Biochemical Society Symposium Number 48. Biochemical Society, London.

STABA E.J. (ed.) (1980) *Plant Tissue Culture as a Source of Biochemicals*. CRC Press, Boca Raton, Florida.

Future prospects

BROWN G.B., MANECKE G. & WINGARD L.B. (eds.) (1978) *Enzyme Engineering*, Vol. 4. Plenum Press, New York.

KLIBANOV A.M. (1983) Unconventional catalytic properties of conventional enzymes: applications in organic chemistry. *Basic Life Sci.* **25**, 497–518.

MAUGH T.H. (1984) Semisynthetic enzymes are new catalysts. *Science*, **223**, 154–156.

MAUGH T.H. (1984) Need a catalyst? Design an enzyme. *Science*, **223**, 269–271.

NEIDLEMAN S.L. & GEIGERT J. (1983) Biological halogenation and epoxidation. In *Biotechnology*, (eds. Phelps C.F. & Clarke P.H.), pp. 39–52. Biochemical Society Symposium Number 48. Biochemical Society, London.

OMATA T., IIDA T., TANAKA A. & FUKUI S. (1979) Transformation of steroids by gel-entrapped *Nocardia rhodochrous* cells in organic solvents. *Eur. J. appl. Microbiol. Biotechnol.* **8**, 143–155

PYE E.K. & WEETALL H.H. (eds.) (1978) Enzyme Engineering, Vol. 3. Plenum Press, New York.

ULMER K.M. (1983) Protein engineering. *Science*, **219**, 666–671.

WINGARD L.B., BEREZIN I.V. & KLYOSOV A.A. (eds.) (1980) *Enzyme Engineering, Future Directions*. Plenum Press, New York.

WISEMAN A. (ed.) (1977–1983) *Topics in Enzyme and Fermentation Biotechnology*, Vols. 1–7. Ellis Horwood, Chichester.

require acid conditions, and ores and waste materials which consume excessive amounts of acid are, therefore, unsuitable for processing.

There has been little application of bacteria in leaching metals *in situ*. Factors which must be considered when leaching under deep solution mining conditions include the effects of elevated hydrostatic pressure and hyperbaric oxygenation on bacterial activity. Hydrostatic pressure arises as a result of the introduction of leach solutions under pressure and the weight of the water column. Hyperbaric oxygenation results when high-pressure oxygen is injected into the *in situ* ore body to regenerate the oxidizing agent. However, oxygen at this concentration would not be necessary if bacteria were used, since the organisms would regenerate the oxidizing agent. Only oxygen at appropriate concentrations for microbial activity would be required. *In situ* leaching technology has primarily been employed in sandstone-type material and fracturing of the rock has not been necessary because the permeability is great enough to allow solution flow. Bacteria have not been used in this type of leaching. However, if bacteria were used, the question remains whether the growth of the bacteria in the unfractured formation may limit permeability and restrict solution flow. This is not anticipated to be a problem in *in situ* ore bodies which have been fractured by explosives or by other means.

When a microbial process is established to leach metal(s) from a specific ore body, it is highly probable that the process will have to be altered before obtaining optimum application at another ore body of the same type of metal. Although the metal values may be the same, the type of ore mineral and the host rock can be substantially dissimilar. The leaching bacteria respond differently to each mineral. For example, some chalcopyrite is more refractory than other chalcopyrite; the refractory ore is more resistant to direct attack by microorganisms or by bacterial products.

Due to the enormous volumes of material to be processed, leaching reactions take place in the natural environment rather than in highly controlled chambers. Therefore, the microorganisms are subject to the vagaries of weather and dramatic changes in mineralization and pH conditions. Neither the system nor the ore will be sterile, so naturally occurring bacteria will always be present. If specially adapted or genetically altered strains of leaching bacteria are used, these organisms must be able to compete favourably with the natural flora. Development of strains with enhanced ability to oxidize iron or minerals and to tolerate high concentrations of metal or acid could undoubtedly be achieved by genetic manipulation. The limitations to such developments are our current incomplete knowledge of all the microorganisms involved and of the detailed mechanism of sulphide mineral degradation, and our virtually total lack of knowledge of genetic mechanisms (e.g. chromosome mapping, presence and function of plasmids, amenability to transformation or plasmid transfer) in the leaching organisms. This is a rich field for research of potentially very great biotechnological importance.

5.3   Metal transformation, accumulation and immobilization by microorganisms

Prompted by strict environmental regulations, a genuine desire to recover valuable metals, and a need to renovate process waters for recycling purposes, the mining industry has increasingly applied new physicochemical technologies for wastewater renovation. Too often these technologies are expensive and ineffectual. Many mining firms are now discovering that biological processes can be used for waste-water renovation, and that these processes can be more economical and more effective than conventional techniques. Several industries now routinely use biological processes to remove contaminating inorganic ions from mine wastewaters. These systems usually consist of large impoundments or slowly meandering streams in which growth of algae and microorganisms is encouraged. These organisms either accumulate dissolved and particulate metals or produce by-products which render the contaminant insoluble. Biological wastewater renovation processes currently in use generally lack process development, and the technology can be considered unsophisticated. Research in recent years indicates that many microorganisms are capable of accumulating large concentrations of metals and that many organisms possess structural components which can select and bind specific ions. Selection of microorganisms capable of metal accumulation and development of more highly engineered systems for using these organisms to remove all or specific trace contaminant ions from large volumes of process waters would have considerable application to the mining and other industries producing contaminated waste streams.

All microorganisms accumulate metals from their environments, since metals such as iron, magnesium, zinc, copper, molybdenum and many others are essential components of enzymes or pigments such as cytochromes and chlorophylls. In some cases the amounts of metals accumulated are large; the cell water of bacteria may contain a 0.2 M concentration of potassium ions even when potassium is present only at 0.0001 M or less in the habitat. Microorganisms have evolved metal uptake systems that are specific for particular metals and capable of concentrating metals against a tremendous concentration gradient. Microbial metabolism may cause chemical transformation of metals: excreted metabolic products may complex or precipitate metals in solution; some metals can be converted into volatile forms and be lost from solution; and metals may be oxidized or reduced. Iron oxidation $(Fe(II) \rightarrow Fe(III) + e^-)$ has already been considered in metal leaching, but iron (III) may also be reduced to iron (II) by the respiration of some bacteria. Selenate, selenite, tellurate and tellurite may be reduced to the free metalloid by using reducing equivalents from respiration, thus immobilizing these metalloids.

The main mechanisms by which microorganisms immobilize, complex or otherwise remove metals from solution are as follows:

(1) volatilization;
(2) extracellular precipitation;
(3) extracellular complexing and subsequent accumulation;
(4) binding to the cell surface;
(5) intracellular accumulation.

The interactions are summarized in Fig. 5.3. Clearly, there can be overlap among these processes, since one process may be a prelude to or component of another.

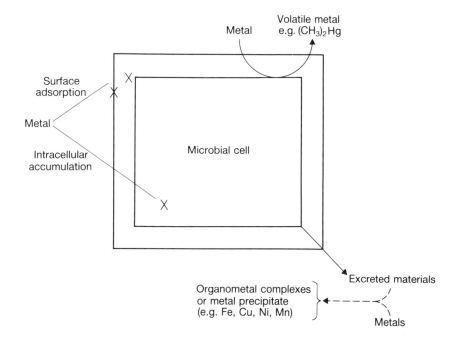

Fig. 5.3. Schematic representation of the interactions which can occur between metals and the microbial cell. Illustrated are surface adsorption, intracellular accumulation, volatilization, complexation with excreted organic material, and extracellular precipitation.

5.3.1 *Volatilization*

It is now well-established that mercury can be methylated by a number of microorganisms. This results in the conversion of Hg(II) ions in sediments or solution into methylmercury compounds (e.g. dimethylmercury), which escape into the atmosphere. This conversion may be an important phase in the natural cycle of mercury. Other metals, such as arsenic, tellurium and selenium may also be microbially methylated and thereby eliminated from soils and water. Such processes may be important in the natural turnover of the metals, and could be significant, e.g. in producing selenium deficient soils or eliminating toxic metal during wastewater treatment. In either case there is a possibility for biotechnological research into respectively decreasing or increasing such activities for man's benefit.

5.3.2 *Extracellular precipitation*

Metals may be immobilized and accumulated in soils or sediments by binding to metabolic products of microbes or to dead organic material that accumulates.

These processes have long been exploited by man in the treatment of sewage and wastewater. Sludge from a normal sewage treatment plant will contain the bulk of pollutant metals removed from the water. Living cells and dead or non-living cellular tissues, which occur in streams and ponds, will accumulate metals and eventually deposit in the sediments. Ponds developing algal 'blooms' (massive growth stimulated by organic or mineral nutrients in the water) are used as a way of removing polluting metals from industrial effluents or from wastewaters of mining operations.

One of the best examples of extracellular precipitation is metal deposition by precipitation with bacterially generated hydrogen sulphide. The sulphate-reducing bacteria are globally distributed in anaerobic environments (lake, ocean and some river sediments, anoxic soils, marshlands, etc). These organisms couple the oxidation of organic matter to reduction of sulphate to sulphide:

$$H_2SO_4 + 8(H) = H_2S + 4H_2O \tag{10}$$

This reaction can result in metal precipitation in soils and sediments leading, for example, to formation of pyrite (FeS_2). Over geological timescales, vast depositions of metal sulphide minerals such as CuS may form. Degradation of algal biomass in sediments leads to sulphide formation. Such a process is shown schematically in Fig. 5.4. Both

Stage 1 Growth and death of algae and sedimentation of biomass

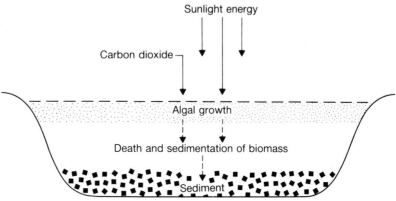

Stage 2 Deposition and accumulation of metal sulphides in sediments

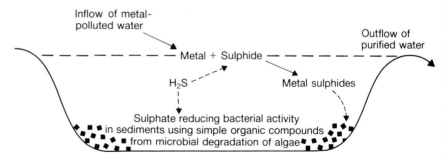

Fig. 5.4. A naturally occurring metal precipitation system in which dead algal material settles to the bottom of an impoundment (Stage 1) and microbial degradation of the organic matter produces hydrogen sulphide which precipitates metals (Stage 2).

processes occur naturally and have been exploited by man on a small scale relative to the total quantity of metal-polluted water that might be treated on a global basis. There is clearly a biotechnological potential in bacterial sulphate reduction as a means of both purifying waters of metals and for the recovery of metals from dilute solutions. Two 'high technology' systems that could be employed are shown in Fig. 5.5. In a one-stage system, a continuous culture of sulphate-reducing bacteria is fed a metal-polluted solution, sulphate and a nutrient such as lactate or, depending on the type of bacteria used, carbohydrate or organic acid waste. To dispose simultaneously of a complex organic waste, mixed cultures of fermentative bacteria, all operating under anaerobic conditions, might be employed. In a more precisely controlled operation, a separate culture of sulphate-reducing bacteria could be used to generate hydrogen sulphide, and the hydrogen sulphide could be pumped through a precipitation vat into which metal contaminated water flowed. Metal sulphide would precipitate, and a flow of metal-free (or at

One-stage system

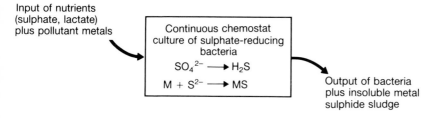

Two-stage system

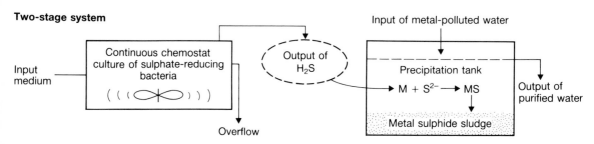

Fig. 5.5. Two 'high technology' systems for removal of metals from solution as metal sulphides. A one-stage system employs a continous culture of sulphate-reducing bacteria; a two-stage system would entail generation of hydrogen sulphide which would be used to precipitate metals in a separate tank.

least substantially decontaminated) water would leave the tank. Surplus sulphide escape could be controlled by balancing the sulphate and metal supply; however, some surplus production of soluble sulphide would be inevitable. Dissolved sulphide could be eliminated by spontaneous chemical oxidation, if dissolved sulphate levels were maintained at permissible levels. Alternatively, sulphide could be oxidized by sulphur-producing photosynthetic bacteria in a third-stage process. Elemental sulphur may possibly be economically recovered. Surplus hydrogen sulphide gas could readily be recycled.

5.3.3   *Extracellular complexation*

Certain microorganisms produce specific chemicals that form very high affinity complexes with some metals. Best known are compounds forming complexes with iron. Molybdenum, vanadium and other trace metals required in metabolism may also be taken into the cell in the form of extracellularly formed complexes. Although these complexing agents do not precipitate metals, their use may provide a novel technology for extraction of specific metals from solution. Complexing agents, isolated from microbial cultures, or chemically synthesized compounds, which imitate biological complexing agents, may be used for immobilizing metals in solution. Metal complexation by these compounds could be enhanced by improving the metal capacity, increasing the reaction rate and developing metal specificity in these complexing agents. Such improvements could be achieved by chemically synthesizing complexing agents using natural products as models or by genetically manipulating microorganisms to bulk produce the desired compounds.

5.3.4 *Intracellular and extracellular accumulation of metals by microorganisms*

Direct accumulation of metals by organisms has been alluded to in the previous sections, and now we shall examine the underlying biochemistry and exploitation possibilities in applied microbiology. The idea of using microorganisms to accumulate metals from solution has long been an attractive one, not only for water purification but for the recovery of valuable or economically important metals. The feasibility of such concentration as an exploitable process is indisputable because of the well-known capacity of living organisms to accumulate metals from dilute solutions. Many plants and animals can concentrate elements up to a million-fold from their environments.

The potential biotechnological application of microbial metal accumulation is shown schematically in Fig. 5.6. Two alternative process streams are apparent: accumulation by organisms in growing culture or accumulation by suspensions of non-growing organisms. Of course, fixed films of organisms in flow-through columns or other biofilter devices are possible. When growing cultures of organisms are considered for use, a rather complex technology is implied. A balance must be achieved among the following factors: (1) organism growth

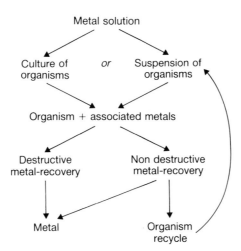

Fig. 5.6. Proposed plans for removal of metals from solution using microorganisms as biosorbents.

(continuous culture); (2) capacity for metal accumulation and hence metal input rate; (3) possible toxicity of the metal to the organisms; (4) competition for the metal to be recovered by the organism and possible complexing agents (organic and inorganic) in the medium; and (5) competition for uptake of the metal by other metals required by the organism. As a 'high technology' process for metal recovery, the use of growing cultures may introduce too many variables and problems. Accumulation by non-growing suspensions or immobilized micro-organisms appears to be a more attractive possibility. The physicochemical condition of the cells and their capacity for metal uptake can be better defined and controlled. Similarly, the metal can possibly be supplied in solution relatively free of complexing agents and at concentrations not lethal to living cells. Having obtained a metal-enriched microbial product, the next problem is metal recovery. Two choices for metal retrieval are the non-destructive recovery of the metal from the organisms or metal recovery by organism destruction. The latter can be accomplished by pyrometallurgical treatment of the organisms or degradation of the organisms with strong acid or base. The method selected would be determined by the ease with which the metal and organism could be separated and the value of the organism as a reusable reagent. Clearly, if the organisms were costly to produce, recycling would be desirable. If, on the other hand, the organisms were available as an inexpensive by-product of another industry (e.g. yeast) and/or the metals to be recovered were exceedingly valuable (e.g. platinum-group metals), a destructive process would be economically acceptable. The choice of organism and process in any biotechnological application must be governed by the basic properties of microbial metal-accumulating systems; a clear understanding of the microbial biochemical process is essential.

Microbial cells may concentrate metals in one or more ways:

(1) extracellular accumulation of metabolic and non-metabolic metals by binding or precipitation on cell walls or membranes;

(2) intracellular accumulation of metabolically essential metals (e.g. K, Fe, Mg, Mo, traces of Cu, Ni);

(3) intracellular accumulation of relatively large amounts of non-metabolic metals (e.g. Co, Ni, Cu, Cd, Ag), primarily by pathways existing for the uptake of metabolically necessary metals.

Uptake of some metals by yeasts and bacteria is almost exclusively by surface binding, as in the case of uranium accumulation by yeast and lead accumulation by *Micrococcus*. Intracellular accumulation may be accompanied by little surface accumulation; potassium uptake is an example. Metal uptake frequently exhibits two-stage kinetics. This is illustrated by Fig. 5.7. Immediately following addition of a metal to an organism, the metal is bound rapidly to the cell surface by energy-independent processes; slower transport into the cell cytoplasm then occurs. The latter process is frequently energy-requiring and depends on the organism actively respiring. This uptake can be blocked by respiratory poisons and by anaerobiosis which inhibit aerobic respiration or energy conservation. Bacteria show a greater initial binding of metals to the cell surface than do most yeasts; thus, surface binding of metals is much lower by *Saccharomyces cerevisiae* than by *Escherichia coli* or *Bacillus* (Fig. 5.8). Surface-bound metals are easily removed from the bacterial wall by a chelating agent or dilute acid; this is illustrated by the easy removal of surface-bound cobalt from *Bacillus* using EDTA (Fig. 5.8). EDTA does not affect intracellular accumulation. For easy removal of metals following rapid binding, bacterial systems seem preferable; however, energy-dependent uptake of metals by yeast often exceeds the total quantity of metals accumulated by bacteria. The extent of surface binding varies considerably among different strains of related bacteria. For example, *Bacillus megaterium* KM (1 g dry weight organism per litre) at 20 °C bind 43 mg cadmium per gram dry weight of organism from a solution of 112 mg Cd/l, as compared to only 10 mg cadmium per gram dry weight by *B. polymyxa*. The high-binding strain, *B. megaterium* (1 g dry weight of cells per litre) binds 38 and 68% of cadmium, respectively, from solutions containing 112 and 11 mg Cd/l.

Subsequent accumulation of metals into a cell generally requires specific transport systems. It is well established that uptake of nickel or cobalt is affected by the magnesium transport system, whereas rubidium uptake is probably influenced by the potassium transport system.

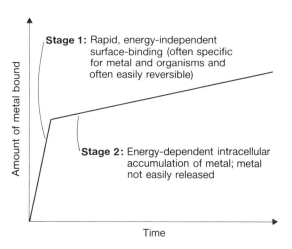

Fig. 5.7. Energy-independent (Stage 1) and energy-dependent (Stage 2) kinetics frequently noted when bacteria accumulate metals.

Stage 1: Rapid, energy-independent surface-binding (often specific for metal and organisms and often easily reversible)

Stage 2: Energy-dependent intracellular accumulation of metal; metal not easily released

Amount of metal bound

Time

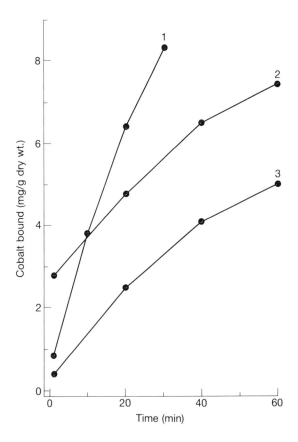

Fig. 5.8. Uptake of cobalt by a yeast and a bacillus from 0.002 M-PIPES buffer solution: (1) *Saccharomyces cerevisiae*; (2) *Bacillus megaterium*; and (3) cobalt remaining with *B. megaterium* following EDTA (1 mM) wash (based on data of Norris & Kelly, 1977, and Norris & Kelly, 1979).

Different metals may compete either for the carboxyl, hydroxyl and other binding sites on the cell surface or for the transport systems. When a metal ion is taken into a cell, ions of an equivalent charge are released from the cytoplasm. Depending on the organism, the cation displaced may be a proton, magnesium or potassium (Fig. 5.9).

The biotechnological application of microorganisms as specific biosorbents for metals is yet to be developed. It has recently been shown that uranium could be absorbed from seawater or from uranium solutions by algae, yeast and *Pseudomonas*. *Pseudomonas* accumulate the uranium intracellularly (Fig. 5.10), while *Saccharomyces* bind the uranium on their surfaces (Fig. 5.11). This bound uranium is easily removed, and the yeast can be reused in the metal recovery process. The potential for development of a useful technology for microbial metal recovery depends on numerous factors:

(1) the specificity of binding must be sufficient to remove a specific metal or mixture of metals from dilute solution;

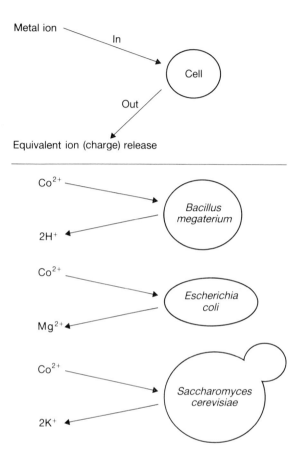

Fig. 5.9. Schematic representation of ionic balance maintained by microorganisms accumulating metals intracellularly.

(2) the metal specificity of the bioaccumulator must be equivalent to physicochemical procedures such as ion exchange, solvent extraction, etc.;

(3) the bioaccumulator must be capable of recovering large quantities of metal;

(4) for the bioprocess to be cost effective, the value of the metals recovered must be equivalent to, or exceed, the cost of producing, and possibly recycling, the organisms; and

(5) the bioaccumulators must not be adversely affected by other materials in the environment.

The extreme versatility of microbial systems and the possibility of significant changes in microorganisms by genetic manipulation suggest that a biotechnology can be developed which meets these specifications. However, speculation in development of this microbial technology must be viewed with caution, since the recovery of metals by microbial processes are governed by the same chemical laws that control metal

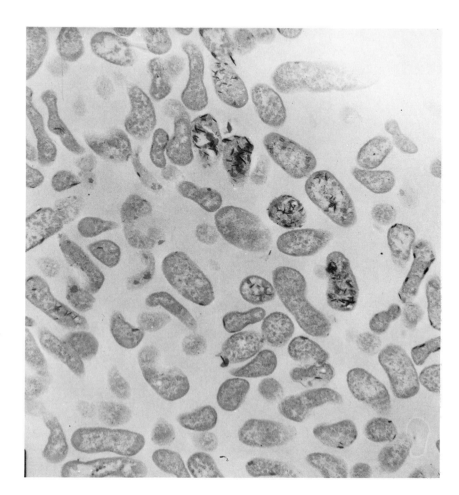

Fig. 5.10. Electron micrograph of *Pseudomonas aeruginosa* showing intracellular accumulation of uranium (magnification: 27 000 ×). Photograph courtesy of G. W. Strandberg, Oak Ridge National Laboratory, Oak Ridge, TN.

recovery by physical and chemical methods. As more is learned about the microorganism's ability to accumulate metals, researchers may discover that microbial metal uptake systems may only serve as models for the development of artificial recovery technologies.

5.4   Biopolymers

5.4.1   *Introduction*

The term 'biopolymers' includes the many large molecular weight compounds, such as nucleic acids, polysaccharides and lipids, that are produced from an extremely wide variety of biological sources. Discussion in this section will be specifically focused upon the production of polysaccharides and poly-β-hydroxybutyrate by microorganisms. These

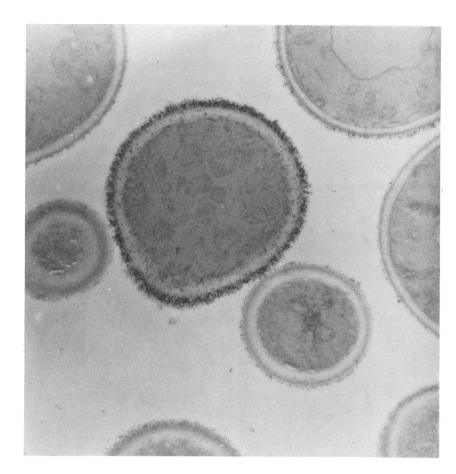

Fig. 5.11. Electron micrograph of *Saccharomyces cerevisiae* NRRL Y2574 showing surface accumulation of uranium (magnification: 35 000 ×). Photograph courtesy of G. W. Strandberg, Oak Ridge National Laboratory, Oak Ridge, TN.

biopolymers are often produced in response to specific environmental conditions when carbon is not the limiting growth nutrient and may, therefore, function as carbon and/or energy reserves (see section 5.4.7).

5.4.2   *Polysaccharides*

Polysaccharides occur as energy reserves and as structural materials in cell walls and extracellular capsules. A large number of such polymers, that have gained commercial importance as industrial gums, have been obtained from plant tissues (seed and seaweed extracts, tree exudates, etc.). The ability of these polysaccharides to alter the rheological properties of water, by causing gel formation and altering the flow characteristics of aqueous solutions, has resulted in their application in a wide variety of industrial situations. Polysaccharide hydrocolloids are used extensively in the food, pharmaceutical, cosmetic, oil, paper and textile industries. For example, carrageenan and agar are produced commercially from red algae, whilst alginates are derived from brown algae. However, there are inherent disadvantages in the production of polysaccharides from plant and algal sources.

(1) The chemical composition of polysaccharides varies in response to metabolic requirements, which, in turn, reflect changes in the environment, e.g. seasonal variations, ageing cycles, time of harvesting, etc. The manufacturer, therefore, has little control over the raw material.

(2) Modification and degradation of the product occurs during processing, as such processing conditions are often harsh treatments (alkali extraction resulting in β elimination and degradation, acid precipitation causing hydrolysis, hot-water leaching, bleaching). Such treatments may also introduce undesirable odour or colour to the finished product.

(3) Plant products are subject to variable supply, dependent upon harvesting, over-harvesting, climatic conditions, disease or pollution.

The production of polysaccharides from microbial sources offers controllable polymer synthesis in constant supply. In addition, microbial polysaccharides often possess unique physical and consistent chemical properties, improved functional characteristics and low biological oxygen demand. Microorganisms synthesize a large range of polysaccharides as extracellular capsules and slimes, unbound to the cell wall. Their composition is limited, in general, to a small number of monosaccharides (neutral hexoses, methyl pentoses, oxo-sugars, amino sugars, uronic acids) but, in combination, they produce polymers with diverse physical properties. However, microbial polysaccharides are still relatively expensive to produce, demanding high capital and energy expenditure and skilled operatives. It is probable that microbial polymers will not displace starch and its derivatives from all their uses. In assessing the feasibility of commercial production of a particular microbial polysaccharide account has to be taken of the following parameters:

(1) the potential annual production capacity and its relationship to the market size, both present and future;

(2) the unique qualities of that product in relation to other microbial and plant polysaccharides;

(3) the economics of production and the prospective product life. Polysaccharide-producing microorganisms are found in a wide variety of environments but there appears to be no definitive isolation strategy. Polymer synthesis has been described in both aerobic and anaerobic cultures. Psychrophilic microorganisms do excrete polysaccharides but

there is little information available about similar activity in thermophiles.

5.4.3   *Polysaccharide production by fermentation*

Bulk polymer production requires a cheap, plentiful carbon source.
Fermentation permits cultivation of the producer organism under
precisely defined environmental conditions, thus controlling the biosynthetic process and allowing manipulation of product type and
product specification. Specific changes in growth conditions can
modulate the molecular weight and structure of the resultant polymer.
In some cases, the maximum rate of polysaccharide synthesis is
achieved during logarithmic growth whilst, in others, it may be a
feature of late logarithmic or early stationary phase. Glucose and
sucrose are commonly used carbohydrate substrates, although polysaccharides can be produced during microbial growth on n-alkanes (C_{12-16}),
kerosene, methane, methanol, ethanol, glycerol and ethylene glycol.
Batch fermentation is disadvantageous in that the fermentation broth
often becomes very viscous and, consequently, the culture quickly
becomes oxygen-limited. Calculation of the relationship between agitation rate and oxygen transfer rate in liquids exhibiting non-Newtonian behaviour is still not possible. Rapid pH changes also have to be
controlled but the method does allow rapid production of the polymer
for purposes of physical characterization and the opportunity to
optimize the medium, especially with regard to the efficacy of different
carbohydrate substrates. Nitrogen is often used as the limiting nutrient
(carbon to nitrogen ratios of 10:1) but other limiting nutrients, such as
sulphur, magnesium, potassium and phosphorus, may be used. The
particular nutrient limitation employed may determine the character of
the polysaccharide, in terms of its viscosity characteristics and degree
of acylation. For example, many fungal polysaccharides are phosphory-
lated. Production under phosphorus limitation may reduce or abolish
the extent of phosphorylation and even alter the ratio of monosaccharides in the final polymer. Potassium limitation tends to reduce polysaccharide synthesis due to an inhibition of nutrient uptake.

   Product recovery from the fermentation broth, after removal of the
microbial cells by centrifugation, is a relatively mild and simple process
involving precipitation with organic solvents (alcohols, acetone), thus
reducing polymer degradation or modification. Due to their greater size,
fungal cells are removed more easily by centrifugation. On the large
scale it often is not feasible to effectively remove the producer organism
and the culture broth is pasteurized, either by heating or homogeniza-

tion. Ease of recovery requires a balance between the optimum conversion of the carbohydrate substrate and the resultant broth viscosity. The addition of an electrolyte frequently assists polymer precipitation in the presence of water-miscible organic solvents. Another variant of this method involves the addition of solvent to a level below that required for polymer precipitation. After solids have been removed by filtration at 100 °C, further solvent is added to precipitate the polysaccharide. Alternatively, culture filtrates may be spray dried, precipitated by the addition of multivalent cations or removed using quaternary ammonium compounds (cetyltrimethylammonium chloride) in conjunction with methanol. This latter method is not often used.

The inherent drawbacks involved in batch fermentation, with the continuous variation in concentration of nutrients, microorganisms and products, may be solved with the introduction of continuous culture. This has been primarily used to date to study the physiology of polysaccharide synthesis but offers the promise of continuous, high productivity. Growth may be controlled by a single limiting nutrient, whilst other parameters are monitored continuously. Thus, the effects of different limiting nutrients and different growth rates may be examined under otherwise constant environmental conditions (e.g. control of dissolved oxygen tension by impeller speed and gassing rate). In certain microorganisms, such as *Xanthamonas campestris* and *Azotobacter vinelandii*, dilution rate does appear to control the rate of polysaccharide production, lower dilution rates favouring higher polymer yields. In others, synthesis is independent of dilution rate. Continuous culture may present problems in terms of strain stability, with the occurrence of non-polysaccharide producing variants. However, this may also be dependent upon the organism concerned or the conditions under which it is cultured. *Xanthamonas juglandis*, a host specific variant of *X. campestris*, exhibited no deterioration in nitrogen-limited culture over 900 h of continuous polysaccharide production. High polymer yields were associated with excellent culture stability. Carbon limitation may well be the nutritional parameter that yields mutants altered in polysaccharide-producing capability in continuous culture. Polysaccharide fermentation may be controlled in a number of ways but constant product yield can be achieved through the regulation of culture viscosity. Culture growth in baffled fermentation vessels may result, however, in localized areas of increased viscosity, due to wall growth or regions of localized polysaccharide accumulation, and viscosity control may not be applicable to such non-uniform growth conditions.

Bioengineering problems that remain to be resolved include the high energy demand for agitation of viscous broths, and the concomitant oxygen limitation due to non-ideal mixing, and the continuous processing of the broth, involving the removal of cells and the use of large amounts of recovery solvents. Due to the low product concentration (often less than 5%, w/v) large fermenters (50–200 m³) will be required. Pilot plant studies help to determine the rheological characteristics during culture scale-up in an attempt to solve these problems and may often involve redesign of fermenters and impellers to deal efficiently with pseudoplastic broths.

Post-production problems in polysaccharide synthesis are primarily associated with the removal of contaminating organisms, which is of extreme importance if the product has applications in the food industry. Lytic and proteolytic enzymes are used to degrade the bacteria but this process in itself may lead to further contamination, as it represents the addition of a potential growth substrate. Polymer treatment with proteolytic enzymes at alkaline pH, followed by addition of a silaceous adsorbent, is used to remove bacterial debris by filtration. More acid conditions may be used to avoid polymer degradation at alkaline pH. Polysaccharide solutions may be clarified subsequently by passage over immobilized endoglucanases, which results in a slight drop in viscosity due to hydrolysis of glycosidic bonds.

Several microbial polysaccharides are now produced commercially (dextran, xanthan, gellan gum, Zanflo and Polytran) and many others are under development. The following section summarizes the major properties and applications, both actual and potential, of microbially derived polysaccharides.

5.4.4 *Microbial polysaccharides: properties, applications and commercial potential*

Xanthan (Keltrol, Kelzan, Rhodogel)

Xanthan is synthesized by *Xanthamonas campestris* during growth on glucose, sucrose, starch, corn sugar or distillers' solubles. The use of industrial wastes, such as acid whey from cottage cheese production, as carbon sources has also been demonstrated. This polymer is a five-sugar repeat unit (D-glucose, D-mannose, D-glucuronic acid) with two types of carboxyl groups (acetate and pyruvate) and an allulose backbone. Molecular weight determinations vary between 2 and 15×10^6. The exact tertiary structure of the polymer is still in dispute but conformation analysis suggests that it exists as a helix. It remains to be established whether this is single or double. Xanthan was the first microbial polysaccharide to be put into commercial production in 1967 and has found a wide variety of applications in industry. This gum produces high viscosity in low concentration and low shear rates. Viscosity is stable over a wide range of pH values, is independent of temperature and has improved resistance in the presence of salts such as potassium chloride. Aqueous solutions of the gum form stable gels in combination with the plant polysaccharide, locust bean gum. It is produced in a conventional, carefully controlled batch process, using commercial D-glucose as a carbon source. The cells are not normally removed before the polymer is precipitated with methanol or iso-propanol in the presence of potassium chloride. Continuous production of xanthan has been investigated but not commercially implemented.

The unique properties of xanthan gum have led to an extraordinary versatility in its commercial application, as a stabilizer and agent for suspension, gelling and viscosity control. The pseudoplastic flow behaviour of the polymer, in combination with its stability towards heat, acid and alkaline conditions and the presence of cations, confers advantages over other lubricants in bentonite muds and xanthan is widely used in oil production. It also finds application in enhanced oil recovery (EOR) where, in combination with surfactants and hydrocarbons, it serves as a mobility control agent in oil field flooding. For EOR, the polymer must be small enough to enter the rock pores to yield high viscosity solutions that are not affected by high salt concentrations, high temperatures and pressures and the shear forces experienced during the pumping operation. The microbial polymer, which must be free of bacterial debris, exhibits a lower degree of retention than the more conventionally used polyacrylamide. Aqueous solutions of the sulphated polysaccharide are also used in oil recovery as they are subject to less shear thinning. In borax complexes, xanthan is exploited as a gelling agent in explosives.

Xanthan was approved for use in the food industry in 1969 by imparting improved 'mouthfeel' to products such as canned and frozen foods, relishes, french dressings, instant foods, toppings, whips and fruit flavoured beverages. Due to the synergistic interaction between xanthan and plant galactomannans, it finds application, in conjunction with locust bean gum, in animal feed such as canned pet food, where it is competitive with agar, and as a suspending agent in low-solids animal feed. Ether and ester xanthan derivatives are used in cosmetics and textiles.

Dextran

Dextran is an α-D-glucan produced by a wide variety of Gram-positive and Gram-negative bacteria, such as *Aerobacter* spp., *Streptococcus bovis* and *S. viridans* and *Leuconostoc mesenteroides*. This polymer is produced commercially by growth of the latter species on sucrose. Most polysaccharides are products of intracellular synthesis but, in the production of dextran, the substrate does not enter the cell. Dextrans are classified on the proportion of three types of linkage class (α-1→3, α-1→4 and α-1→6) present and their water solubility. The high molecular weight polymer is solvent precipitated and subsequently degraded, either enzymatically (using *exo-* or *endo*dextranases), by mild acid hydrolysis or by heat, to generate a product with the correct molecular weight range. The extent of this degradation is controlled by changes occurring in viscosity. Alternatively, low molecular weight primer polysaccharides may be used to initiate polymerization to produce dextrans with the required degree of polymerization. Bacterial seed cultures are often washed several times in saline to remove adhering high molecular weight polymers that might otherwise act as sites of polymerization. In addition to the nature of the glucose acceptor molecule, the molecular weight of dextran is also determined by the concentration of sucrose and the reaction temperature. High sucrose concentrations (70% w/v) favour the formation of a uniform range of low molecular weight dextrans. Using a low molecular weight dextran primer (10 000–25 000) and sucrose (10% w/v, pH 5, 15 °C) a large proportion of the product (50%) has a molecular weight range of 50 000–100 000, which does not require further fractionation for clinical use.

The extracellular enzyme responsible for dextran synthesis, dextran-sucrase (α-1,6-glucan:D-fructose 2-glucosyltransferase, E.C.2.4.1.5.) is inducible in *Leuconostoc mesenteroides* but not in *Streptococcus mutans* or *S. sanguis*, where it is produced during logarithmic phase. The enzyme releases fructose and transfers a glucose residue on to the enzyme-bound acceptor molecule. The growing dextran chain also remains firmly bound to the enzyme. Studies with immobilized dextransucrase have demonstrated that a narrow range of molecular weight products is formed, whereas the soluble system forms very high molecular weight dextrans. Large amounts of this enzyme are produced by dextran synthesizers, which may either be in the soluble form or cell bound. The production of microbially derived dextrans is, however, limited despite the development of such cell-free systems.

Dextrans are primarily used as plasma substitutes (blood expanders) and other medical applications might include its use as a hydrophilic layer on the surface of burns, to absorb fluid exudates. Cross-linked dextran derivatives, to which functional groups (such as carboxymethyl or diethylamino groups) are attached to glucose residues by ether linkages, are widely used for the separation and purification of biological molecules. Dextrans are also sulphated and used as polyelectrolytes.

Microbial alginate

Alginates have been traditionally produced from seaweeds (*Laminaria* spp., for example) but this source of supply is subject to considerable variability. Bacteria such as *Pseudomonas aeruginosa* and *Azotobacter vinelandii* produce alginate-like heteropolysaccharides of D-mannuronic and L-guluronic acids and the process has been commercialized using *Azotobacter* under carbon excess conditions. Microbial alginate differs from the algal product in possessing *O*-acetyl groups associated with D-mannuronic acid residues. The type of alginate produced may be manipulated by altering various parameters. For instance, low phosphate favours the production of a high molecular weight polymer in good yield, whereas the ratio of mannuronic to guluronic acid may be determined by the concentration of calcium ions, which itself affects the epimerization of one acid to the other. Microbial alginate production in continuous culture has also been studied and the yield from sucrose was raised to 50%, compared to 25% in batch culture. Large losses in carbon substrate, through conversion to carbon dioxide, can result from high respiration rates. The highest rates of polysaccharide synthesis in continuous culture have been obtained under molybdenum limitation. Under phosphorus limitation and at low dilution rates there is additional carbon loss through the synthesis of poly-β-hydroxybutyrate by cultures of *Azotobacter vinelandii*.

Alginates from plant sources are predominantly used commercially at present as thickening or gelling agents in dairy products. They are used to stabilize yoghurts, control ice crystal formation in ice creams and, as propylene glycol derivatives, are used in acidic foods such as salad dressings, because they do not gel until the pH is below 3. Alginate polysaccharides gel in the presence of multivalent cations (Ca^{2+}, Sr^{2+}) and have, therefore, been used as a gentle technique for the immobilization of microorganisms. In other circumstances this may cause problems unless the polymer is derivatized to prevent such gelling.

Gellan gum

Gellan, a polysaccharide of glucose, rhamnose and glucuronic acid residues with 3–4.5% O-acetyl groups, is produced by the aerobic fermentation of *Pseudomonas elodea* ATCC 31461 with a carbohydrate carbon source. It occurs in three forms (native, low acetyl and low acetyl/clarified) and the low acetyl form, which is readily produced from the native polymer by heating at pH 10, forms firm, brittle gels upon heating and cooling. Gel strength is a function of gum and salt concentration and the nature of the cations present. Maximum gel strength requires lower concentrations of divalent (Ca^{2+}, Mg^{2+}) than monovalent cations. Although this polymer has not yet been approved for the food industry, it has the potential to replace carrageenan and agar. It has already found application in microbiological media, under the trade name 'Gelrite', where its advantages over agar include improved clarity, equivalent gel strength at half the polymer concentration, reduced toxicity and high resistance to enzymatic degradation.

Zanflo

'Zanflo', produced from *Erwinia tahitica*, is similar in properties to xanthan with the exception that solution viscosities undergo thermally reversible changes (above 60 °C). This high viscosity polysaccharide is composed of fucose, galactose, glucose and uronic acid residues with some esterified O-acetyl groups. Lactose, hydrolysed starch or mixtures of the two are used as the carbohydrate source. With its freeze–thaw stability, resistance to enzymatic degradation and flow and levelling qualities, this polymer, originally developed for carpet printing applications, may find uses in the paint industry. *Erwinia tahitica* does produce a cellulase, which may cause problems if the polymer is used in conjunction with cellulose viscosifiers in paint.

Polytran (Scleroglucan)

Polytran is a linear β-1,3-glucan isolated from the fungus *Sclerotium glucanicum* and related genera, grown in submerged culture on cornsteep liquor. A single D-glucopyranosyl group is linked β-1$\rightarrow$6 to every third residue in the chain. It is pseudoplastic over a broad range of pH and temperature and unaffected by various salts. It finds application in the stabilization of bentonite drilling muds and enhanced oil recovery, as well as ceramic glazes, latex paints, printing inks and seed coatings. This neutral polysaccharide is susceptible, however, to degradation by exoglucanases to glucose and gentobiose.

Many other microbial polysaccharides are now reaching the stage of commercial development and the last five years has seen rapid advances in the isolation and development of various polymers.

Microbial polysaccharides from Alcaligenes *spp.*

The Kelco company of the US has recently developed several polysaccharides, derived from various *Alcaligenes* spp. to the commercialization stage.

The polymer S130, of unknown structure, is produced at high conversion rates in submerged culture. It exhibits high viscosity at low concentration, excellent solubility and viscosity in seawater and brines and has extreme stability at temperatures as high as 149 °C. Most polysaccharides lose their viscosity at temperatures in excess of 93 °C but S130 is not only as viscous at 149 °C as it is at room temperature but is also stable for long periods at these elevated temperatures, especially in the presence of low level oxygen scavengers. Although very viscous at low shear rates, it becomes almost water-thin at high shear rates. With such properties this polymer has wide potential in the oil industry as a good suspending agent and viscosifier for low solids drilling mud.

S194 is similar in properties to S130 but is extremely shear stable, which, in combination with good salt compatibility, allows application as a suspending agent in flowable pesticide suspensions (concentrated suspensions of water-insoluble material) and agricultural suspension fertilizers.

The polysaccharide S198 contains *O*-acetyl and *O*-succinyl substituents and is produced by aerobic fermentation using glucose as a carbon source. It too is highly viscous at low concentration, with good stability over a wide pH and temperature range, but it exhibits a shear stability that confers excellent suspending properties. With its stability to the presence of heavy metal contamination its application as a thickener for water-soluble lubricants, in a move away from petroleum-based hydraulic lubricants, is foreseen.

Curdlan

Curdlan, an α-1,3-glucan, is produced by *Alcaligenes faecalis* var. *myxogenes* strain 10C3 and heating aqueous solutions of this polymer above 54 °C results in non-reversible gel formation. The resultant gel strength depends on the temperature used, being constant from 60 to

80 °C and increasing over the range 80–100 °C. Above 120 °C the molecular structure of this polysaccharide changes from a single- to a triple-stranded helix. Curdlan is insoluble in cold water and gels may also be formed by dialysis of alkaline solutions against water. This polysaccharide has potential as a gelling agent in cooked foods, a molecular sieve, a support for the immobilization of enzymes and a binding agent.

Pullulan

Pullulan, an α-D-glucan polysaccharide composed of α-1→6 maltotriose or occasionally maltotetraose units, is produced by *Aureobasidium pullulans* and forms strong, resilient films and fibres that can be moulded. These films have a low permeability to oxygen, compared to cellophane and polypropylene. Pullulan could find use in packaging or as a flocculating agent in clay suspensions in mining operations. It is resistant to amylases but is degraded by the enzyme pullulanase.

Polysaccharide production has also been described in species of *Arthrobacter*, *Beijerinckia* and methane and methanol-utilizing bacteria.

5.4.5   *Polysaccharide biosynthesis*

Although some bacteria synthesize polysaccharides extracellularly, as in the case of dextrans, most syntheses occur within the microbial cell, thus requiring membrane penetration by the substrate. Nearly all the work carried out on polysaccharide biosynthesis has used organisms that are of little or no commercial importance but results can be extrapolated to these other commercial organisms.

Substrate uptake occurs by facilitated diffusion, active transport (the substrate entering unaltered) or by group translocation (the substrate undergoes phosphorylation). The rate of growth may be dependent on the rate of substrate uptake, which may thus represent the first limitation in polysaccharide synthesis. However, carbon conversion rates are often very high and this may not, therefore, be the limiting factor in the synthetic process. Specific carbon requirements for polysaccharide formation may be associated with specific uptake mechanisms.

Substrate entering the cell is committed to anabolic or catabolic processes, the former involving conversion to polysaccharides, lipopolysaccharides or glycogen. Glycogen is rarely synthesized in proliferating bacteria but would be a possible source of carbon drain in a non-proliferating environment, which might be found in a two-stage production process where microbial growth is followed by polysaccharide synthesis. Control of glycogen synthesis is by allosteric regulation of ADP-glucose production, as there is no involvement of isoprenoid lipids. Exopolysaccharide synthesis is dependent on nucleotide diphosphate monosaccharides, such as UDP-glucose. UDP-glucose pyrophosphorylase assumes a key role, producing precursors for the synthesis of wall polymers, such as teichoic acids and lipopolysaccharides, as well as exopolysaccharides, and strict control over these activating enzymes may allow channelling of intermediates into the synthesis of a particular polymer. Some sugar nucleotides (UDP-galacturonic acid, GDP-mannuronic acid) are precursors of exopolysaccharide synthesis only, whereas others are also activated precursors of monosaccharide formation (e.g. UDP-glucose to D-galactose, D-glucuronic acid). If a monosaccharide occurs within a polysaccharide, then the organism is required to form the activated nucleotide precursor. The one exception is L-guluronic acid that occurs in bacterial alginate. It is formed from GDP-mannose and GDP-mannuronic acid but is not itself activated. The polymer forms using activated mannuronosyl residues which are subsequently acted upon by an epimerase *in situ*. The degree of specificity of this epimerase is uncertain but it acts on non-acylated residues only. L-Guluronic acid has been found in some polysaccharides which do not possess mannuronic acid residues. However, a number of monosaccharide monomers are found in polysaccharides, for which no equivalent activated sugar derivative has been identified (amino- and methyl-uronic acids, methyl hexoses).

Further control over the pool of activated monosaccharides may be exerted through the action of UDP-sugar hydrolases, although these enzymes have been found to be periplasmic. In addition, some enzymes involved in nucleotide sugar synthesis are membrane bound and it is not known whether the products of these reactions will occur freely in the cytoplasm.

Sugars are subsequently transferred sequentially from the nucleotide to a lipid carrier, usually an isoprenoid alcohol phosphate (C_{50}–C_{60}), which has been activated by the initial transfer of a sugar phosphate. In this way sugar repeating units are formed (tetra- to octa-saccharides), from which multiples of the repeat unit are synthesized by transfer to the reducing end of the chain. The exact mechanism of chain elongation and polymer extrusion are still not known. However, the involvement of isoprenoid lipids may not be universal, as they are not found, for instance, in alginate-producing *Azotobacter*. In those organisms where they are found, control of exopolysaccharide synthesis may also reside in the availability of such lipid carriers, which may, in turn, be controlled through the balance between the free alcohol and the phosphate. Control through dephosphorylation would require an ATP-dependent kinase to reactivate the free alcohol. The antibiotic moenomycin inhibits C_{55} isoprenoid alcohol kinase. Most cells are thought to possess sufficient isoprenoid lipid to allow the necessary simultaneous syntheses to occur, but if the organism is limited then polysaccharide synthesis will only be expressed in late log phase or at low temperature. The antibiotic bacitracin binds effectively to isoprenoid lipids, rendering them unavailable for biosynthesis. Mutants with increased resistance to this antibiotic often demonstrate increased polysaccharide synthesis. In addition, mutants unable to synthesize peptidoglycans may make more lipids available for exopolysaccharide synthesis. There appears to be an order of priority in the biosynthetic demand for isoprenoid lipids: peptidoglycan > lipopolysaccharide > exopolysaccharide.

Different types of lipid carrier may be involved in exopolysaccharide synthesis and heteropolysaccharides may require more than one species of carrier. Modifications of the polymer, such as O-acetylation or the addition of pyruvate ketals, may occur whilst it is still attached to the lipid carrier. Acetylation is not necessary for polymer formation and failure to acetylate might be due to the absence of the appropriate acetylase enzymes or of sufficient precursor, presumably acetyl CoA. The introduction of acetyl groups in bacterial alginate, where they are associated with D-mannuronic acid residues, must occur before epimerization takes place and it has been suggested that the acetyl groups may protect the latter from epimerization. The pyruvate/acetate content of xanthan gums varies considerably and the process of polymer modification must be able to encompass these differences. It is not known whether pyruvylation is a single or multi-step process or whether pyruvylation is dependent on the intracellular concentration of phosphoenolpyruvate. The xanthan polymer may also be composed of differing compositions of pyruvylated and non-pyruvylated strands.

The release of the lipid carrier from the polymer has yet to be characterized but may involve a ligase reaction, in which the polymer is removed and attached to the cell surface. Polysaccharides are normally bound to the cell surface upon excretion and the attachment site may be an outer membrane protein. A definite number of attachment sites may exist, which become saturated and cause the excretion of excess polysaccharide as slime. Alternatively, the attachment sites may only accommodate a polymer of a particular size. Slime mutants of capsulate bacteria may either lack the ligase enzyme or the attachment sites themselves. The exact location of the attachment site is unknown but the process of excretion will involve the transfer of the polymer from its site of synthesis, the cytoplasmic membrane, to its final extracellular location, necessitating the passage of a hydrophilic molecule through a hydrophobic membrane. It has been postulated that the adhesion (Bayer) sites of Gram-negative organisms, where the inner and outer membranes associate with each other, may function as export locations but the mechanism and its control are still far from clear.

5.4.6   *Approaches to the improvement of microbial polysaccharide production*

Advances in the use of microorganisms to produce industrially useful polysaccharides may be made by effecting the following improvements:
    (1) increasing the rate and extent of polysaccharide formation;
    (2) modifying the polysaccharide produced;
    (3) altering the surface properties of the producer microorganism to simplify cell separation in subsequent downstream processing;
    (4) eliminating enzyme activities that may make unwanted modifications to the polysaccharide;
    (5) transferring the genetic determinants of polysaccharide synthesis to more amenable host process organisms.

Increase in the rate or extent of conversion of the carbohydrate substrate to the polymer product will require increase in the specific activities of the synthetic enzymes involved, alteration of the control mechanisms of the synthetic process or increased availability of polysaccharide precursors. The number of enzyme steps involved in the synthesis will depend on the complexity of the particular polymer, but any attempt to increase polymer production will require a detailed knowledge of the synthetic pathway and of its metabolic control. At present, improvements in yield are brought about by random mutational events. The rate of substrate uptake may be improved by the duplication of genes involved in uptake mechanisms, but this may not be necessary if the organism possesses several transport routes for each substrate.

Modification of the polysaccharide, for the purpose of enhancing one or more particular characteristics, also requires some prior knowledge of the synthetic pathway. Xanthan forms microgels in aqueous solution due to the interaction of the pyruvate ketals of the polymer with cations. These groups may be removed chemically by treatment with oxalic or trifluoroacetic acid. Alternatively, mutants can be isolated which produce non-pyruvated or non-acetylated xanthans that are otherwise unaltered in carbohydrate structure and have lower viscosity. Screening of such mutants is, however, difficult as the products are often indistinguishable from the parent polysaccharides, although the colonies may differ in appearance and have a tendency to adhere to the surface of the growth media. Other solutions to this problem involve the growth of the wild-type bacteria in the presence of acylation inhibitors or under potassium or magnesium limitation. Similarly, mutants of polysaccharide-producing microorganisms can be isolated that synthesize polymers of increased molecular weight and, therefore, increased viscosity but that are otherwise identical in chemical composition. Polymer chain lengths may also alter with growth conditions. For example, *Xanthamonas juglandis* in continuous culture produces longer unbranched molecules at lower dilution rates.

Alteration of the surface properties of the producer microorganism, e.g. by loss of surface polymeric material such as lipopolysaccharides, eases polysaccharide harvesting. Such mutant cultures will autoagglutinate, spontaneously flocculate and reduce the amount of centrifugation required. However, care must be exercised to ensure that such mutants do not 'leak', losing cell material, such as proteins from the periplasmic space, or lyse to contaminate the final product. Other surface alterations involve the mutation of capsular organisms to stable, slime-forming bacteria or the isolation of phage-resistant mutants, to reduce the risk of phage contamination during the production run.

Some microorganisms produce exopolysaccharides that are subsequently degraded by hydrolytic enzymes. *Azotobacter vinelandii*, for example, synthesizes alginate and an alginase (alginate lyase). Xanthan producers often secrete an active cellulase that may be the cause of unwanted degradation if xanthan is subsequently added to cellulose-based products. Control of such unwanted enzymic activity may be achieved either by careful manipulation of culture conditions or by the use of mutants unable to produce such hydrolytic enzymes.

Polysaccharide-producing strains also synthesize other polymeric products, such as other polysaccharides, poly-β-hydroxybutyrate or glycogen, which may represent a considerable carbon drain from the desired synthetic process. Mutants defective in these alternative synthetic pathways should be sought. Glycogen formation in prokaryotes is a distinct process from other pathways of polysaccharide synthesis and isolation of mutants defective in the key enzymes of this pathway (ADP-glucose pyrophosphorylase and glycogen synthetase) effectively solves this problem. However, prior knowledge of the regulatory mechanisms of both pathways is required for this approach.

Some species of curdlan producers also synthesize considerable quantities of succinoglucan. Mutants of *Alcaligenes* spp., *Agrobacterium radiobacter* and *Rhizobium trifolii*, devoid of succinoglucan, have been isolated. Four types of mutants have been isolated: succinoglucan producers, curdlan producers, producers of both polymers, and mutants that synthesize no curdlan and little succinoglucan. Succinoglucan is a heteropolysaccharide of glucose and galactose and both polymers are acetylated and pyruvylated. The mechanism for the alteration in polymer production patterns is not known.

The transfer of genetic determinants of polysaccharide synthesis may be advantageous under a number of circumstances. For example, transfer of alginate synthesis from a strain of *Pseudomonas aeruginosa*, originally isolated from a patient with cystic fibrosis, to more harmless, non-pathogenic *Pseudomonas* spp. would allow this synthetic capacity to be employed commercially. A chromosomal locus has been postulated for the genes of alginate synthesis. Transfer of the genes for xanthan synthesis to a host that is not pathogenic to plants would also be advantageous. Alternatively, polysaccharide synthesis could be introduced into bacterial strains that possess faster growth rates. It has been found that strains unable to synthesize exopolysaccharides could be converted into polymer-producing strains by selection for carbenicillin resistance. This technique has been successfully applied to a variety of *Pseudomonas* spp. to produce *muc* mutants, which excrete polysaccharides. It would appear that, in such cases, alginate synthesis in the wild-type organisms has been suppressed and mutation of the relevant genes has caused subsequent expression.

5.4.7   *Poly-β-hydroxybutyrate (PHB)*

Poly-β-hydroxybutyrate (PHB) is a thermoplastic polyester consisting of repeat units of the formula $-CH(CH_3)-CH_2-CO-O-$. For over fifty years, PHB has been recognized as an energy reserve material accumulated by a wide variety of microorganisms (e.g. *Alcaligenes*, *Azotobacter*, *Bacillus*, *Nocardia*, *Pseudomonas* and *Rhizobium* spp.). Under certain conditions, some species, such as *Alcaligenes eutrophus* and *Azotobacter beijerinckii*, can accumulate up to 70% of their dry weight as this polymeric material. The intracellular accumulation of PHB during growth on carbohydrates, or other carbon sources such as methanol, is promoted by phosphorus or nitrogen limitation. Production of the polymer may, therefore, be a one-stage process, in which the nutrient limitation is imposed from the onset of growth, or a two-stage process, where the organism is first grown under no limitation, to produce sufficiently high concentrations of biomass, followed by nutrient limitation, other than carbon, to promote high levels of intracellular PHB. Although many species are capable of PHB accumulation, both the extent of that accumulation and the molecular weight of the particular polymer are too low to render them useful for commercial polymer production. PHB content of the biomass needs to be at least 35–40% of the dry weight and the molecular weight of the polymer should be of the order 200 000–300 000. As PHB is produced as intracellular granules, the polymer is extracted from the microorganisms after cell breakage. Solvent extraction into halogenated hydrocarbons, such as 1,2-dichloroethane, is the preferred method of polymer purification. An alternative approach involves the moulding or extrusion of the dried cells of the microorganism itself, containing at least 50% by weight of the PHB polymer. In order to make the composition melt processable for the production of shaped articles, the cells are subject to cell breakage and, if necessary, mixed with the necessary amount of extracted polymer. In addition, the polymeric properties of PHB may be engineered by altering the composition of the homopolymer *in situ*. Although this energy reserve material has always been thought of as a simple polyester of β-hydroxybutyrate monomers, it is now evident that in many cases the polymer is a more complex heteropolymer of various β-hydroxy fatty acid monomers. Therefore the composition of the polymer may be manipulated during growth, e.g. by the addition of propionic acid to the growth medium, which gives rise to β-hydroxyvalerate units in the resultant copolymer. This, in turn, will alter the rheological properties of the polymer.

The advent of microbially produced PHB heteropolymers will not have a significant affect on the burgeoning plastics industry, but the special properties of the polymer, especially with regard to its biodegradability, may find application in certain specialized areas.

5.4.8   *Other microbially derived polymers*

The biopolymers described so far are all synthesized totally by the microorganism during growth upon a particular carbon source. There are alternative routes to novel polymeric materials, which involve the intervention of the microorganism at one stage of the synthesis. In this mode the microorganism is used merely as a microbiological catalyst to perform an otherwise difficult chemical transformation. The use of microbially mediated hydroxylation reactions to produce phenols or dihydrodiols for further chemical polymerization has already been mentioned in Chapter 4. A case in point is the biotechnological route proposed for the production of a new polymer, polyphenylene, which is of interest to materials scientists because of its thermal stability and electrical conductivity. The polymer could be synthesized from benzene and oxygen, using genetically modified *Pseudomonas putida* to effect the initial conversion of benzene to the dihydrodiol. This dioxygenated product is excreted from the cell, isolated by solvent extraction and then derivatized for the chemical synthesis of polyphenylene. Other possibilities of using such technology exist for the production of polymers from naphthalene and biphenyl.

5.5   The biodeterioration of materials

5.5.1   *Introduction*

Although the word 'biodeterioration' has only recently been officially accepted as part of the English language, it has been with man as a process ever since he began to fashion natural raw products and store his food. The organisms responsible for the biodeterioration process have almost certainly been with us for a much longer period, providing a valuable link in the recycling of elements through the biosphere. Man soon recognized the need to prevent his materials and foodstuffs from being recycled by the use of preservation or delaying techniques in order to retard the activities of the organisms. Preservation as an empirical process has spanned many thousands of years and there are ample records which describe its use.

However, it required the advent of the microscope and the ability to selectively grow and identify the agents of biodeterioration, for the principles of the process to be more fully understood and studied in any proper scientific fashion.

It took the Second World War to initiate studies on the process of biodeterioration, as a result of the influx of biologically susceptible materials into theatres of war, where the high temperature and humidity accelerated the breakdown of these materials on a large scale. Since that time, there has been an increasing awareness of the problems that biodeterioration can cause to products fashioned from natural raw materials and, more latterly, fashioned from synthetic raw materials. Ecological, physiological and biochemical disciplines need to combine with non-biological areas such as those of engineering and materials science in order to give us a complete understanding of the process involved.

5.5.2   *Definition of biodeterioration*

Biodeterioration is defined as 'any undesirable change in the properties
of a material brought about by the vital activities of organisms'. It is a
process which, broadly speaking, results in the lessening of the value of
a material. Hence the change in the properties of a material refers to
those properties which make that material suitable for a particular
in-service use. The change may be mechanical, physical, chemical or
aesthetic in nature and need not result in chemical breakdown of the
material. This latter point is an important distinction between biodeterioration and biodegradation. The former is a broader term, whereas
biodegradation is restricted to the breakdown of a product, often a
waste, released into the environment, such as oils, pesticides or
detergents.

In general terms, biodeterioration is an undesirable process, whilst
biodegradation is generally regarded as desirable.

The use of the word 'organism' suggests a cross-section of the animal
and plant world to be involved and indeed this is so. Although the
microorganisms have been extensively studied and are well represented
in the literature as biodeteriogens, the effect of insects, rodents, green
plants (including the algae) and even birds should not be underestimated.

5.5.3   *Classification of biodeterioration processes*

The use of the word 'materials' in the definition of biodeterioration
means that it is 'dead' and thus distinct from the study of living
matter—pathology. The dividing line is often very fine and can overlap,
in that organisms found in the living or dying matter remain in the dead
material and may continue their activities, reducing the value of the
product during storage. However, in many cases once the host has died
the nutrient conditions and changes in the cell constituents will alter
the type of organism colonizing the material. Thus, we talk of 'field
fungi' and 'storage fungi' in connection with the fungal attack on
cereals.

The processes of biodeterioration can be conveniently but artifi-
cially divided into three types (Table 5.1). There is often considerable
overlap but it is very valuable to know that these exist when deciding
upon a line of investigation or the recommendation of a preservative
regime.

Process	Examples
(1) Mechanical	The gnawing and boring of non-nutritive materials, such as lead pipe and plastic cables, by rodents and insects. Damage to road surfaces and walls due to growth of plants
(2) Chemical	
(i) Assimilatory	The utilization of substrates in the material as a nutrient source, e.g. cellulose in wood and keratin in wool
(ii) Dissimilatory	The organism produces a metabolic product, such as an acid or a toxin, which may corrode or render the material unfit for use
(3) Fouling and soiling	Certain organisms in particular situations can block up pipes or foul the hulls of ships. The staining of decorative finishes and plastic shower curtains may be the result of fungi growing, not on the material but on contamination on the surface

Table 5.1. Classification of biodeterioration processes

5.5.4   *Materials subject to biodeterioration*

In any discussion on biodeterioration, it is easier to classify the topics according to product types. This becomes difficult, however, when dealing with composites such as paint, where there may be a combination of raw materials, such as cellulose and a synthetic polymer, which cut across the product classification. It has been shown, however, that the environment in which the product is stored and used often has a significant effect on the organisms which colonize it and the activities of that organism. The ensuing sections will thus briefly review the products that have been recorded as being subject to biodeterioration. For a fuller appreciation, the reader should consult the recommended reading list at the end of this chapter.

Foodstuffs

The post-harvest decay of raw foods is an area which is much neglected in those parts of the world where food is in greatest demand. In the developed countries, food is protected in a number of ways to prevent fungal, insect and rodent attack such that losses are minimized. During storage of grain it is necessary to use chemical and physical barriers such as pesticides and drying procedures. There has been much concern over the presence of mycotoxins in food which has been contaminated by fungi, often at an early stage in its storage life. This may lead to the condemning of large shipments of grain, particularly those used for animal feed.

Processed foods in many cases require more protection from contamination. Packaging can often act to both discourage and encourage the growth of microorganisms, whilst chemical preservatives are limited in chemical composition, and thus type, by legislation.

Cellulose

Cellulose, in its native form as various fibres and as wood, has provided us with a range of materials and products over the centuries. Products made from cellulose-based materials have probably received most attention from research workers in the field of materials protection, the wood and textile industries being supported by a significant preservative industry. Although cellulolytic species represent only a small percentage of the total number of fungal and bacterial species known, the breakdown of cellulose-based products is widespread and can be rapid under appropriate conditions. Cotton textile buried in the soil will completely lose its textile strength within 10 d at 25 °C. The fungi appear, from our present knowledge, to be the main agents of cellulose attack. The environmental conditions (relative humidity less than 90%, very low nitrogen content, acid pH) in cellulosic materials often favour the fungi. The ramifying nature of fungal hyphae penetrating cell walls to get close to the cellulose, which is often bound up in a matrix of lignin and hemicellulose, may also be an advantage. The bacteria undoubtedly have a role to play, certainly in utilization of the pectin layers and degradation of the pit regions in softwoods, allowing movement of water and hence bacteria into the wood.

The distribution of nutrients in wood will determine its susceptibility to biodeterioration. Sapwood contains soluble nutrients more easily available to microorganisms and if the wood is not seasoned properly and the outer sapwood planed off to remove the migrating nutrients, typical blue stain and superficial decay may commence if the wood is subsequently wetted.

Cellulose may be chemically modified to improve or extend its properties. Cellulose acetate, rayon and the various substituted celluloses such as carboxy- and ethoxycellulose all increase microbiological resistance, probably by reducing hydrolysis reactions. The substituted celluloses act as extenders and viscosity agents when incorporated into emulsion systems.

Animal products

Most of the animal products susceptible to biodeterioration are proteinaceous in origin. They include animal hides, wool, and glues. The quality of hides and wool is often adversely affected by bacteria and fungi during the early processing stages. Whilst still at the slaughterhouse or in the shearing shed the warm hide or wool, with its high attendant contamination, can begin to deteriorate within 48 h. Tanning of hides and degreasing of wool can help preserve the products. Some tanning procedures, however, involve an initial soak in water and, if this is prolonged at elevated temperatures, bacterial growth occurs. Leathers used for book covers are often high in sugar content and this encourages mould growth in damp storage conditions.

Wool is generally most susceptible during its processing, whilst glues may suffer from both 'in-can' storage problems and dry film attack.

Surface coatings

Surface coatings such as paints, lacquers and varnishes have a dual function. They act as decorative finishes, and also protect a surface against environmental ravages, including microbiological attack. The phasing out of lead in paints and the influx of emulsion paints has led to biodeterioration problems with the paint itself. This can occur both in the can and as a dry film. Most of the research in this area revolves around designing efficacious preservative systems that will remain active over the service life of the paint. Paint contains pigments, binders, emulsifiers, oils, resins and wetting agents and may be water or solvent thinned. Although some of these ingredients, such as casein, starch, cellulose and plasticizers, are susceptible to microbial utilization, alternative non-utilizable components are often not possible. The colonization of paint films by microorganisms is very much dependent upon environmental factors such as temperature, humidity, the weathering of the film and the settling of nutrients such as wind-blown agricultural fertilizers. In-can preservation problems are often due to bacteria whilst deterioration results from fungal colonization. In addition, extracellular enzymes, such as those which form the cellulase system, are able to persist in liquid emulsion paints and cause losses in viscosity.

Rubbers and plastics

Rubber and plastic products are formulations which contain rubber or a particular synthetic polymer. Up to 50% of the formulation may be made up of additives put in to confer plasticizing, antioxidant, antihydrolysis and UV stabilizing properties on the product. In addition, there may be fillers and pigments. Many of these additives are more susceptible than the polymer backbone itself. Thus polyvinyl chloride (PVC) in its unplasticized form is extremely resistant to biodeterioration. However, it is a rigid material, which limits its uses. The addition of a plasticizer, which is often an organic acid ester, confers flexibility on the product but also increases susceptibility and resultant embrittlement.

Natural rubbers consist of regularly repeating isoprenoid units which are subject to oxidation, resulting in an increased tendency for microbiological attack to occur. Cross-linking and the addition of antioxidants slow down this process. Synthetic rubbers have been developed with superior properties, including biodeterioration resistance. Silicone, nitrile and neoprene rubbers are extremely resistant to microbial attack.

Although, in general, the synthetic polymers used as plastics, such as polyethylene, polystyrene and PVC, are resistant in themselves, some polymers which are finding increasing usage are susceptible. These are the polyurethanes, an ill-defined group of polymers based loosely upon the presence of an urethane group, but containing substituted urea, biuret, amide and allophanate groups within the final molecule. Ethers and esters are also chemically locked into the molecule to confer elasticity on the product. Most of these groups are susceptible to catalytic hydrolysis and probably enzyme hydrolysis under suitable conditions. The configuration of the molecule probably controls the extent to which the sites are available for attack, but it is known that the ester bonds are particularly prone to microbiologically mediated hydrolysis. The use of antihydrolysis agents has been shown in at least one instance to reduce microbiological susceptibility but there remains a large gap in our knowledge of the breakdown pathways of polyurethanes. Their superior abrasion and hard wearing properties have guaranteed them a large niche in the plastics market and it is unlikely that they will be replaced in the near future.

Fuels and lubricants

The products under this heading are primarily petroleum based fractions which are essentially hydrophobic. However, when in contact with water a number of phenomena can occur involving microorganisms. Like plastics, they contain a number of additives to improve performance and these may enhance the susceptibility of the product to biodeterioration.

Kerosene is widely used as an aviation fuel in jet engines. In the presence of water, which is inevitable in storage tanks and occurs in fuel tanks on aircraft, fungal growth takes place both at the fuel–water and the water–metal interfaces. If unchecked, the growth eventually becomes detached and moves into the fuel supply system for the engines. Filters and fuel lines may become blocked, and gauges give faulty readings. One particular fungus, *Cladosporium resinae*, is repeatedly isolated from this type of environment. The movement of fuel around the aircraft increases aeration and in supersonic aircraft the fuel acts as a heat sink. Temperatures within some of the fuel tanks reach up to 55 °C and it has been suggested that this will encourage thermophilic fungi.

The heavier fraction of petroleum consists of oils, which are used in industry in hydraulic systems as straight lubricants, and to facilitate the machining of metals on lathes and grinding operations. The ingress of water into these systems has resulted in failure of the systems due to either blockages, the breakdown of the efficiency of the lubricant or the initiation of corrosion due to acidic microbial by-products. In the engineering industry metal working fluids exhibit the greatest problems of microbial infection, resulting in emulsion breakdown, corrosion of machined parts, and a shortening of lathe tool-life. The relative roles of the microorganisms in deteriorating these types of oils are not fully appreciated, but those capable of directly utilizing the oil, degrading the emulsifier and capable of producing hydrogen sulphide appear to be readily isolated under suitable conditions.

Metals and stone

There is now strong evidence which links certain microbiological activities with corrosion processes. There are three possible mechanisms: production of corrosive products such as acids, hydrogen sulphide or ammonia, production of differential aeration cells, and cathodic depolarization. These activities result in the pitting and eventual perforation of fuel storage tanks with water bottoms, aluminium fuel tanks on aircraft, and the formation of tubercles in iron water pipes which impede water flow. Conventional techniques for corrosion protection are employed, such as the use of chemical inhibitors, surface coatings or the use of sacrificial anodes.

Stone is subject to a number of chemical and environmental factors which contribute to its erosion. The problem of biodeterioration of stone arises when the stone is fashioned into buildings and monuments and it becomes necessary to conserve the structure in its original form. It is extremely difficult to demonstrate the role of microorganisms in the deterioration process, as with metals. However, several mechanisms have been put forward. The first is a mechanical effect in that the growth of the organism on the structure encourages the retention of water, which on freezing and thawing results in flaking of the surfaces. The second mechanism involves expansion and contraction of the microbial cell or filament, and the third is the result of chelation of insoluble minerals in the stone by excreted organic acids. Bacteria have been demonstrated to be able to release insoluble phosphates and silicates by production of 2-ketogluconic acid. The lichens, so far not mentioned, appear to play an important part in stone deterioration. Their ability to contract and expand (between 15 and 300% moisture in 2–3 h) in response to wet and dry conditions and the way in which they are able to penetrate rocks demonstrate their potential in this area. Recent work, using sectioned rock samples colonized by lichens, has shown that deep penetration of the fungal symbiont can occur and that there is selective solubilization of the mineral constituents.

5.6   Recommended reading

Microbial leaching

BRIERLEY C.L. (1978) Bacterial leaching. *Crit. Rev. Microbiol.* **6,** 207–262.
BRIERLEY C.L. (1982) Microbiological mining. *Scient. Am.* **247,** 42–51.
FENCHEL T. & BLACKBURN T.H. (1979) *Bacteria and Mineral Cycling.* Academic Press, London.
MURR L.E., TORMA A.E. & BRIERLEY J.A. (eds.) (1978) *Metallurgical Applications of*

Bacterial Leaching and Related Microbiological Phenomena. Academic Press, New York.

POTTER G.M. (1981) Design factors for heap leaching operations. *Mining Eng.* **33**, 277–281.

Metal accumulation

KELLY D.P., NORRIS P.R. & BRIERLEY C.L. (1979) Microbiological methods for the extraction and recovery of metals. In *Microbial Technology: Current State, Future Prospects*, (eds. Bull A.T., Ellwood D.C. & Ratledge C.). Cambridge University Press, Cambridge.

NORRIS P.R. & KELLY D.P. (1979) Accumulation of metals by bacteria and yeast. *Devs ind. Microbiol.* **20**, 299–308.

NORRIS P.R. & KELLY D.P. (1977) Accumulation of cadmium and cobalt by *Saccharomyces cerevisiae*. *J. gen. Microbiol.* **99**, 317–324.

NORRIS P.R. & KELLY D.P. (1982) The use of mixed microbial cultures in metal recovery. In *Microbial Interactions and Communities*, Vol. 1, (eds. Bull A.T. & Slater J.H.). Academic Press, London.

STRANDBERG G.W., SHUMATE S.E. & PARROT J.R. (1981) Microbial cells as biosorbents for heavy metals: accumulation of uranium by *Saccharomyces cerevisiae* and *Pseudomonas aeruginosa*. *Appl. Environ. Microbiol.* **41**, 237–245.

TUOVINEN O.H. & KELLY D.P. (1974) Use of microorganisms for the recovery of metals. *Int. Metall. Revs* **19**, 21–31.

Biopolymers

BAIRD J.K., SANDFORD P.A. & COTTRELL I.W. (1983) Industrial applications of some new microbial polysaccharides. *Biotechnology*, **1**, 778–783.

BERKELY R.C.W., GOODAY G.W. & ELLWOOD D.C. (eds.) (1979) *Microbial Polysaccharides and Polysaccharases.* Society for General Microbiology and Academic Press, London.

DAWES E.A. & SENIOR P.J. (1973) Energy reserve polymers in microorganisms. *Adv. Microbial Physiol.* **10**, 136–266.

SUTHERLAND I.W. (1982) Biosynthesis of microbial exopolysaccharides. *Adv. Microbial Physiol.* **24**, 79.

SUTHERLAND I.W. (1983) Extracellular polysaccharides. In *Biotechnology*, Vol. 3: Biomass, Microorganisms, Products 1, Energy, (ed. Dellweg H.). Verlag Chemie, Weinheim.

The biodeterioration of materials

EGGINS H.O.W. & OXLEY T.A. (1980) Biodeterioration and Biodegradation. *Int. Biodeterioration Bull.*, **16**, 53–56.

GILBERT R.J. & LOVELOCK D.W. (eds.) (1975) *Microbial Aspects of the Deterioration of Materials.* Academic Press, London.

ONIONS A.H.S., ALLSOPP D. & EGGINS H.O.W. (1981) *Smith's Introduction to Industrial Mycology*, 7th edn. Edward Arnold, London.

ROSE A.H. (ed.) (1981) *Microbial Biodeterioration. Economic Microbiology*, Volume 6. Academic Press, London.

SEAL K.J. & EGGINS H.O.W. (1981) The biodeterioration of materials. In *Essays in Applied Microbiology*, (eds. Norris J.R. & Richmond M.H.). John Wiley & Sons, Chichester.

6   The Environment and Biotechnology

D. J. BEST, J. JONES & D. STAFFORD

6.1   Introduction

The maintenance of environmental quality has been a continuing problem since the dawn of civilization. Man, through his industrial, agricultural and domestic activities, has bought about physical, chemical and biological modifications to his environment, many of which have had a deleterious effect. Biotechnology will exert an increasingly important impact on the control and improvement of environmental quality in a variety of ways. This is most obviously witnessed through improved or innovative waste treatment procedures, but there is also a fundamental influence, which will alleviate many problems at source, shown in the increasing use of biotechnology in the chemical and agricultural industries.

This chapter examines the treatment of waste products, giving both a historical perspective and consideration of the effect biotechnology has already and will continue to make. The role biotechnology plays in more recent threats to the environment in the treatment of xenobiotics and oil pollution is also covered.

Industries are rapidly developing which use the activity of microorganisms to produce recycling systems, effluent control, alternative energy resources and chemicals for use in industry, as well as a new approach to agriculture. In the Third World, this activity, known as 'appropriate technology', could bring about a substantial increase in the standard of living and, more importantly, the quality of life for many millions of people. The scale of some of these new biotechnological processes as they apply to environmental treatment can be quite staggering. Large fermentation tanks of capacity 4000 or 5000 m^3 have been built for waste treatment processes and, with a bacterial concentration of 10^8–10^9/ml within the reactor, a considerable amount of microbial power is available to the biotechnologist.

The biological processing of waste matter draws upon a number of scientific disciplines, including biochemistry, genetics, chemistry, microbiology, chemical engineering and computing. All of these disciplines have been brought together in three main areas:

(1) the degradation of toxic wastes, both organic and inorganic;

(2) the recovery of resources, to recycle carbon, nitrogen, phosphorus and sulphur;

(3) the production of valuable organic fuels.

6.2   The processing of waste

Man's wastes have been dealt with for thousands of years by natural biological processes, controlled through contained biological systems. Most publicly owned sewage treatment plants have four basic operations (Fig. 6.1).

(1) Primary processing removes solids, which are either disposed of or sent to a sludge digester.

(2) Secondary processing degrades the dissolved organic compounds. This is effected by natural aerobic microorganisms. The resulting sludge (mainly microbial cells) is either disposed of or sent to a digester. In the activated sludge process, some is returned to the aeration tank.

(3) Tertiary processing (optional) involves chemical precipitation and separation of phosphorus and nitrogen.

(4) Digestion processing is used to treat the sludge from the primary and secondary stages, and is conventionally an anaerobic process. It reduces the solid volume, the odour and the number of pathogens and, in addition, generates the valuable organic fuel, methane.

The treatment of industrial wastewater, particularly from the chemical, petroleum, food, pulp and paper industries, uses processes similar to those described above. Therefore, any biotechnological improvements to these processes are likely to have immediate industrial application. The scope for biotechnological improvements is in an

Fig. 6.1. Stages of sewage treatment in a complex incorporating anaerobic digestion.

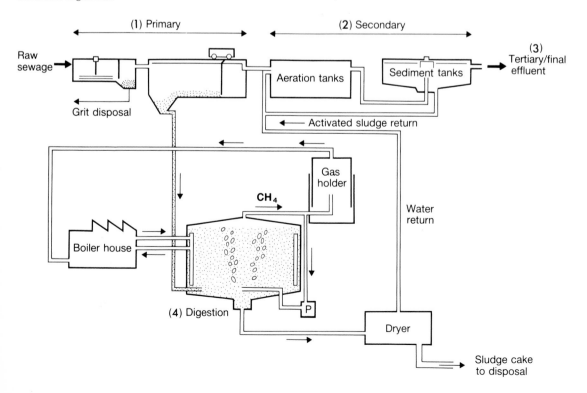

increase in the capacity of the treatment plants, increased recovery of useful by-products, replacement of currently used synthetic chemical additives and the removal of odour, metals and recalcitrant compounds.

6.2.1.   *Aerobic processing of waste* (see also Chapter 9)

Aerobic effluent treatment is the largest controlled use of microorganisms in the biotechnological industries. The steps involved are as follows:

(1) substrate adsorption to the biological surface;

(2) adsorbed solid breakdown by extracellular enzymes;

(3) dissolved material absorption into cells;

(4) growth and endogenous respiration;

(5) release of excretory products;

(6) ingestion of primary population by secondary grazers.

This ideally should result in the complete mineralization of the waste to simple salts, gases and water. The efficiency of treatment is proportional to the amount of biomass and the contact time of the waste with the biomass.

Aerobic treatment systems can be divided into the percolating (or trickling) filter system, and the activated sludge process.

Percolating (trickling) filters

The percolating filter was the earliest system used in the biological treatment of waste, and the basic design used today is virtually unchanged since its introduction in 1890. Seventy per cent of treatment plants in Europe and America use this system due to its advantages of easy maintenance, reliability, low running costs, small surplus biomass production and a plant life of thirty to fifty years.

A major limitation of the percolating filter is excessive growth of the micoorganisms in the filter, which restricts ventilation and flow, eventually causing blockage and failure. A recent modification to plant design is the use of alternating double filtration (ADF), in which the order of the filters first receiving the effluent is periodically reversed. ADF is of particular benefit to the processing of industrial waste. In addition to ADF, recirculation and intermittent dosing are used to dissipate the load deeper into the filter. This improves the total biological oxygen demand (BOD) removed, but reduces the nitrifying activity. Other modifications to plant design and operation are the slowing down of the distributor to even the spread of biomass, and the use of direct double filtration in which a larger size of medium is used for the first filter, allowing higher loading.

In 1970, plastic media were introduced to replace the clinker, stone or gravel in percolating filter systems, which extended the range of these systems to the relatively high concentrations of wastes found in some industrial effluents. Another major advantage is that the lightness of plastics allows taller, space-saving plants to be built. The plastics are moulded to provide optimum surface area, ventilation and voidage.

A major modification to plant design was introduced in the UK in 1973 in the form of the Rotating Biological Contactor (RBC). This is a rotating honeycomb of plastic sheets alternately in contact with the waste and air, thus providing a large surface area for the biomass and good aeration (section 6.6).

The environment of the percolating filter is not truly aquatic as there is only a thin film of water over the surface of the biomass. Studies of the organisms present are hampered by the complexity and hetero-geneity of the biomass. The Zoogloea are thought to be the bacterial group primarily active in the wastewater treatment process, although a number of other bacterial species are involved. The active growth of several species of filamentous bacteria and fungi in treatment systems has also been demonstrated. The most common algae present are the blue-green (*Cyanophyceae*) and *Chorophyceae*. A wide range of Metazoa are present, including earthworms, insects and crustaceans. Flies and worms are very important in controlling the film accumulation.

Activated sludge

This is a completely mixed process, first introduced to treat wastes in 1914. It is a more powerful process than filtration, effecting the treatment of ten times the quantity of effluent per volume of reactor, but it has several disadvantages. It incurs higher running costs due to the requirement for mixing and aeration, it is more difficult to operate and maintain, and it produces large amounts of surplus biomass. In spite of these drawbacks, the activated sludge process remains the preferred method of waste treatment for high density populations, as the area requirement is less than for the equivalent filtration system.

As is the case for filtration systems, a number of modifications have been introduced into the activated sludge system. Those associated with aeration include the following:

(1) tapered aeration, which relates aeration capacity to the oxygen demand, this being less at the outlet than the inlet;

(2) step aeration, which introduces the waste at intervals through-out the length of the tank;

(3) contact stabilization, in which the returned sludge is aerated to encourage organisms to utilize any stored nutrient, resulting in an increased ability to assimilate greater amounts of waste on return to the main treatment tanks—the sludge volume is thereby reduced through the aerobic digester stage and is thus similar in principle to extended aeration treatment;

(4) the use of pure oxygen in closed tanks, which can then operate at higher biomass concentrations, thus reducing residence times and additionally solving the 'bulking' problem (excessive growth of filamen-tous bacteria and fungi which seriously inhibits settling out of the sludge);

(5) the development of the deep-shaft air-lift fermenter, by ICI in 1974 (Fig. 6.2), which is more economic than the conventional process through reduced residence times and low running costs.

Activated sludge is a truly aquatic environment and, as in the case of percolating filters, the major group of bacteria involved in the treatment process is the Zoogloea. It is assumed that only a small proportion of the sludge floc is actively growing. There is less ecological diversity in activated sludge compared with percolating filters. The growth of algae is restricted by light availability, and the species and diversity of protozoa present is dependent upon the degree of treatment (Fig. 6.3).

Fig. 6.2. The deep-shaft air-lift fermenter.

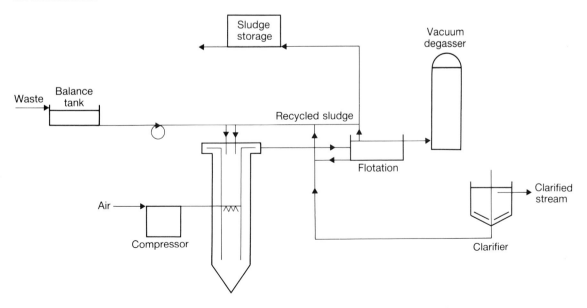

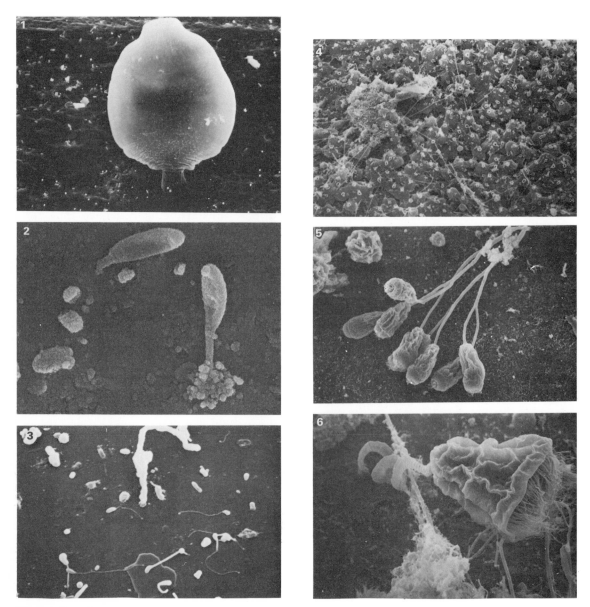

Fig. 6.3. Scanning electron micrographs of microorganisms present in activated sludge (courtesy of Mr G.P. Morris). (1) *Epistylis* (×1000); (2) stalked bacteria (×16 000); (3) flagellated bacteria (×5000); (4) *Vaginocolidae* (×750); (5) *Epistylis* sp. colony (×350); (6) *Vorticella* sp. (×1000).

The successful treatment of domestic and trade wastes requires an accurate knowledge of the composition and strength of an effluent. This provides a guide to the treatment as once the components are known, both qualitatively and quantitatively, estimates of the type of 'microbial seed' required to initiate a system may be made. It is often difficult to demonstrate that specific microbes isolated from biological treatment systems are those effecting the oxidation of specific components.

A study of the microbiology of an activated sludge system involves:

(1) identification of the microorganisms and their enumeration; (2) estimation of microbial activity with respect to both the total population and the individual species; (3) the relationship of 1 and 2 to the feed input and the treated output.

Microbial activity of activated sludges can be estimated in terms of growth or of general metabolism, where the latter includes changes brought about in the suspending medium. Measurements can also be related to the population of selected sub-strains of microorganisms. It is possible to demonstrate that certain bacterial types are associated with activated sludge activities and these bacteria can be enumerated and their various metabolic activities assessed. It remains to be shown whether specific activities of the sludge can be attributed to particular bacterial species, once their properties are known, and how they can be affected by adverse conditions produced either by the composition of the input feed or the metabolic products of the reactant microorganisms. In sewage fed to an activated sludge unit there are high concentrations of organic nutrients and, therefore, a large number of chemo-organotrophic species such as *Achromobacter*, *Flavobacterium*, *Pseudomonas* and *Moraxella*, in addition to many other bacterial species. Where the incoming waste contains a high concentration of inorganic compounds, *Thiobacillus*, *Nitrosomonas*, *Nitrobacter* and *Ferrobacillus* spp., oxidizing sulphur, ammonia and iron respectively, are found. These organisms have been isolated and identified from waste treatment systems using elective culture techniques. During this procedure, it is important to establish whether any particular species plays a dominant role in the changes found in the activated sludge system. This aspect is often overlooked, especially by the non-biologist. It is frequently difficult to assign an unambiguous role to a specific microorganism. For example, *Thiobacilli*, which oxidize sulphur compounds, can be isolated from treatment systems in which this activity occurs, but it is also possible to demonstrate partial oxidation of some sulphur compounds by *Pseudomonas* spp.

The interrelationship between the various organisms involved in the catabolism of organic and inorganic substrates is an important consideration in determining the control of the activated sludge process. Intermediates from metabolic pathways of one organism can influence the degradative process of another. For example, phenol is now known to affect the activity of those organisms oxidizing ammonia and concentrations as low as 3 or 4 p.p.m. can inhibit this oxidative process.

During the breakdown of benzoic acid to catechol, succinate and acetate, intermediates inhibit the production of enzymes involved in the initial degradative steps. Catechol and succinate repress synthesis of enzymes degrading benzoyl formate and benzaldehyde by feedback regulation. Acetate inhibition, for example, is known as catabolite repression; the presence of a simple organic compound represses the breakdown of a more complex molecule until the simpler compound is metabolized. Derepression allows synthesis of the new enzymes to degrade the more complex aromatic structures. In a practical situation, the presence of an homologous series of compounds in a waste will require the production of enzymes that will be able to cope with the breakdown of the most complex molecule of the series. Total breakdown of such compounds must occur within the given minimum hydraulic retention time of the treatment process. It is possible, therefore, to predict the treatment required for the oxidation of phenolic compounds; e.g. the more complicated the side chain in the molecule, the longer the time required for the enzymic degradation of these compounds.

The efficiency of the process can be improved through a greater understanding of the metabolic control in the microflora of activated sludge systems. The control of biodegradation is complex, but appreciation of the biochemistry of these pathways may allow manipulation of the control processes. For example, addition at low concentration (2–5 mg/l) of TCA cycle intermediates, glucose, amino acids and vitamins, such as alanine and nicotinic acid, to the sludge can accelerate the rate of oxidation of specific components. Incorporation of these intermediates into biomass produces an energy requirement, resulting in the stimulation of ATP production by increased oxidation of inorganic compounds such as sulphur or ammonia. An understanding of the biochemistry of these pathways may allow manipulation of the metabolic control processes.

The role the microbial biotechnologist can play in developing the wastewater industry is to understand more fully the relationship between microbial activity, sludge flocs and plant performance in treating wastes. In this respect, the activated sludge system can be regarded essentially as a wet oxidation process with concomitant growth of the sludge, which may be viewed as a nuisance, or an advantage if the sludge is regarded as a recyclable resource. Clearly, a wide range of organic and inorganic compounds can be treated biologically and rendered relatively harmless in terms of their toxic effect on the environment.

The 1974 Prevention of Pollution Act, as it is progressively brought into effect, will have a profound influence on the wastewater treatment industry. The term 'biodegradation' is widely used, but there are many different interpretations of its meaning. At one extreme, it refers to the complete mineralization of a compound by microorganisms to carbon dioxide, sulphate, nitrate and water. At the other extreme, it is used to describe a compound which has only been slightly altered and has lost some of its characteristic properties. Standard methods for assessment of degradation have been established, and the term 'biodegradation', can be qualified as follows:

(1) primary degradation, in which the characteristic property of the original compound has disappeared and is no longer detectable by specific chemical tests;

(2) environmentally acceptable biodegradation, in which the minimum alteration of the parent compound necessary to remove properties occurs (both these definitions depend on arbitrary criteria and are, therefore, imprecise);

(3) ultimate biodegradation, which involves the complete conversion of the parent compound to the inorganic end-products associated with the microorganism's normal metabolic processes.

Research into the degradation of a wide range of organics has resulted in the isolation of microorganisms capable of degrading unusual compounds (Figs. 6.4 and 6.5).

The fluidized bed

Introduced in 1980, this is, in many respects, a combination of the percolating filter and activated sludge systems. It effects substantial savings by utilizing high concentrations of microbes and by having no requirement for final settlement. There are two basic designs.

(1) Simon Hartley captor: This was developed by Simon Hartley, based on research at the University of Manchester Institute of Science and Technology. The biomass is grown in spaces inside polyester foam pads which are retained in the reactor by mesh. The pads are periodically removed from the reactor, the thick biomas (up to 15 kg in each cubic meter of fluidized support element) machine squeezed out and the empty pads returned to the reactor.

(2) Dorr–Oliver oxitron: Sand particles are used as the support medium; the sand is allowed to overflow the reactor, is cleaned and then recycled.

A potentially unwanted consequence of the intensification of aerobic treatment is an increase in the surplus sludge to be disposed of. Sludge disposal can represent 50% of effluent treatment costs. The alternatives are to utilize this sludge (see section 6.5), or to uncouple anabolic and catabolic activities to create inefficient conversion of substrate to biomass, which may be achieved by maintaining a deficiency of trace nutrients and intermittent feeding.

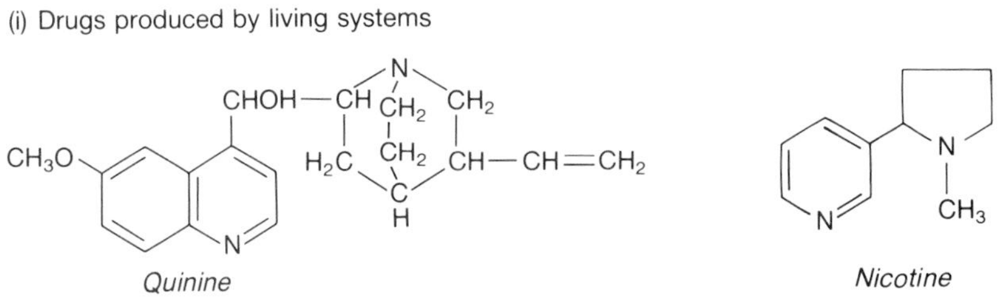

(i) Drugs produced by living systems
CHOH
CH
N
CH2
CH2
CH3O
H2C
CH2
CH
CH
CH2
C
H
N
Quinine
N
N
CH3
Nicotine
(ii) Oxidant and vulcanization accelerator
N
SH
S
2-Mercaptobenzothiazole
(iii) Anticorrosive
O
N
H
Morpholine

Fig. 6.4. Major groups of compounds that may be found after the production of drugs, synthetic herbicides and other petrochemical compounds. All are biodegradable.

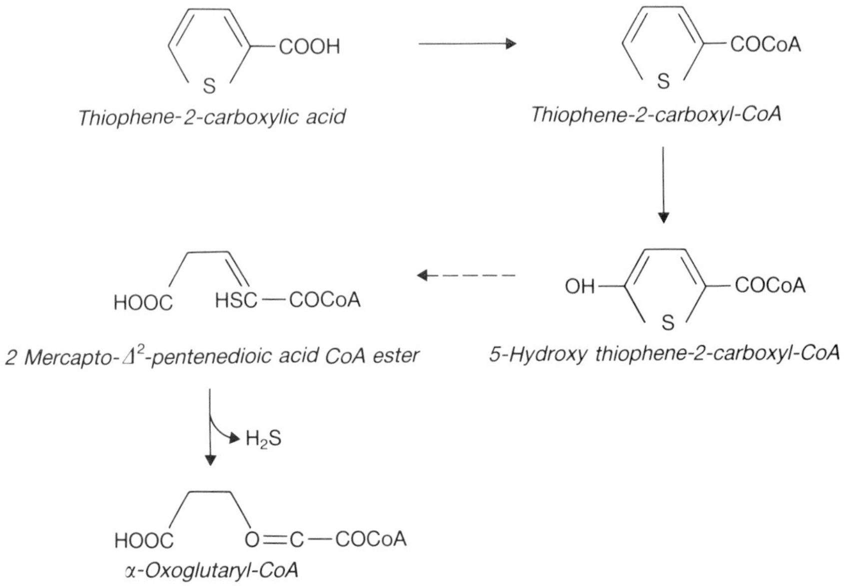

COOH
S
Thiophene-2-carboxylic acid
COCoA
S
Thiophene-2-carboxyl-CoA
OH
COCoA
S
5-Hydroxy thiophene-2-carboxyl-CoA
HOOC
HSC—COCoA
2 Mercapto-Δ2-pentenedioic acid CoA ester
H2S
HOOC
O=C—COCoA
α-Oxoglutaryl-CoA

Fig. 6.5. The intermediate metabolism of a thiophene analogue using a *Flavobacterium* sp.

6.2.2 *The anaerobic digestion process* (see also Chapters 2 and 9)

The rising costs of both aerobic digestion and energy, coupled with new developments in microbiology and engineering, have renewed interest in the anaerobic treatment of waste. The digestion of sewage sludge is the most common anaerobic treatment and is a well established technology, having been used successfully since 1901. There are, however, a number of problems due to the slow growth rate of the obligate anaerobic methanogenic bacteria used in the system, such as susceptibility to interference and low resilience to load changes. The conversion is also relatively slow and, therefore, expensive. Although some problems may be due to poor engineering design and operation, biotechnological advances could be made, e.g. through the addition of enzymes to the waste to improve efficiency, or progress monitoring of treatment through biological parameters (section 6.3).

The anaerobic fermentation of either waste materials or crops grown specifically for energy production offers tremendous potential for producing a fuel gas economically at mesophylic temperatures (30–35 °C). This new biotechnology has been developed by microbiologists in conjunction with chemical and mechanical engineers, agriculturalists and economists.

When a mixture of organic compounds is fed to a consortium of different bacteria, complex biochemical reactions take place (Fig. 6.6).

In the case of methane production, methanogenic bacteria can produce the required energy source directly from hydrogen and carbon dioxide. Cellulose degrading organisms produce fatty acids, which can

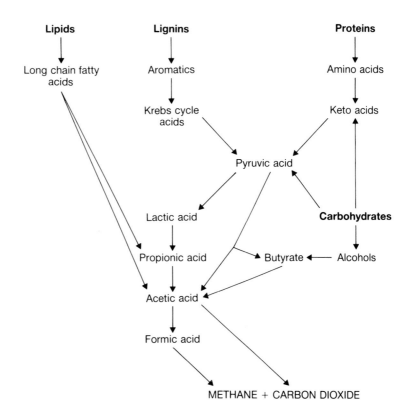

Fig. 6.6. Biochemical sequences for the breakdown of individual compounds to methane and carbon dioxide during the anaerobic digestion of waste materials.

be reductively cleaved to methane and carbon dioxide, and even molecular hydrogen can be produced by some bacteria. There is a complex, interdependent microbial community, in which three different groups of bacteria have been described: the hydrolytic, fermentative group, the hydrogenic, acetogenic group and the hydrogenotrophic, methanogenic group. The methanogens are slow growing and susceptible to stress from shock loads and hydrogen accumulation. Developments in reactor design and control of the process will incorporate the need to avoid large fluctuations in the loads delivered to the reactor, and the control of the delivery rate through monitoring the concentration of hydrogen and intermediates such as propionic and butyric acids. Overload problems, particularly associated with industrial wastes, can be overcome by using high recycle rates and buffering with chemicals or domestic sewage. There remains the possibility of increasing methanogenic activity by conventional strain selection or genetic manipulation techniques. Further research into the physiology and ecology of the microorganisms involved would enable assessment of reaction to a combination of wastes, characterization of nutritional requirements, and possibly an improvement in the start-up procedure by reducing the amount of inoculum needed.

Waste materials, derived from a number of sources, can produce a net energy gain plus a useful residue. Crops grown specifically for the purpose of conversion of fuel gas energy include cassava, the end-fuel being methanol or ethanol. Some countries of the world, e.g. Brazil, Australia and New Zealand, are intending to rely on biological fuel production for their major fuel sources by the year 2000. Some European countries also have this potential, e.g. Finland, Sweden and Ireland.

The Department of Energy's Solar Energy Programme in the UK is funding work on energy bioconversions and in Europe, the solar energy programme is funding EEC projects for producing biological energy. In the US, many schemes are under way and one landfill site produces enough gas for twelve thousand homes from the biological conversion of domestic refuse. The first International Conference on Anaerobic Digestion held in Cardiff (1979) identified the basic microbiological and engineering problems and considered future applications in Third World countries. Eventually, anaerobic fermenters may be used to produce intermediates for the chemical industry such as acetic acid, lactic acid and acrylic acid for chemical feedstocks (see Chapter 4).

However, there are obstacles to implementing anaerobic digesters for gas fuel production. Traditionally, reactor designs have comprised stirred tanks with long residence times. In order to reduce residence time, reactors have been developed which retain the biomass or recycle it after separation of the processed waste. Low cost, non-fouling, easily maintained heat transfer equipment should be tailor-made for the digester business. For future research and development, a major line of investigation in anaerobic digestion will be into the rate limiting steps where celluloses and starches are hydrolysed during the first stage to produce soluble organic acids and alcohol. A second potential rate limiting step is the production of methane from these short-chain fatty acids. Modelling of the digestion process is complicated due to the difficulties in associating each step with a particular microorganism, and identifying which of these steps is rate limiting. Other steps may have limitations placed upon them by reactor conditions. Measurement of dissolved gases produced by the microorganisms, particularly hydrogen, carbon dioxide and hydrogen sulphide, is essential; the latter inhibits the activity of methanogenic bacteria. Recent studies have investigated the anaerobic conversion of defined substrates by known cultures of microorganisms. The process is a very complex one, from which many new types of bacteria have been isolated.

The commercial application of anaerobic digester systems is currently increasing for the treatment of farm, industrial and food manufacturing wastes, in addition to the processing of energy crops. Figs. 6.7, 6.8 and 6.9 indicate some of the designs available on the market. Great improvements have been made in digester design, with increases in efficiency of up to 300%. Although many of the new models are at the laboratory or pilot plant stage, some full-scale systems are operational and commercially available.

In Third World countries, energy production from wastes has obvious advantages in that energy can be produced from natural resources. Co-operation between developing and developed countries is increasing, and institutes are being set up in the Third World countries for the practical application of the technology developed particularly in Europe and America. Some members of the Third World, however, are themselves pioneering and forging ahead with basic research into this renewable energy source.

6.3   Biological control of microbial waste treatment systems

A major requirement for the biological treatment of effluents is the assessment of any toxic effect of the effluent input, to avoid severe disruption or even failure of the system. Warning of shock loads is an important aspect of running a sewage works or industrial effluent treatment plant. The usual methods of measurement, such as oxygen demand and pH, are often neither sufficiently rapid nor sensitive. There

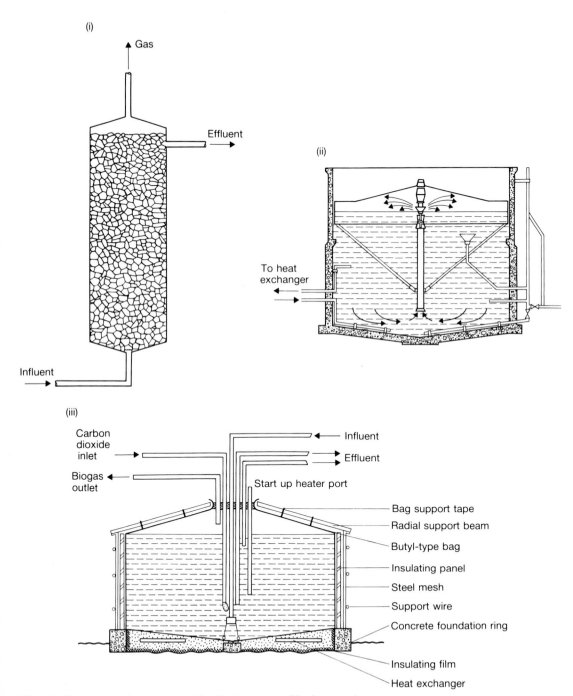

(i)

Gas

Effluent

(ii)

To heat
exchanger

Influent

(iii)

Carbon
dioxide
inlet

Influent

Effluent

Biogas
outlet

Start up heater port

Bag support tape

Radial support beam

Butyl-type bag

Insulating panel

Steel mesh

Support wire

Concrete foundation ring

Insulating film

Heat exchanger

Fig. 6.7. Three types of reactors used for the treatment of food-processing wastewaters.
(i) Anaerobic filter. (ii) Simplified layout showing mixing by screw pump and draught
tube. Scum formation is also controlled by scattering the digester contents over the scum
surface. (iii) Coulthard-type high-rate digester.

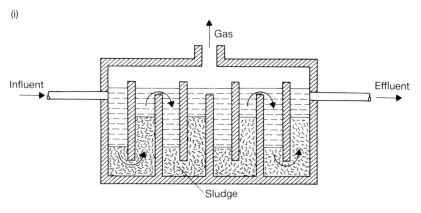

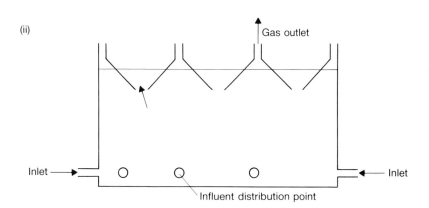

Fig. 6.8. Reactor types for treating farm wastes (i) and individual effluents (ii).

are, however, key points in the metabolic activity of microorganisms which might allow sensitive monitoring of their status. For example, the ATP level is usually maintained in a population of microorganisms by intracellular metabolic control at a relatively constant level of approximately 2 μg per milligram dry weight of cells. Changes in substrate availability or toxin exposure produce rapid changes in the concentration of ATP within the cells. Death leads to a complete loss of ATP through autolysis in seconds. The turnover time for ATP is usually less than 1 s. Thus, the status of the population, in terms of stress or dormancy for example, can be assessed from changes in intracellular ATP levels. The measurement of ATP concentration in activated sludge in a normally operating system enables the determination of the 'active biomass' within the system. Since oxygen is involved in energy

(i)

Wastewater flows through
sedimentation chambers only

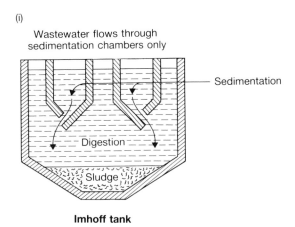

Imhoff tank

(ii)

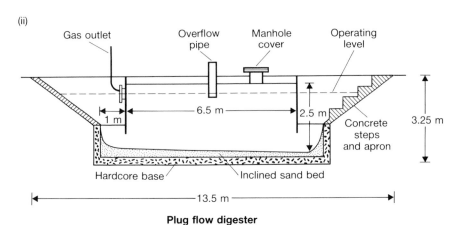

Plug flow digester

Fig. 6.9. Anaerobic digester treatment of sewage sludge (i) and farm wastes (ii).

production in the cell, the relationship between oxygen supply and demand will be reflected in the ATP levels when oxygen becomes limiting (Fig. 6.10).

In the activated sludge process, most of the sludge is recirculated from the final settlement tank to the aeration tank, while a proportion is wasted into a digester (either aerobic or anaerobic). The proportion and the rate of sludge recycling determines the contact time between the incoming wastewater and the sludge microorganisms and, therefore, the rate at which the wastewater is purified. These factors affect the age of the sludge, which in turn influences its ability to sediment when it leaves the aeration tank (Fig. 6.11). The performance of the activated sludge is largely determined by the specific growth rate of the new biomass in the aeration tank, which is affected by the wide variations in the rate of flow and composition of the incoming waste. Therefore,

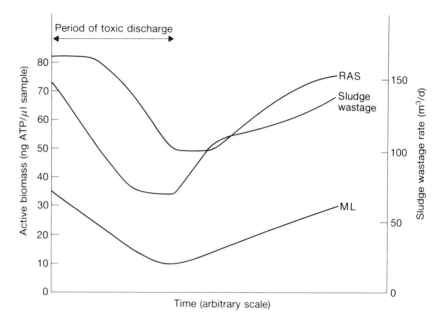

Fig. 6.10. Effect of addition of toxic discharge to sewage input on active biomass, represented by ATP measurements. The activity of the returned activated sludge (RAS) and mixed liquor suspended solids (ML) is also shown.

control of the process to operate optimally and to produce a final effluent which always meets the required discharge standards is fraught with difficulties. This may be possible using ATP level determination with feedback control and, in fact, the control of activated sludge treatment of sewage has improved using ATP concentration measurements. ATP levels can be determined manually, but automatic measuring devices are required for continuous control of the activated sludge process. Samples of the mixed liquor leaving the treatment lane are taken at predetermined time intervals during a 24 h cycle. Analysis of the ATP levels in these samples allows the ratio of the mixed liquor suspended solids to ATP to be determined, thereby indicating the amount of activated sludge–ATP to be recycled and the appropriate wastage rate. ATP levels can be determined within 10 min of sampling, and thus time intervals as short as 15 min may be used for updating the control of the wastage rate. Using programmed microprocessors, the wastage rate value can be maintained automatically by valves or pumps, which feed the surplus activated sludge to the digester. In this way, the rate of wastage of sludge may be changed throughout the day in response to changes in the environmental conditions.

This method of control produces a much clearer effluent with lower suspended solids and BOD discharge value. The advantages of controlling the treatment of industrial waste waters and sewage by this method

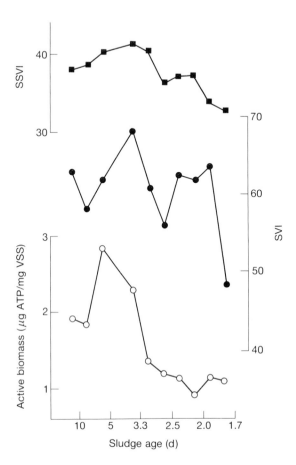

Fig. 6.11. Changes in active biomass, sludge volume index (SVI) and stirred specific volume index (SSVI) of aeration tank treated sewage as a function of sludge age.

are enormous and, with the addition of the oxygen control input system, make the process far more economic.

6.4   Pathogen control

One of the major benefits of the microbial anaerobic digestion process is the elimination of pathogens, particularly the causative agents of food poisoning, principally *Salmonella*. In studying the fate of these organisms, and also the *E. coli* group, a sensitive Most Probable Number technique (MPN) has been selected to study the viability of these organisms in laboratory-scale systems. One of the major end-products of non-methanogenic digestion is saturated fatty-acids, and there is evidence that the fatty acid chain is hydrogenated prior to degradation by β oxidation. Octanoic acid which is formed has been shown to be particularly effective in killing pathogenic species. Fig. 6.12 shows the

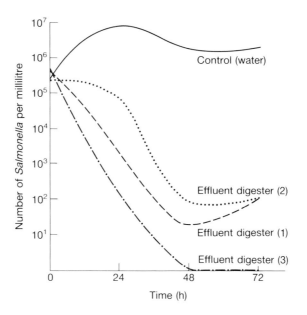

Fig. 6.12. Survival curve of *Salmonella*/ml when exposed to anaerobic digest supernatant.

survival of *Salmonella* during anaerobic digestion. Fatty acids have a toxic inhibitory effect on growth and kill these bacteria.

Digestion is also effective against plant pathogens, and a study recently completed has shown that *Fusarium*, *Corynebacterium* and *Globodera* species are reduced in number during the digestion process. These pathogens are almost completely destroyed in a digester with a 10 d hydraulic retention time (Figs. 6.13, 6.14 and 6.15). This provides encouraging prospects for the use of digesters in the treatment of diseased plant material. (*Fusarium oxysporum* is a dykaryotic fungus which causes wilt and root rot of tomatoes, carnations and rice. *Corynebacterium michiganese* causes vascular wilt, canker and leaf spot on tomatoes, potatoes and tobacco. *Globodera pallida* is a nematode, commonly known as potato root eel worm, which destroys the root tissues of its host plant.)

A further study of other pathogens, perhaps those found in sewage in more tropical climes, would necessitate the inclusion of these organisms in studies of pathogen removal by digestion. They include *Entamoeba hystolitica*, which causes amoebic dysentry. The benefits of these biotechnological processes for improving public health need to be developed on a large scale.

6.5   Resource recovery

One of the broader aims of environmental technology is the conserva-

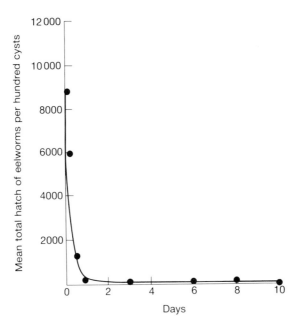

Fig. 6.13. Mean total hatch of *Globodera* eelworm) per hundred cysts after exposure to an anaerobic digester at 35 °C. The control cysts showed a recovery of 8739 ± 444 in water after 7 d.

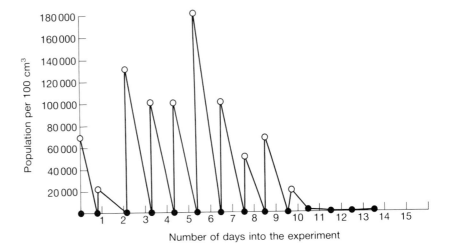

Fig. 6.14. Graph showing the population before (●) and after (○) introduction of *Fusarium oxysporum* into an anaerobic digester for a 10 d period.

tion of resources via the recycling of waste materials. Pressure to achieve this aim has arisen in some cases from government subsidies, but at present the recovery and marketing costs of large-scale biomass recycling are not justified by the value of the product. The recovery of more valuable products, however, such as oils, metals, vitamins and peptides may produce an expanding area of this technology.

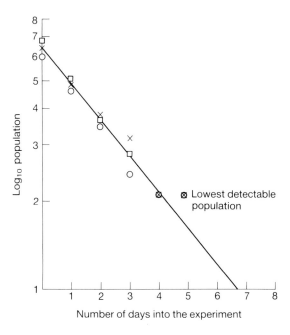

Fig. 6.15. The most probable number of *Corynebacterium michiganense* at 35 °C for a 5 d period after introduction to an anaerobic digester. The counts were obtained from three experiments □, ○, ×.

There are two aspects of recovering useful materials from wastes: (1) extraction and/or concentration of useful materials from the waste; and (2) transformation of the waste into useful materials.

6.5.1 *Water*

Water can be viewed as a recyclable resource. However, reclaiming organically polluted water is generally regarded as being uneconomic when the cost of the equipment required is compared with that of mains water. Similarly, recycling industrial wastewater is economic only in heavy industries such as power, steel and coal, where an inferior quality is acceptable and wastewater treatment can be kept to a minimum. Recalcitrant compounds are an obvious source of problems to this form of supply. A possible solution is the application of microbes, which have acquired the ability to degrade recalcitrant compounds (see section 6.7).

6.5.2 *Fertilizer*

There is an increasing demand for cheaper and more plentiful high quality animal protein, but there has been a concomitant reduction in the agricultural labour force to effect this increasing need. The solution to these changes—the use of increasingly intensive farming methods—

should theoretically produce a higher volume and concentration of waste usable as fertilizers. However, over the last hundred years, the use of animal waste as fertilizer has diminished, to be replaced by imported phosphate and nitrogen fertilizers, manufactured using fossil fuels. A significant proportion (40%) of agricultural costs in the UK are attributable to fertilizer application. In view of the increasing prices of chemical fertilizers, it may well prove economic to reverse this trend and return to the use of animal waste as organic fertilizers.

Table 6.1 shows the production potential of manure from poultry, pigs and dairy cattle. A full analysis of each form of waste is shown in Table 6.2. The content of the waste determines its potential usefulness

Table 6.1. Manure production from farm animals (faeces and urine)

Parameters	Poultry	Pigs	Dairy cattle
Kg/day	0.1	4.5	45
% Live weight	5–6	8–9	9–11
% Dry matter	20–30	15–20	10–11
Ave. dry wt. (Kg/day)	0.03	0.7	4.5

Table 6.2. Nutrient content of farm animal manure

Manure	Dry matter %	Nutrients (kg/t fresh manure) N	P	K	Mg
Cattle	4–23	2.4–6.5	0.4–1.8	2.0–5.8	0.2–0.6
Pig	5–25	1.6–6.8	0.6–2.1	1.7–3.6	0.3–0.7
Poultry	23–68	9.6–23	2.4–12	3.8–11.6	1.2–2.2

as a fertilizer and its route for utilization. In addition, it is possible to derive energy from manure by anaerobic digestion before its discharge on to the land (section 6.2.2) and, whilst it is difficult to determine changes in the value of the fertilizer during digestion, there is considerable retention of fertilizer value in digested animal manures.

6.5.3   *Animal feed*

Humans produce about 25×10^9 kg waste per year in the UK and this, coupled with 180×10^9 kg produced from intensive animal farming, provides a high tonnage of biological sludge in waste treatment plants. Treatment processes using microorganisms produce a large amount of microbial protein which could potentially be recycled as an animal feed, since 30–40% of the dry weight of the cells produced is crude protein.

The method for extraction of protein from activated sludge is shown in Fig. 6.16, and the composition of single cell protein (SCP) from the same source is outlined in Table 6.3. The heavy metals found in sewage sludges (e.g. copper in pig waste due to use of copper concentrates in animal feeds) would require removal. The ideal result from this upgrading of sewage sludges by extracting the protein would be an inoffensive, pure, economic animal feed. Additionally, other active biomolecules of value, such as amino acids and vitamins, are produced in the sludge. Table 6.4 shows the relative concentrations of amino acids in activated sludge protein fed to fish. Domestic and agricultural wastes, however, are likely to prove unsuitable for large scale SCP production due to problems of consistent nutritive value and material free from toxins and pathogens. In addition, there are economic problems due to the cost of fermentation and drying equipment required. Non-contaminated wastes from the food, drinks and fermentation industry may be utilized as an additive to animal feed, but its economic viability has yet to be demonstrated.

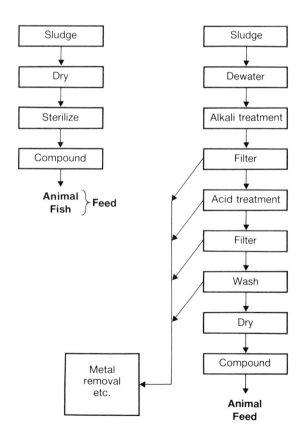

Fig. 6.16. Protein recovery process using sewage activated sludge as primary substrate.

Component	% By weight
Water	7.85
Crude protein (N × 6.25)	37.80
True Protein	31.25
Fat	1.73
Ash	25.16
Acid-insoluble ash	12.33
Carbohydrate (as glucose)	9.80
Fibre	10.30
Gross energy value (kg/g)	16.20

Table 6.3. Composition of activated sludge single cell protein

Amino acid	g/100 g sample
Aspartic acid	2.86
Serine	1.4
Glutamic acid	3.62
Proline	1.28
Glycine	2.26
Alanine	2.95
Cysteine	0.25
Tyrosine	0.72
Threonine	1.86
Valine	2.19
Methionine	0.25
Isoleucine	1.42
Phenylalanine	2.07
Histidine	0.69
Lysine	1.68
Arginine	1.59
Tryptophan*	—

Table 6.4. Relative concentration of amino acids in activated sludge protein

* Destroyed during acid hydrolysis.

A consultative document has been drawn up by the Ministry of Agriculture, Fisheries and Food (*Protein Processing Order 1981*), concerning the reuse of sludge proteins for animal feed. In this particular case, the Government is restricted to prescribing a standard for enforcement of the order, and the proposals would incorporate the Diseases of Animals Act 1950 and 1975. This particularly covers the control of diseases such as *Salmonella*.

6.6   Biological processing of industrial waste

Industrial wastes can be classified broadly into two categories: those arising from biologically based industry, such as food, drink and fermentation, and those arising from the chemical industry. Wastes from the former are of variable composition and have been treated usually by biological oxidation, as used traditionally for domestic waste. However, this solution is expensive and there is now great interest in the reduction of volume of these dilute wastes or in their direct utilization, either by transformation (to produce biomass or other valuable products) or by the recovery of valuable components of waste.

The many diverse activities of the chemical industry generate a plethora of waste compounds, many of which are recalcitrant and, therefore, persistent in the environment. Often they may require chemical or physical pretreatment before conventional biological effluent treatment can be used. However, there are as yet few practical industrial applications in which specific microorganisms are used for the breakdown of xenobiotics in the treatment of waste water. However, such biological treatment does offer great potential for the following:

(1) *in situ* degradation of specific wastes using specialized cultures or consortia;

(2) inoculation of conventional waste treatment systems with adapted, specialized cultures;

(3) decontamination and detoxification of spillages;

(4) metal removal (see Chapter 5);

(5) bioscrubbing of odour and waste noxious gases (mercaptans, hydrogen sulphide, cyanide, chlorinated hydrocarbons, etc.);

(6) generation of biomass from wastes;

(7) conversion of wastes to methane (biogasification) (see Chapter 2). The types of xenobiotic compounds introduced into the environment by the chemical activities of man include plasticizers, explosives, additives, polymers, dyestuffs, surfactants, pesticides and petroleum-derived organics. Conventional activated sludge and trickling filter waste treatment systems were developed to process domestic sewage. Waste streams from the chemical industry rarely fall within the operating limits for which these systems were designed. Conventional processes are often limited in their ability to transfer sufficient oxygen to support the maximum rates of oxidation of the indigenous microflora. They are also sensitive to fluctuations in waste loading, especially when inhibitory and toxic wastes are introduced in high, intermittent concentrations.

Several processes have been developed to overcome the oxygen limitation of conventional activated sludge systems in the treatment of chemical waste. Two types (the bubbling distributor and the Unox system) use pure oxygen to increase the rates of gaseous transfer. One innovative waste treatment system, the deep-shaft air-lift fermenter developed by ICI, develops high levels of dissolved oxygen (Fig. 6.1). The fermenter shaft is partitioned into a downflow section, into which the incoming waste, the returning activated sludge and the process air are injected. As the mixture rises through the inner upflow section, the pressure in the system decreases, releasing bubbles of gas. High levels of dissolved oxygen and turbulence maintain a very active biomass, resulting in increased resistance to shock loads and reduced aeration and residence times, especially for very concentrated wastes.

These high aeration processes are resistant to shock loads that are neither toxic nor inhibitory. Waste treatment systems based on film growth processes are more resistant to toxic wastes. The population cannot be washed out of the system, even though their growth and metabolism may be adversely affected by the incoming waste. In addition, concentration gradients are established within the film due to diffusion limitations. This results in lower concentrations and thus higher rates of uptake and oxidation of toxic wastes within the film. The film also provides an environmental niche for those organisms whose growth rate may be very slow in the presence of high concentrations of the shock load. The simplest form of film system is the trickling filter (see section 6.2.1) but these films tend towards instability when they are of a thickness such that the concentration of substrate at the support surface falls to zero. The cells at the support surface die, lyse and the film falls off, blocking filters within the waste treatment system. Growth at the surface of the film is rapid at high substrate concentrations, leading to the formation of a thick film and establishing a cyclic sloughing process. Dilution of the incoming feed with the clarified effluent may reduce the extent to which this occurs. However, there is still a great deal of interest in establishing methods in which a fixed film thickness can be maintained. By slowly rotating a polystyrene disc through an effluent, film thickness is maintained by the hydraulic shear forces and thorough aeration occurs when the film is out of the water. This efficient and stable system has been proposed for effluents of low BOD. Alternatively, the development of fluidized bed reactors, in which the organisms are grown on the surface of small, inert particles (sand, glass, anthracite, etc.) and through which the effluent and air are passed at a controlled rate, may provide an efficient method of toxic waste treatment *in situ*.

Wastes containing neither nitrogen nor phosphate will not support growth and it may be possible, in these circumstances, to use resting cells to oxidize the toxic components to carbon dioxide, provided the hydrolytic and oxidative enzymes remain stable. The fluctuating environment found in the deep-shaft waste treatment process exposes the organisms to periodic starvation, during which growth ceases. Upon exposure to the carbon source, uncoupled metabolism will proceed for a short time, the organisms respiring but not growing. This has the advantage of reducing the overall yield of biomass (sludge) in the process.

The methods by which industrial wastes are treated biologically are described in the following examples from the dairy, pulp and dye industries.

6.6.1   *Wastes from the dairy industry . . . whey* (see also Chapter 3)

Whey is a by-product of cheese making and varies in composition, depending on the type of milk used and cheese produced. It has been utilized in a dried or concentrated form as an animal feed but is nutritionally unbalanced, due to the high mineral and lactose content. Processes have been developed to recover proteins, by ultrafiltration, precipitation or separation by ion exchange, which could be used in enzyme reactors to produce protein hydrolysates. The filtrates, or permeates, from the recovery procedures are produced in large volumes, contain high concentrations of lactose (35–50 g/l), minerals, vitamins and lactic acid, and present disposal problems. Transformation of lactose to lactic acid by the lactic acid bacteria would make this carbon source available for yeast fermentation (e.g. mixed cultures of *Lactobacillus bulgaricus* and *Candida krusei*). Alternatively, lactose may be fermented directly by the yeasts *Kluyveromyces fragilis* or *Candica intermedia*. After such fermentations, it may not be necessary to separate the microorganisms from the medium, which may be reduced in volume and sold as a protein-enriched whey.

In addition to the generation of useful protein from whey, it is also possible to produce chemical feedstocks, such as ethanol, by fermentation. Galactose can be produced by the chemical hydrolysis of lactose, followed by the removal of glucose from the liquor by fermentation. An alternative biological route would use yeast mutants lacking β-galactosidase. Such mutants would retain the ability to hydrolyse lactose and use the resultant glucose as a carbon source. Hydrolysis of permeate lactose improves the sweetness of whey and has been achieved on a pilot scale using immobilized β-galactosidase (see Chapter 4). Not only does the hydrolysed permeate find uses in the food industry but it also helps to overcome the enzyme deficiency problems found in some weaning animals and overcome lactose intolerance in humans. Other chemical products from whey include lactose, lactulose, lactitol and lactobionic acid.

6.6.2   *Wastes from the pulp industry*

Fibrous material used in the manufacture of paper commodities is produced from wood and annual plants by the chemical dissolution of lignin. However, this process also results in the dissolution of a large percentage of the woody material and the annual generation of tremendous volumes of wood waste, which has theoretical potential for the generation of an alternative chemical technology.

Two pulping processes are in operation today. The predominant one is alkaline kraft cooking (sulphate process), which generates kraft black liquor. This waste contains refractory aromatic material from the degradation of lignin and low molecular weight organic acids (glucoisosaccharinic, lactic, acetic and formic). The pulping of resinous pine woods generates tall oil and terpenes, which are used extensively in industry. There is no industrially applicable biological waste treatment for kraft black liquor and it is far more economic to evaporate the liquor and burn it, thus recovering energy from the waste.

The sulphite pulping process is less extensively used but generates a waste of the following composition: lignosulphonates with aromatic elements (60%), sugars (mannose, galactose, glucose, xylose, arabinose; 36%), acetic acid, methanol and furfural. This waste liquor is an attractive raw material for fermentation, because of its high sugar content, and the large-scale fermentation of these liquors was first introduced in 1909. Yeast fermentation is now the traditional method of removing pentoses, hexoses and acetic acid from the wastes. New processes for converting the wastes into fungal protein, using species such as *Paecilomyces variotii*, *Sporotrichum pulverulentum* and *Chaetomicum cellulolyticum*, will supplement the traditional methods. The refractory compounds in the treated waste can then be concentrated and burned. Lignosulphonates are used as binders and oil drilling aids and can be converted to vanillin by alkaline pressure oxidation. In general, energy independence has assumed priority over chemical production in the pulping industry.

A major environmental problem produced by the pulping industry involves the treatment of wastewaters that arise from the above processes and as condensates from evaporators and digesters. These wastewater streams are clarified by neutralization and sedimentation, oxidation in one- or two-stage activated sludge plants, in aerated lagoons or by a combination of biological and chemical oxidation procedures. These measures are effective in the elimination of biodegradable compounds and toxic phenolics but are expensive to run and ineffective in the removal of recalcitrant lignin-derived compounds. Bleachery effluents, containing chlorinated biphenyls, can be decolourized using several species of white rot fungi.

Chemically altered lignins are the predominant by-products in the sulphite pulping process and these compounds result from the many reactions of active sulphite with a complex natural polymer. There is, therefore, no detailed knowlege of the structure of lignosulphonates, a heterogenous mixture of compounds covering a broad spectrum of molecular weights (300–100 000), whose composition is dependent upon the nature of the wood being treated. Sulphonation does result in the partial solubilization of these lignin fragments. Given the complexity of the lignosulphonates, studies on the microbial degradation of these compounds are equally as complex and have usually been simplified with the use of model compounds, such as dehydropolymers of coniferyl alcohol and other small model compounds. Low molecular weight lignosulphonates are more susceptible to microbial degradation, although lignin derivatives appear to be more resistant to degradation than lignin itself. Sulphonation, therefore, increases recalcitrance. Soil fungi and bacteria are more effective than white rot fungi in this co-oxidative degradation process, in which another carbon source is also required. The decay of lignosulphonates is often accompanied by polymerization reactions, with a resultant shift in the distribution of polymer molecular weights. These changes can be correlated with the presence of extracellular phenol oxidases (e.g. laccase), whose physiological function is still unknown. Phenols are converted to the respective quinones and phenoxy radicals, which spontaneously polymerize. Polymerization and degradation thus occur concomitantly. With some fungi these polymerization reactions do not occur in the presence of cellulose. Cellulose is degraded to cellobiose, the substrate for cellobiose:quinone oxidoreductase, which simultaneously oxidizes cellobiose and reduces quinones and phenoxy radicals. Other oxidoreductase systems may exist, linking readily available carbon sources with quinone reduction. One possible role for biological polymerization of this type would be in facilitating the precipitation of lignosulphonates. Lignosulphonates have found application as binders in particle boards, using laccase-containing culture filtrates as the polymerization catalyst. The environmental role of phenol oxidases in the polymerization of phenols and production of organic soil polymers may be important in determining the fate of many xenobiotics.

Chemical and physical pretreatment of lignosulphonates may increase their biodegradability. UV irradiation and ozonation fragment the molecules, and removal of sulphonic acid groups, whilst decreasing solubility, will also decrease the recalcitrance of these molecules. Microbial desulphonation has been attempted with some success using anaerobic sulphate-reducing bacteria and *Pseudomonas* spp. The application of mixed cultures to this process has excellent potential. The use of mixed cultures in the degradation of lignosulphonates will be more efficient than the use of pure strains, as it should be possible to construct communities with varying abilities, such as desulphonation, cleavage of resistant bonds, demethylation and depolymerization. Such an approach may lead to the development of a high efficiency bio-oxidation system. One pilot-scale process uses *Candida utilis* in the production of SCP from the carbohydrates of pulp industry waste and a mixed culture to degrade the remaining lignosulphonates. As there is no market for the biomass from the second stage of this process, it is recycled, after thermolysis, to enhance the growth of the *Candida* species.

6.6.3   *Waste from the dye industry*

The textile and dyestuffs industries produce a bewildering array of dyes and pigments, whose common structural feature is merely that they possess a chromophore. They are released into the environment in effluent streams, although they are not major pollutants in terms of amounts released. In addition, these wastes are not generally considered to be toxic or carcinogenic to fish or mammals, with the exception of benzidine and cationic dyes.

Chemical methods for the treatment of coloured effluents have proved successful and the microbial removal of dyes and pigments is very limited. In activated sludge, the degree of elimination has been attributed to adsorption rather than degradation. The extent to which these compounds are subject to microbial degradation is determined by their solubility, ionic character and degree and type of substitution. Microorganisms have been found to degrade dyes but only after adaptation to much higher concentrations than those normally found in effluent streams. Interest in the use of microorganisms for the control of this type of environmental pollution has involved screening for the degradation of model compounds, such as *p*-aminobenzenes, and screening and selection of organisms from the drainage ditches of dyestuffs factories. Attempts have been made to correlate degradability with chemical structures.

Azo dye compounds can be degraded aerobically or anaerobically. Anaerobic decolourization is relatively easy to achieve microbiologically, as a number of organisms possess non-specific enzymes which catalyse the reductive fission of the azo group (Fig. 6.17). However, no further degradation is observed under these conditions, necessitating a two-step process in which the resultant products, amines which may be potentially hazardous, are subject to oxidative degradation. Azoreductases isolated from aerobic systems appear to be more substrate specific and may not be applicable generally to waste treatment. The process of azo dye degradation has been examined using the simple model compound 4,4′-dicarboxyazobenzene (DCAB) and organisms have been isolated on this compound as sole source of carbon, nitrogen and energy. It was commonly believed that the azo linkage was not synthesized in nature, but an azo compound has been identified as an excretory product of an insect fungal pathogen and bacteria will have been exposed to such a chemical linkage *in vivo*. A *Flavobacterium* species isolated in this manner degrades only *trans*-DCAB and the inclusion of inhibitors of the ring-opening enzymes results in the accumulation of aminobenzoate. However, this organism will not

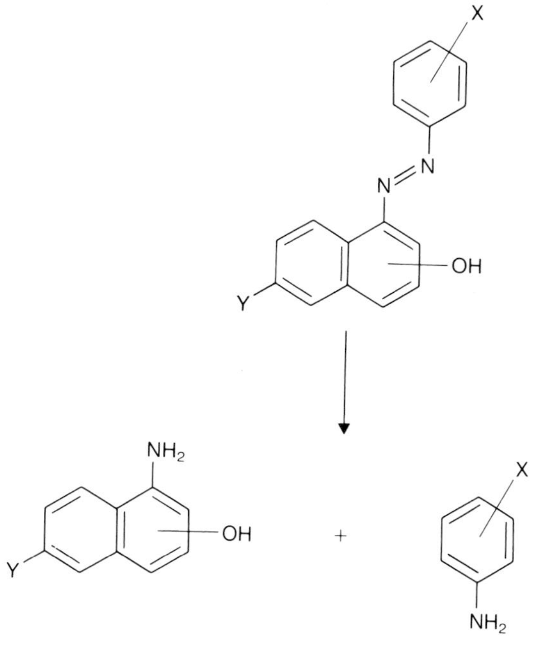

Fig. 6.17. Anaerobic reductive fission of model azo dyes.

degrade commercial dyes and attempts to evolve the organism in continuous culture, in a long-term adaption procedure, were unsuccessful. The *Flavobacterium* was outcompeted by another organism, a pseudomonad, which could grow on and, therefore, remove high concentrations (750 mg/l) of the commercial dye used. The fact that the ability was lost by growth at higher temperatures may indicate that this function was plasmid-encoded. However, the strains evolved were highly substrate specific and unlikely to survive in wastewater treatment plants.

6.6.4 *Bioscrubbing*

The discharge of noxious, toxic and odorous gases is a serious environmental problem. Reduced sulphur compounds (thiosulphate, hydrogen sulphide, methyl mercaptans, dimethyl sulphide) are generated from a variety of industrial processes (in the photographic and pulp industries, in oil refining and the purification of natural gas) and as by-products from the anaerobic digestion of animal wastes high in organic content. Most inorganic reduced sulphur compounds can be utilized, either aerobically or anaerobically, as an energy source by a range of different microorganisms (Fig. 6.18) and microbiological treatment systems can be devised based on *Thiobacilli*, for example, where anaerobic desulphurization is coupled to denitrification. One technique of dealing with a hydrogen sulphide effluent is to pass the gas through an appropriate salt solution, such as copper sulphate, thus precipitating an insoluble metal sulphide, which can then be oxidized microbiologically (see also Chapter 5 for the microbiological processing of mining waste). The increase in efficiency of animal husbandry has

Aerobic process

$$H_2S + 2O_2 \longrightarrow H_2SO_4$$

$$(CH_3)_2S + 5O_2 \longrightarrow 2CO_2 + H_2SO_4 + 2H_2O$$

Anaerobic process

$$5H_2S + 8NaNO_3 \longrightarrow 4Na_2SO_4 + H_2SO_4 + 4H_2O + 4N_2$$

Fig. 6.18. Degradation of reduced sulphur compounds.

$$(CH_3)_2S + 4NaNO_3 \longrightarrow 2CO_2 + Na_2SO_4 + 2NaOH + 2H_2O + 2N_2$$

resulted in the increased production of malodorous animal wastes. Deodorization of such wastes involves the removal of reduced sulphur compounds and this may be achieved either with or without a loss of nitrogen, through the formation of ammonia, depending on the type of microorganism used in the process. Organic sulphides are often toxic to microorganisms. For example, it is difficult to enrich for microorganisms capable of utilizing dimethyl sulphide, although it is possible to isolate a community on the closely related dimethyl sulphoxide. The dominant member of this community, a *Hyphomicrobium* sp., is capable of the rapid oxidation of dimethyl sulphide and it is now possible to envisage relatively simple microbiological solutions to the treatment of such noxious wastes.

Various biological methods have been described for the detoxification of industrial waste cyanide. These range from the use of activated sludge to the application of specific cyanide-metabolizing enzymes. Rhodanese, found in *Bacillus stereaothermophilus*, catalyses the conversion of cyanide to thiocyanate. An alternative immobilized system is based on the hydrolysis of cyanide to formamide catalysed by the inducible enzyme, cyanide hydratase, found in fungal pathogens of cyanogenic plants. Microbiological methods for the scrubbing of waste effluents are, therefore, available although, in general, their performance at the point of waste discharge remains to be determined.

6.7   The degradation of xenobiotic compounds in the environment

The biodegradation of organic pollutants in the environment can either be beneficial, resulting in complete mineralization, decontamination and detoxification, or deleterious, where biological modification of the pollutant molecule increases the biological effects, through magnified toxicity or environmental persistence. Detoxification of pollutants may involve as little as a single modification of structure to render a potentially hazardous chemical innocuous. The fate of xenobiotic compounds will depend upon a complex array of interacting factors, both intrinsic (stability, water solubility, molecular size and charge, volatility) and extrinsic (pH, photooxidation, weathering), which will determine the degree of recalcitrance of that compound and the rate and extent of its degradation. Once available to the microbial community, transformation will be determined by the entry of the xenobiotic into the cell and the degree of structural analogy between the synthetic compound and the natural compound for which the biodegradative machinery exists. Fortuitous metabolism and co-metabolism play important roles in the removal of xenobiotic compounds from the environment.

Most biodegradation studies have centred on the traditional approach of the selection and characterization of pure isolates from the environment. However, the heterogeneity of microenvironments provides habitats for a vast range of microorganisms with many different metabolic capacities and interaction between these environments must occur. Xenobiotics will encounter mixed populations of microorganisms, communities that possess co-operative, commensal and mutualistic relationships.

6.7.1   *The involvement of microbial communities in the biodegradation of xenobiotics*

It is possible to isolate stable communities, in which the advantageous relationships between the community members make the association more succesful than the individual species would have been on their own. The classification of microbial communities is based on the interrelationships of the individual species.

(1) Communities in which one or more members are unable to synthesize a particular growth requirement(s) and this deficiency is supplemented by the metabolic activities of other community members. This type of relationship may vary in the complexity of the various interactions but is probably important in the degradation of many compounds. For instance, a two-membered community growing on cyclohexane consisted of a *Nocardia* sp., which could oxidize but not grow on cyclohexane alone, and a *Pseudomonas* sp., which provided the necessary growth requirement, probably biotin, for the other species. These relationships may be obligatory for the growth of a particular species or may provide additional nutrients for the improved growth and rates of degradation of the primary oxidizing organism. This strongly suggests that the true rate of xenobiotic degradation in the environment can only be assessed by consideration of microbial communities and not individual species.

(2) Communities in which metabolites that are rather inhibitory to the producer organism or other organisms in the same environment are removed by other community members.

(3) Communities in which the growth parameters of the individual organism(s) have been altered to produce a more competitive community, stable to environmental perturbations. Thus, the community may be less susceptible to substrate inhibition and able to cope with shock loads that are encountered in waste treatment systems.

(4) An extremely important role of microbial communities in the degradation of xenobiotic compounds is the ability to achieve a co-operative metabolic attack upon the substrate. Individual members of the community do not possess the necessary metabolic capacity to totally degrade the compound, although this capacity is present within the community as a whole. Thus, failure to demonstrate the mineralization of a particular compound may result from undue emphasis on the degradation by single microbial species. It has also been noted that this combined metabolic approach can be expressed through the synthesis of different enzyme components by individual species, with expression of enzyme activity only within the community (e.g. expression of lecithinase activity by two pseudomonads).

(5) The importance of co-metabolism in the context of xenobiotic breakdown has been stressed. Microorganisms growing at the expense of one substrate transform another in one or a sequence of enzymic reactions. These reactions are not growth associated in that they do not generate intermediates that can be used for growth by that organism. However, other interacting organisms may use these intermediates as carbon sources. The isolation or construction of such communities to deal with the problems of xenobiotic pollution may be possible, although the stability of such associations in waste treatment situations is unknown.

(6) Communities have been identified in which the movement of reducing equivalents between populations has occurred. Under anaerobic conditions, the classic fermentative solution to the generation of excess reducing equivalents has been to reduce the end-products of metabolism. However, these communities use a second, acceptor organism and many such anaerobic communities have now been isolated. The anaerobic degradation of aromatic compounds by these communities may be a process of environmental significance.

(7) Communities in which more than one primary population are able to utilize completely the growth limiting substrate have been isolated in continuous culture, despite the expectation that the most competitive organism would dominate. Communal interactions that occur must, therefore, stabilize the free competition between such community members.

Stable microbial communities provide the opportunity for the exchange of genetic information between populations. Plasmid transfer (see Chapter 7) between microbial communities may be an important evolutionary mechanism for the *in vivo* development of novel strains able to deal with environmentally novel compounds. The frequency with which such events occur in the natural habitat is not known, but such evolution has been demonstrated in the laboratory. A mixed culture of *Pseudomonas* spp. was grown in continuous culture for many generations. One species was capable of growth on chlorocatechols and the other possessed the TOL plasmid, which coded for the enzyme benzene dioxygenase. Neither strain could grow on 4-chlorobenzoate but enrichment resulted in the isolation of a mutant which could grow upon this chlorinated aromatic acid. It appeared that plasmid transfer had occurred, enabling the mutant strain to oxidatively decarboxylate 4-chlorobenzoic acid, using the acquired broad specificity dioxygenase, and hence grow on the resultant chlorocatechol.

In the light of these type of interactions, all of which are environmentally important, specific examples of xenobiotic degradation will be considered. The potential for the application of biotechnological treatment of industrial wastes is discussed in section 6.6. Wastewaters from industry can be expected to contain stable reaction by-products, often of known composition, and the technology for the control of these limited point discharges is reasonably well advanced. The environment, however, will be contaminated by the complex formulated commercial product (by spillages, etc.) from innumerable unspecified non-point discharges. It is from that perspective that xenobiotic degradation will be examined.

6.7.2   *Chlorinated hydrocarbons*

Chlorinated C-1 and C-2 hydrocarbons are employed extensively as solvents and represent significant environmental contaminants. However, knowledge of the microbial degradation of these compounds is not well advanced. Organisms that are able to utilize dichloromethane have been isolated but the pathway of degradation remains to be fully elucidated. It is likely that primary dehalogenation is catalysed by a halidohydrolase to form chloromethanol, which spontaneously decomposes to formaldehyde.

The dehalogenation of halogenated aromatic compounds proceeds primarily after the aromaticity of the system has been abolished by the action of dioxygenases. The carbon–halogen bond is labilized by the introduction of an oxygen atom on to the halogen-carrying carbon with the formation of halocatechols. The relaxed specificities of the initial catabolic enzymes may allow a relatively fast co-metabolic turnover of the haloaromatic, although they have to cope with the steric and negative inductive effects of the substituent groups. This is an example of the recruitment of ordinary catabolic enzymes for the initial attack upon xenobiotic compounds. Halocatechols are critical metabolites in the degradation of aryl halides. They are degraded via *ortho*-cleavage pathways, as *meta*-cleavage enzymes are irreversibly inactivated by lethal metabolites of the cleavage reaction. Ordinary pyrocatechases are relatively inefficient in the cleavage of halocatechols, and ring cleavage thus assumes fundamental importance in the degradation of these compounds. Subsequent metabolic steps involving cycloisomerization serve to eliminate the remaining ring substituents. Conventional waste treatment plants are severely disturbed by shock loads of wastes containing chloroaromatics: *meta*-cleavage pathways are destroyed and the inefficiency of the *ortho* pathways causes an accumulation of toxic phenols and their black autooxidation products.

6.7.3   *Other substituted simple aromatic compounds*

In the degradation of aryl halides, elimination of the substituent groups is often achieved in the later catabolic steps after the aromaticity of the system has been destroyed. In the case of sulphonated aromatics, the carbon–substituent bond is highly polar and must be labilized at the earliest possible opportunity; otherwise, the whole sequence of catablic enzymes would have to be able to deal with the effect of this substitution. Sulphonated naphthalenes are widely used as emulsifiers, wetting agents and in the manufacture of azo dyes and, in conventional sewage treatment plants, these compounds appear to be inert. However, in continuous culture, a naphthalene-utilizing community from sewage acquired the catabolic capability to degrade naphthalene sulphonic acids. The first catabolic steps involved dioxygenation, substituent elimination and re-aromatization, the key step requiring a dioxygenase of broad specificity, capable of hydroxylating regiospecifically in the 1,2 position of the ring system (Fig. 6.19).

Nitrotoluenes are toxic effluents produced by munitions factories and disappear quickly from the environment. They are not completely mineralized, however, as the products of degradation (aromatic amines) condense with the carboxylic groups present in cellular macromolecules to form polyamides. These appear to be resistant to further microbial attack.

6.7.4   *Polyaromatic hydrocarbons*

Polychlorinated biphenyls (PCB) are extremely recalcitrant compounds, which are very persistent in the environment, absorbing strongly to biological material and sediments. They are considered to be practically immobile in soil and strong microbial degradation is

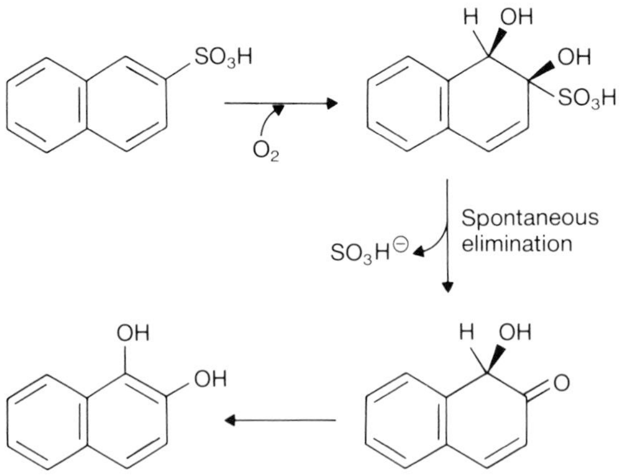

Fig. 6.19. Elimination of the sulphonic acid group from naphthalene sulphonic acid.

unlikely to occur. The composition of PCBs in environmental samples differs from that of commercial samples due to this retention in environmental systems. The microbial degradation of biphenyl has been characterized and is consistent with the catabolic systems described for other aromatic hydrocarbons. As the degree of chlorination increases, the rate of metabolism decreases. The microbial degradation of commercial PCBs to uncharacterized low molecular weight hydrocarbons in mixed cultures has been reported. Molecules with lower degrees of chlorination are preferentially attacked and substitutions greater than tetrachloro-PCB appear to render the molecule completely inert. Most microbiological studies have examined the metabolic fate of pure PCB isomers.

Benzopyrene is a recalcitrant polyaromatic, whose transformation can give rise to carcinogenic hydroxy and epoxy derivatives. It is not mineralized by activated sludge treatment, although several microbial cultures which can partially metabolize this compound have been described, involving inducible and non-inducible hydroxylase systems to produce a complex mixture of conjugated derivatives.

Polystyrene is extremely resistant to microbial degradation. One process has been described in which finely pulverized car tyres, containing a styrene-butadiene rubber, were partially degraded biologically following the addition of a surfactant. The process was retarded by the presence of antiozonates, antioxidants and vulcanizer accelerators, but the final end-product could be used as a soil conditioner. The growth of a microbial community on styrene, which resulted in the removal of the polymerization inhibitor, 4-*tert*-butyl-catechol, has been described. Free radical polymerization of the styrene resulted in the precipitation of polystyrene but it was subsequently noted that the polymer was removed, indicating the ability of the community to degrade the polymer.

6.7.5   *Microbial treatment of oil pollution*

It is pertinent at this point to consider the microbiological degradation of complex mixtures of hydrocarbons and their derivatives, charactersitic of oil pollution, both in the form of wastewaters from the petroleum industry and from accidental and deliberate discharges of oil to the environment. These discharges arise from a variety of sources, such as the washings of ships' bunkers, major environmental pollution from tanker accidents at sea, leaks in oil storage systems and dumping of waste oil.

Wastewaters from the oil industry are usually treated biologically, after the removal of most of the oil by physical processes or the use of coagulants. The toxic effects of the constituents of the wastewaters on the activated sludge systems can be minimized by gradual acclimatization of the treatment system with an increasing throughput of wastewater and, thereafter, ensuring a constant flow and composition. However, the waste loadings of these systems may vary significantly and may be better dealt with by employing the improved technologies of pure oxygen aerated sludge systems or the deep-shaft process.

The greatest environmental input of oil occurs in the sea, where the oil undergoes a series of physical changes, known as weathering. The total abiotic losses arising from these processes, which include dissolution, evaporation and photooxidation, amounts to 24–40%, depending upon the type of oil and the meteorological conditions. Many of the short-chain alkanes are removed at this stage. The extent of microbial degradation of the weathered spillages will be governed by a number of factors. The composition of the oil is an important consideration, as the relative proportions of saturates, aromatic compounds, nitrogen-, sulphur- and oxygen-containing compounds and asphaltenes varies from oil to oil. Branched alkanes, sulphur-containing aromatics and asphaltenes will confer recalcitrance to some degree. In addition, the availability of nutrients such as nitrogen and phosphorus will determine the rate of microbial growth and, therefore, degradation. It has been found that the addition of such nutrients does cause a stimulation in the rate of biodegradation. The variety of organisms capable of growth upon oil constituents is related to the degree of hydrocarbon pollution. For example, the highest number of oil-degrading microorganisms are found near large ports, following chronic oil pollution, or around oil platforms. Total degradation of oil is often not achieved, even in diverse microbial communities, and the more biologically inert components, such as asphaltenes, are found in sediments and oil deposits. The main physical conditions influencing the rate of oil degradation are temperature, oxygen concentration, hydrostatic pressure and the degree of oil dispersal. An oil-in-water emulsion is the best phase for enhanced microbial attack on the oil.

The dumping or accidental spillage of oil in soil presents a particular problem, as this may eventually lead to the contamination of ground water and drinking water supplies. Microorganisms capable of hydrocarbon degradation are widespread in the soil environment, but even their activities may be problematical in the production of soluble derivatives or surface active agents, which may serve to increase the mobility of the remaining oil.

6.7.6 *Pesticides*

The discharge of pesticide wastes and residues from the sites of manufacture is now under stringent legal control and the wastewater treatment technologies for such point source control are now well advanced, although complex and diversified. This primarily involves solvent extraction, followed by conventional biological treatment. No technology exists for non-point discharges, which occur with spillages or the cleaning and disposal of pesticide containers. In addition, pesticides contribute to environmental pollution subsequent to their intended use through water run-off from crops. The majority of pesticides can be degraded by bacteria and fungi. The transformation of the parent pesticide to less complex products is often achieved by the concerted effort of microbial communities. Various extents of degradation of DDT have been reported, for example, including the involvement of co-metabolic processes for the complete mineralization of this recalcitrant pesticide. These communities can often be isolated from the environment with the xenobiotic serving not as a primary carbon source but as phosphorus, sulphur or nitrogen sources. The extreme toxicity of the pesticide is often lost in the first transformation step and the potential exists, therefore, for the development of simple microbiological treatments for pesticide detoxification. Hydrolysis, for example, may reduce the toxicity significantly or greatly enhance the biodegradation potential of the transformed pesticide. The requirement exists for cell-free enzymes able to operate without coenzymes or special requirements that are capable of detoxifying various classes of pesticides. Hydrolases such as esterases, acylamidases and phosphoesterases could be considered. For *in situ* application, the chosen enzyme should possess favourable kinetics over a broad range of temperature and pH, must be insensitive to low levels of solvents or heavy metals, exhibit no substrate inhibition at the levels encountered in the waste treatment environment and also have good storage stability. Parathion hydrolase, isolated from a *Pseudomonas* sp., has been tested as a biological detoxifying agent in a number of situations. It has been used to remove 94–98% of residual parathion (approximately 75 g) from a pesticide container in 16 h at substrate levels of 1% (w/v). Buffered solutions of parathion hydrolase have also been applied to soil spillages of parathion, where pesticide concentrations would be expected to be high. In this instance, soil type, moisture content, buffer strength and enzyme concentration were all critical factors determining the rate of detoxification but substantial levels of the pesticide were removed after only 8 h. A further application of the immobilized enzyme, which has been demonstrated in the laboratory, is the treatment of wastewater effluent. Hydrolases for other pesticides have been reported. Many possess a broad substrate specificity, which offers great potential for the development of other simple pesticide detoxification systems. Future applications of such systems may include the cleaning of chemical plants and reactors, the use of enzyme sprays to remove applied pesticides and pesticide–enzyme formulations for rapid decay after pesticide application.

6.7.7 *The microbial degradation of surfactants*

Synthetic surface-active agents, employed domestically and industrially for cleaning purposes, can be classified as either 'hard' or 'soft' according to their susceptibility to biodegradation. Anionic surfactants, such as alkylbenzene sulphonates, claimed public attention as environmental pollutants in the late 1950s, causing foam formation in waterways. Initially, 'hard' detergents were marketed, rendered recalcitrant by their branched alkyl side chains. Biodegradable, linear, unbranched alkylbenzene sulphonates (LAS) were introduced voluntarily by the detergent industry to counteract this environmental persistence. Degradation of these surfactants occurs initially by terminal methyl group oxidation followed by β oxidation of the linear side chain. Ring degradation does not occur usually until the side chain has been virtually eliminated. As the initial oxidative step requires oxygen, this process only occurs aerobically. Branching does not necessarily confer recalcitrance, although the process of β oxidation is inhibited. The mechanisms of branched chain degradation are still unclear. The carbon–sulphide bond is stable, increasing the biological resistance of the detergent molecule. Desulphonation reactions have not been fully elucidated but could be effected by hydroxylase or monooxygenase mechanisms. The ultimate fate of the sulphonate group is thought to be conversion to sulphate, probably via intermediate sulphite. The metabolic fate of aryl sulphonates has already been described (see section 6.7.3) and there is evidence for plasmid-encoded desulphonation and *meta*-cleavage of the aromatic ring.

Alkyl sulphates are still used industrially in laundry formulation and cosmetic products, the majority being primary alkyl sulphates. Linear sulphates are readily degraded but this is retarded by the presence of branch points along the molecule. Sulphatase enzymes are involved in the initial attack upon these molecules to produce the corresponding alcohols, which are then further metabolized. This process does not require molecular oxygen and can, therefore, occur anaerobically. Alkyl sulphatase enzymes are unusual amongst sulphatase enzymes in that they attack the C–O bond of the C–O–S linkage. Primary and secondary alkyl sulphates induce a complex pattern of sulphatase enzymes.

Non-ionic detergents are used domestically and their wetting and emulsifying properties are used to advantage in the formulation of agricultural sprays and cosmetics. Linear primary alcohol ethoxylates are rapidly and completely mineralized but higher molecular homologues are increasingly recalcitrant. Degradation proceeds by terminal methyl group oxidation, followed by β oxidation to produce short-chain alkanoate ethoxylates with no surface-active properties. In secondary alcohol ethoxylates, the hydrophobic alkyl chain is degraded from both ends by ω- and β oxidation. The ether linkages found in these compounds contribute to their biological recalcitrance. Possible modes of cleavage of this linkage have been suggested and include monooxygenative cleavage, hydrolysis, the involvement of a carbon-oxygen lyase and the oxidation of the carbon atom *alpha* to the ether linkage, followed by ester hydrolysis.

Commercial detergent formulations seldom contain more than 30% by weight of the surface-active agent. Other constituents include optical brighteners, oxidizing bleaches, foam boosters, anti-corrosion agents and, in some cases, enzymes with the bulk of the formulation being composed of builder compounds. The purpose of such builders is (1) to reduce the free concentrations of calcium and magnesium to prevent inorganic deposition, which would otherwise cause the precipitation of insoluble alkaline earth salts of the detergent anion, and (2) to deflocculate particles of soil and stabilize the soil suspension. Builders should possess a good buffering capacity and not interact unfavourably with other components of the formulation. For many years, sodium triphosphate was used for this purpose but it contributed to the accelerating eutrophication on inland waters and alternatives have been sought. However, any biodegradable nitrogen or phosphorus compound undergoing mineralization can contribute to the process of eutrophication. The examination of nitrogen- and phosphorus-free builders led to the identification of synthetic carboxylate esters as candidates (CMOS, carboxymethylsuccinate; ODA, oxydiacetate; EGDA, ethylene glycol diacetate; Fig. 6.20). The study of the metabolism of CMOS and its analogues has provided a valuable insight into the way in which microorganisms acquire the potential to degrade novel xenobiotic compounds. CMOS itself is rapidly degraded in the

$COOH.CH_2.O.CH(COOH).CH_2COOH$	CMOS
$COOH.CH_2.O.CH_2.COOH$	ODA
$COOH.CH_2.O.CH_2.CH_2.O.CH_2.COOH$	EGDA
$N(CH_2COOH)_3$	NTA

Fig. 6.20. Structure of some organic builder compounds for use in detergent formulations.

environment, the initial reaction involving an inducible lyase, which cleaves the molecule to produce glycollate and fumarate. Analogues of CMOS prove more recalcitrant, probably due to the specificity of the induced lyases. Uptake of the analogues does not contribute towards their recalcitrance as it has been shown that the compounds are transported on the constitutive citrate transport system. CMOS lyase has been noted to bear a remarkable resemblance to polygalacturonate lyase and organisms that grow on CMOS also grow well on polygalacturonate. This may represent an important step in the evolution of a degradative capacity by the requisition of existing enzymes for a novel function. In addition, it has been suggested that this lyase may be plasmid-encoded. Nitrilotriacetate (NTA) has been widely used as a builder compound as it is subject to rapid biodegradation in activated sludge systems and in river water.

6.8   Recommended reading

CALLELY A.G., FORSTER C.F. & STAFFORD D.A. (eds.) (1977) *Treatment of Industrial Surfactants*, pp. 283–327. Hodder & Stoughton, London.
CHATER K.W.A. & SOMERVILLE M.J. (eds.) (1978) *The Oil Industry and Microbial Ecosystems.* Heyden & Son Ltd. (on behalf of the Institute of Petroleum), London.
GIBSON D.T. (ed.) (1984) *Microbial Degradation of Organic Compounds.* Microbiology Series, Vol. 13. Marcel Dekker, New York.

HIGGINS I.J. & BURNS R.G. (1975) *The Chemistry and Microbiology of Pollution.* Academic Press, London.

LEISINGER T., COOK A.M., HUTTER R. & NUESCH J. (1981) Microbial Degradation of Xenobiotics and Recalcitrant Compounds. *FEMS Symp.* **12.** Academic Press (on behalf of the Federation of European Microbiological Societies), London.

MONLIN G. & GALZY P. (1984) Whey, a Potential Substrate for Biotechnology. In *Biotechnology and Engineering Reviews*, Vol. 1, (ed. Russell G.E.), pp. 347–374. Intercept Ltd., Newcastle-upon-Tyne.

STAFFORD D.A. (1982) Biological Treatment of Organic Compounds and Solvents. In *Safe Use of Solvents*, (eds. Collings A.J. & Luxton S.G.), pp. 293–304. Academic Press, London.

STAFFORD D.A. & ETHERIDGE S.P. (1982) Farm Wastes, Energy Production and the Economics of Farm Anaerobic Digesters. In *Anaerobic Digestion*, pp. 255–268. Elsevier Biomedical Press, Amsterdam.

STAFFORD D.A., HAWKES D.L. & HORTON R. (1980) *Methane Production from Waste Organic Matter.* CRC Press, Boca Raton, Florida.

TURNER J., STAFFORD D.A., HUGHES D.E. & CLARKSON J. (1983) The Reduction of Three Plant Pathogens (*Fusarium*, *Corynebacterium* and *Globodera*) in Anaerobic Digestion. *Agric. Wastes*, **6,** 1–11.

WHEATLEY A.D. (1984) Biotechnology and Effluent Treatment. In *Biotechnology and Genetic Engineering Reviews*, Vol. 1, (ed. Russell G.E.), pp. 261–310. Intercept Ltd., Newcastle-upon-Tyne.

7 Genetics and Biotechnology

K. G. HARDY & S. G. OLIVER

7.1 Introduction

The optimization of any industrial-scale process using a living organism will involve a major effort to improve the genetic characteristics of that organism. Traditionally, strain improvement has involved mutagenesis followed by screening and selection of favourable variants. Indeed, in the most ancient fermentation processes (such as brewing and cheese making) selection must have been an unconscious act on the part of practitioners in the era before Pasteur. In more recent times, hybridization has been employed to combine desirable characteristics from different strains in a single organism.

A programme of strain improvement through mutant selection and recombination is more likely to succeed if the biochemistry and physiology of the fermentation process are fully understood. It is important to know which are the rate-limiting steps in a metabolic pathway or what is the thermodynamic limit to improvements in product yield. Usually a process can be improved by both physiological and genetic tricks and the application of the two disciplines should be regarded as complementary rather than competitive.

Today, we have moved from a situation in which strain improvement was the sole application of genetical techniques to one where processes can be innovated by exploiting the tools of recombinant DNA technology ('genetic engineering'). Novel products, such as human proteins and peptides or viral antigens, may now be produced by fermentation routes as a result of these techniques. Much of the current excitement about biotechnology has been due to the advent of genetic engineering. This chapter discusses how both *in vivo* and *in vitro* techniques of genetic manipulation may be used to improve existing biotechnological processes and to create new ones.

7.2 Conventional routes to strain improvement

7.2.1 *Mutagenesis and selection*

The use of mutagenesis and selection has been the traditional route of strain improvement which has increased the yield of antibiotics produced by fungi and actinomycetes. The way in which these techniques were used to increase yields of penicillin is illustrated in Fig. 7.1. By successively selecting over twenty strains synthesizing more penicillin, a 55-fold increase in productivity was achieved. In this example, as in many others, there was no direct selection; it was not possible to design conditions under which only the desired mutants would grow.

(i)

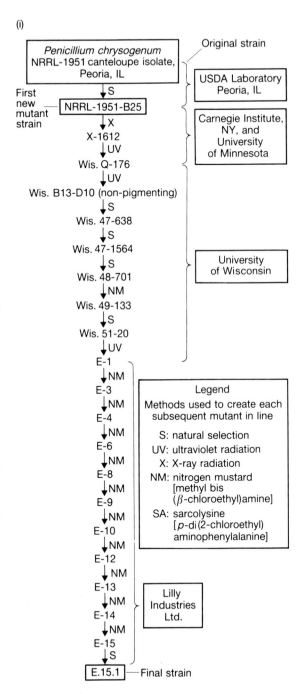

Penicillium chrysogenum
NRRL-1951 canteloupe isolate,
Peoria, IL

—— Original strain

USDA Laboratory
Peoria, IL

First new mutant strain ——

NRRL-1951-B25

↓S

↓X

X-1612

↓UV

Wis. Q-176

↓UV

Wis. B13-D10 (non-pigmenting)

↓S

Wis. 47-638

↓S

Wis. 47-1564

↓S

Wis. 48-701

↓NM

Wis. 49-133

↓S

Wis. 51-20

↓UV

E-1

↓NM

E-3

↓NM

E-4

↓NM

E-6

↓NM

E-8

↓NM

E-9

↓NM

E-10

↓NM

E-12

↓NM

E-13

↓NM

E-14

↓NM

E-15

↓S

E.15.1 ——Final strain

Carnegie Institute,
NY, and
University
of Minnesota

University
of Wisconsin

Lilly
Industries
Ltd.

Legend

Methods used to create each
subsequent mutant in line

S: natural selection
UV: ultraviolet radiation
X: X-ray radiation
NM: nitrogen mustard
[methyl bis
(β-chloroethyl)amine]
SA: sarcolysine
[p-di(2-chloroethyl)
aminophenylalanine]

Fig. 7.1. (i) Development of a high penicillin-producing strain via genetic manipulation. Adapted from *Impacts of Applied Genetics*, Office of Technology Assessment, US Government Printing Office, Washington DC, USA. (ii) Evolution of the penicillin fermentation: graph showing the increase in final concentration of penicillin in industrial fermentations in the period 1940–1975.

(ii)

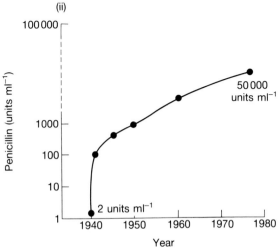

50 000 units ml^{-1}

2 units ml^{-1}

Instead, a screening procedure had to be adopted, in which survivors from intensive mutagenesis were grown in shake flasks and the culture filtrates assayed for the quantity of antibiotic produced.

Such screening procedures are both labour intensive and time consuming and it is not unusual to discover that strains which produce high yields in shake flasks do not maintain that performance in large-scale reactors. Short-cuts to this approach often involve knowledge of arcane facts, e.g. that particular types of morphological variants are often high yielding strains.

Screening on agar plates instead of in shake flasks allows far more mutants to be processed. Bioassays for the production of antibiotics or metabolites are often performed by incorporating indicator organisms into the agar plates. Systems have been developed to automate plate screens by the use of mechanical inoculating devices and time-lapse photography coupled to computer image analysers.

It may be possible to design conditions under which mutants of the desired type will not grow. In these situations an enrichment technique, which kills off the actively growing wild-type cells while permitting the non-growing mutants to survive, may be employed. Examples of such techniques include the use of penicillin or nystatin to kill, respectively, growing bacterial or fungal cells. In the fungi, starvation for the lipid component, inositol, causes the autolysis of growing but not quiescent cells.

True selection, in which conditions are established such that only mutant organisms will grow, is by far the most desirable technique. It has been exploited to isolate mutants which over-produce some metabolite, such as an amino acid or citrate. The mutagenized culture is grown in the presence of some inhibitor, such as an amino acid analogue, and mutants are selected which overcome the metabolic block by over-producing the desired compound.

Agar plate selection techniques require an all-or-none response on the part of the organisms. Mutant colonies will grow and wild-type ones will not. However, in the development of industrial strains, particularly those used in long-established fermentation processes, some small increment in net growth rate under particular conditions is often sought. In such a situation selection in continuous culture is indicated. Exponential growth may be maintained for long periods in a chemostat and this permits mutants with quite small improvements in substrate affinity, maximum specific growth rate or tolerance to the toxic effects of high substrate or product concentrations to be isolated.

7.2.2   *Hybridization by mating*

An obvious way to develop organisms with a desirable combination of
genetic traits is by the use of breeding programmes which exploit the
natural sexual systems of those organisms. Both prokaryotic and
eukaryotic microorganisms have mating systems involving cell-to-cell
contact, which may be used to construct recombinant organisms.

Sexual conjugation in bacteria

In bacteria, the sexual system is known as conjugation and cell-to-cell
contact is mediated by a long tube known as a sex pilus. It is probable
that the pilus represents a channel through which DNA can be
transferred from one organism to another. Many plasmids (see below)
are able to encode this sexual structure and transfer their own genes
and, in some cases, those of their host to a recipient cell. Plasmids that
are able to transfer the genes of the host bacterium are said to have
chromosome mobilization ability or Cma. The F plasmid acts as a sex
factor in *Escherichia coli* and is able to mobilize the chromosome not
only of *E. coli* but also of related enterobacteria such as *Shigella*,
Klebsiella, *Salmonella* and *Erwinia*. The F factor can, therefore, be used
to effect gene transfer between different bacterial genera. Some
plasmids isolated from *Pseudomonas aeruginosa* are even more promis-
cuous. In particular, the plasmid R68.45 can transfer chromosomal
genes between various *Pseudomonas* species, including *Pseudomonas
putida*, and also has Cma for such genera as *Rhizobium*, *Rhodopseudo-
monas*, *Azospirillum*, *Agrobacterium* and *Escherichia*.

    Streptomyces species are among the most important industrial
bacteria since they produce more than 60% of the types of antibiotics in
current use. These organisms have a well developed sexual system
which has been extended to such close relatives as *Nocardia* (which
produces the rifamycin antibiotics). Conjugative ability was first
demonstrated in the species *Streptomyces coelicolor*, which is a laboratory, not an industrial, organism. Two plasmids which have sex factor
activity, SCP1 and SCP2, have been discovered in this organism. SCP2
produces recombinants at very high frequency. SCP1 has many properties in common with the *E. coli* F plasmid. It is able to integrate into the
Streptomyces chromosome and also to incorporate segments of chromosomal DNA into the plasmid molecule. SCP1 is a fairly large plasmid
($Mr = 18 \times 10^6$) and carries all the sixteen genes required for the
synthesis of an antibiotic called methylenomycin.

    Many industrial species of *Streptomyces* appear to lack a mating system of their own. However, *Streptomyces rimosus*, which is used in oxytetracycline production, carries a plasmid (SRP1) which has sex factor activity and produces recombinant organisms at a frequency of 10^{-4}–10^{-3} in SRP1$^+$ × SRP1$^-$ crosses.

Conjugative plasmids are fairly rare among Gram-positive bacteria other than *Streptomyces*, although a mating system has recently been demonstrated for the industrially important genus *Bacillus*. *Bacillus* also has the advantage of being able to take up naked DNA readily and so recombinant organisms may easily be constructed by transformation. This ability has been exploited to produce strains of *Bacillus* in which the starch-hydrolysing enzyme α-amylase constitutes about 50% of the total protein synthesized by the bacterium. Mutations in several genes leading to the over-production of amylase were selected. These different mutations were then combined in a single strain by successively transforming with chromosomal DNA and selecting for an enhanced ability to degrade starch.

Mating systems in fungi

Fungi exhibit a considerable sexual versatility, which may be exploited in genetic studies. Many of the *Ascomycete* and *Basidiomycete* fungi have a highly developed sexual system, which prevents self-fertilization and other forms of in-breeding and encourages outbreeding. The mating process is governed by an incompatibility system. Some fungi show bipolar incompatibility, in which there is a single genetic locus controlling mating behaviour with alternate alleles possible at that locus. This results in two mating types, e.g. *a* and *α* in *Saccharomyces cerevisiae* with only *a*/*α* pairings being permitted. Other fungi (e.g. *Schizophyllum commune*) have tetrapolar incompatibility: there are two mating-type genes, which may each take a large number of allelic forms. For a successful pairing to occur, the two sexual partners must carry different alleles at each of the two mating-type loci (i.e. AxBy × AyBx is a fully compatible pairing while AxBy × AxBx or AxBy × AyBy or AxBy × AxBy are prohibited and arrest at various stages of the mating process).

Fungal parasexual cycle

Many industrially important filamentous fungi have no perfect or sexual stage which can be exploited in breeding programmes for strain improvement. However, the parasexual cycle can be used to effect genetic recombination in such organisms and it has been demonstrated in the following genera of commercial importance: *Aspergillus, Penicillium, Cephalosporium* and *Fusarium*. The standard parasexual cycle is illustrated in Fig. 7.2

Heterokaryons, where genetically distinct nuclei coexist in the same cytoplasm, may be formed either by the anastamosis of hyphae from separate mycelia or mutation of individual nuclei within a mycelium. The fusion of nuclei within a heterokaryon occurs at low frequency (10^{-5}–10^{-4}) but may be promoted by either (+) camphor or

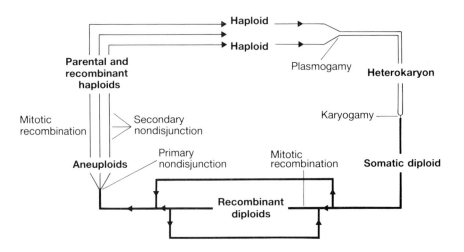

Fig. 7.2. The standard parasexual cycle: ——, haploid or aneuploid nuclei; ▬, diploid nuclei. From C.E. Caten (1981) Parasexual processes in fungi. In *The Fungal Nucleus*, (eds. K. Gull & S.G. Oliver), pp. 191–214. Cambridge University Press.

UV light. Mitotic recombination within diploid nuclei can be induced by a number of chemical and physical agents which are either mutagens or inhibitors of DNA synthesis. The return to the haploid state, following recombination, appears to occur by successive loss of individual chromosomes and is promoted by anti-microtubule agents such as *p*-fluorophenylalanine and the benzimidazole carbamates.

The parasexual cycle is a very inefficient means of forming recombinants when compared with true sexual systems. However, by the use of intense selection and the various promoting agents described it is possible to isolate the desired recombinants. The system has been used to improve antibiotic yields from both *Penicillium* and *Cephalosporium*.

7.3   *In vivo* genetic manipulation

7.3.1   *Plasmids*

Many of the properties of bacteria which make them interesting for the biotechnologist are encoded by genes on bacterial plasmids. Plasmids are circular DNA molecules which can be stably inherited in bacterial cells without being linked to the main bacterial chromosome. Plasmids are also used in genetic engineering to clone genes of interest. Some of the important general properties of plasmids will be summarized to provide a background for a discussion of those properties which are of special relevance to biotechnology.

Replication and structure

Plasmids have molecular weights ranging from about 1×10^6 to about 200×10^6. The smallest plasmids could, therefore, code for one or two average-sized proteins, whereas the larger plasmids could code for 300 or more proteins. The numerous enzymes required for a whole biochemical pathway, such as the degradation of toluene to catechol, are specified by certain of the larger plasmids. Plasmids having molecular weights of more than about 100×10^6 have been found only in Gram-negative bacteria, especially in *Pseudomonas* spp. and *Agrobacterium* spp. Plasmids found in Gram-positive bacteria very rarely have molecular weights of more than 40×10^6.

Plasmids exist in bacterial cells in the form of double-stranded circular DNA molecules ('closed circles') which are twisted to form supercoils. Some plasmids can maintain themselves in only one or two closely related species. Plasmids which have a wide host range include RP4, R68.45, RK2 and closely related drug-resistance plasmids (R plasmids) belonging to the P incompatibility group (members of the same incompatibility group cannot coexist together in a bacterial cell). These plasmids have been transferred to many genera of Gram-negative bacteria and it is likely that all Gram-negative strains are potential hosts for them.

R plasmids belonging to the P group are, therefore, particularly useful for the genetic manipulation of many Gram-negative bacteria which are economically important. Using these plasmids, it is possible to transfer chromosomal genes between unrelated species. Small derivatives of RP4 and RK2 are used as vectors to clone DNA molecules, which can then be transferred to a variety of different species.

The replication of several plasmids has been investigated extensively and many details of the process are now understood, particularly for the small plasmid ColE1. This information is important in the use of plasmids for gene cloning. The plasmids most commonly used for cloning are derivatives of pBR322, which comprises 4362 base pairs and codes for resistance to penicillin and tetracycline. The resistance genes are derived from R plasmids (see Fig. 7.3) and have been attached to part

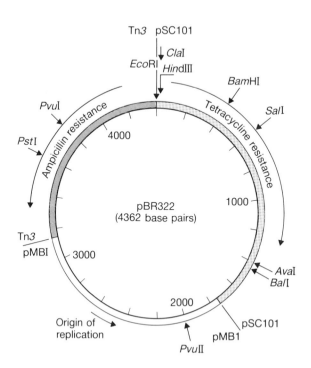

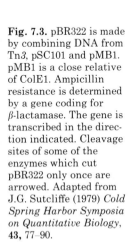

Fig. 7.3. pBR322 is made by combining DNA from Tn*3*, pSC101 and pMB1. pMB1 is a close relative of ColE1. Ampicillin resistance is determined by a gene coding for β-lactamase. The gene is transcribed in the direction indicated. Cleavage sites of some of the enzymes which cut pBR322 only once are arrowed. Adapted from J.G. Sutcliffe (1979) *Cold Spring Harbor Symposia on Quantitative Biology*, **43**, 77–90.

of a small plasmid called pMB1. Plasmid pMB1 is itself very closely related to the most extensively studied small plasmid, ColE1.

pBR322 and ColE1 replicate in *E. coli* and related members of the *Enterobacteriaceae*. They exist in bacteria in multiple copies; ColE1 has a copy number (the number of plasmid molecules per chromosome) of about 20 and pBR322 of about 40. The copy number of a plasmid is an important feature in relation to cloning. A larger number of copies of a cloned gene may result in a higher yield of the protein encoded by that gene. Plasmids with high copy numbers may be less likely to be lost from bacterial cultures. On the other hand, a low copy number plasmid is sometimes useful for cloning genes whose products are lethal to the bacterial cell if they are produced in large amounts.

Much research has focused on the mechanisms which control the replication and the copy numbers of bacterial plasmids. Plasmids are maintained at an approximately constant copy number in cells growing at different growth rates. This implies that plasmid replication keeps in step with bacterial growth. This co-ordination is achieved by mechanisms which control the initiation of replication. Once a round of replication starts, it proceeds at about the same velocity at all bacterial growth rates; at high growth rates, plasmid replication is initiated more often on high copy number plasmids than on those which have low copy numbers.

Initiation of replication of ColE1 and pBR322 is controlled by a small RNA molecule which interacts with a specific region of the plasmid near the origin of replication. Once initiated, replication proceeds in one direction around the plasmid and terminates near the origin. Geneticists interested in maximizing the production of proteins encoded by genes cloned in pBR322 have obtained mutants which specify an altered RNA molecule. Some of these mutants have copy numbers of more than 100 and this can result in higher yields of the products of cloned genes. An interesting feature of replication of ColE1 and pBR322 is that it is entirely dependent on host enzymes; the plasmids themselves do not code for any proteins which are needed for replication.

The replication of pBR322 and of ColE1 differs from that of the *E. coli* chromosome in that it is not inhibited by chloramphenicol, which stops protein synthesis. In the presence of this drug, chromosomal replication stops once the round of replication already in progress is completed. Replication of pBR322 is not inhibited in this way; plasmids accumulate in chloramphenicol-treated cells so that they eventually comprise about 50% of the DNA in the cell. When preparing plasmids, chloramphenicol is sometimes added to cultures several hours before they are harvested to increase the yield of plasmid DNA.

The initiation of replication of the R1 plasmid of *E. coli* is controlled by a repressor protein which is encoded by an R1 gene. Mutants of the plasmid have been isolated which code for an abnormal repressor or are defective in the mechanism which controls its production. Some of these mutants are temperature sensitive such that the repressor is inactive at high temperature (43 °C). These are called 'runaway replication' mutants: at 30 °C they replicate normally and have copy numbers of 1 or 2; at 43 °C replication is initiated much more frequently so that several hundred plasmids accumulate in the cell. The small region of R1 which contains the origin of replication and the associated control elements has been used as the basis for plasmid vectors to clone various genes. The advantage of such vectors is that an elevated temperature can be used to generate many copies of a cloned gene and, hence, many copies of the protein specified by the gene.

Important plasmid genes

Plasmid genes which are particularly interesting to the biotechnologist include those coding for nitrogen fixation, the degradation of organic compounds and virulence factors of pathogenic bacteria.

The use of bacteria, or of bacterial genes, to enhance the ability of plants to acquire nitrogen is often discussed as a long-term benefit of genetic engineering. *Rhizobium* spp. form nodules in the roots of leguminous plants and fix atmospheric nitrogen which is made available to the plants. It was recently discovered that at least some of the bacterial genes required for nodulation and nitrogen fixation are on plasmids in rhizobia. The initial evidence came from experiments in which plasmids were transferred, by conjugation, from one strain of *Rhizobium* to another. A particular species of *Rhizobium* generally forms a symbiotic relationship only with certain leguminous plants. Thus, *R. phaseoli* forms nodules in beans, whereas *R. leguminosarum* nodulates peas. It was found that variants of *R. leguminosarum* that have lost the ability to form nodules were again able to nodulate peas when they received a conjugative plasmid from a nodulating strain. Moreover, when a conjugative plasmid from *R. leguminosarum* was transferred to *R. phaseoli*, it enabled this bacterium to nodulate peas rather than beans. Large plasmids, having molecular weights of at least 90×10^6, which occur in *Rhizobium trifolii* and *R. meliloti* also have at least some of the genes which are necesary for nodulation and for nitrogen fixation.

Plasmids can increase the virulence of pathogenic bacteria by a variety of different mechanisms. In some cases, knowledge of these plasmid-specified virulence factors can help to control the pathogens. The most extensively studied virulence plasmids are those found in pathogenic *E. coli* strains. Enterotoxigenic strains of *E. coli* are an important cause of acute diarrhoea in young animals, and of traveller's diarrhoea in adults. Plasmids in these strains often code for a heat-labile toxin (LT), which is related to the cholera toxin, and a heat-stable toxin (ST). Enterotoxigenic strains usually also harbour plasmids coding for fimbriae (or pili), which are also called 'colonization factors'. These are found on the bacterial surface and attach the bacteria to the wall of the small intestine. Strains which infect a particular animal usually have the same colonization factor. Thus, *E. coli* strains causing diarrhoea in pigs usually have either K88 antigen or 987P antigen as their adhesive pili, whereas strains isolated from calves usually produce K89 antigen. Loss of the plasmids coding for colonization factor antigens from enterotoxigenic strains greatly reduces their virulence, even though the strains may still produce toxins.

One approach to controlling the disease in pigs caused by enterotoxigenic strains has been to vaccinate the animals using a preparation containing K88 antigen. Attempts are being made using recombinant DNA techniques to enhance the production of K88 antigen for use in vaccines. In addition, these techniques are also being used to develop a toxoid which can be safely administered to animals to elicit antibodies against the toxins produced by enterotoxigenic bacteria. A different approach to controlling the problem of bacterial diarrhoea in livestock has been to breed strains of animals which are resistant to the plasmid-encoded colonization factor antigens.

Transposons

Transposons and insertion sequences are related DNA elements which occur in bacterial and bacteriophage chromosomes and on plasmids. Transposons are useful for causing mutations in bacteria and for cloning genes of interest from bacteria. They were discovered by Hedges & Jacob, who found that a small part of the RP4 plasmid which coded for resistance to penicillin could become inserted into other plasmids which had no DNA sequence homology with RP4. The usual mechanism of recombination in bacterial cells, which occurs between homologous DNA molecules and is dependent on the bacterial $recA^+$ gene product, was not responsible for the translocation of the drug resistance gene. Numerous other transposons have since been described. The majority were found on R plasmids from Gram-negative bacteria and in most cases they confer resistance to an antibiotic. The best studied transposon is Tn*3* which is 4957 base pairs long and confers resistance to penicillin. The 38 base pairs at the two ends of Tn*3* are the same, except that they are in an inverted order. The transposon codes for an enzyme, a transposase, which is essential for its transposition. Tn*3* can insert at many different positions in a DNA molecule although it preferentially integrates in A + T-rich regions.

Transposons can be used to provide a marker on a gene which is to be cloned. A difficulty sometimes encountered in cloning a chromosomal gene from a bacterium is that there is no simple method for screening plasmids containing pieces of chromosomal DNA to find out which carries the gene of interest. This difficulty can sometimes be overcome by first isolating a mutant of the gene in question by inserting a transposon within it or close to it.

7.3.2   *Protoplast fusion*

The fusion of protoplasts, cells with their walls removed, has been developed to break down the barriers to genetic exchange imposed by conventional mating systems. Protoplast fusion may be used to produce interspecific or even intergeneric hybrids. It is also used to make hybrids between cells of the same species when they are of incompatible mating types or when the natural mating system is very inefficient at producing genetic recombinants.

Protoplast fusion in bacteria

Protoplast fusion was first used in bacteria for the production of intraspecific hybrids of *Bacillus megaterium* and *Bacillus subtilis*. These organisms do not have a natural conjugation system but are readily transformed by purified DNA. Hybrids can be formed by inducing protoplast fusion with either polyethylene glycol or $CaCl_2$. Recombination frequencies between 10^{-5} and 10^{-4} are obtained in this way and the process is DNase insensitive, which demonstrates that it is not due to a transformation event.

The technique has been used extensively for the hybridization of *Streptomyces* spp. *Streptomyces coelicolor* has a plasmid-determined mating or fertility system (see page 260) but it is not very efficient. Many of the industrially important *Streptomyces* spp. have no natural means of mating. Recombination frequencies are increased by about four orders of magnitude over those achieved by mating when protoplast fusion induced by polyethylene glycol (PEG) is employed. The maximum frequency of recombination obtainable by this technique is 10–20%. This means that it is not necessary for the strains to carry selectable genetic markers since it is practicable to screen for hybrids. This is a great asset in the strain development of industrial organisms, where the introduction of suitable markers is often tedious or unwise.

Protoplast fusion in fungi

The formation of hybrid fungi by protoplast fusion has been the subject of intensive study and has found industrial application in the development of strains of *Cephalosporium acremonium*, which combine high cephalosporin yields with fast growth rates. The first fusion experiments were performed using auxotrophic mutants of *Geotrichum candidum*, which has no known sexual or parasexual mating system. Centrifugation was used to promote protoplast fusion and prototrophic heterokaryons were formed at low frequency. Since then, fusion has usually been promoted by treatment with Ca^{2+} ions and PEG.

Intraspecific fusions have been achieved with a wide range of filamentous fungi and yeasts, including such industrially important organisms as *Penicillium chrysogenum*, *Aspergillus niger*, *Candida tropicalis*, *Yarrowia lipolytica*, *Kluyveromyces lactis* and *Saccharomyces cerevisiae*. Fusion frequencies are generally in the range 0.2–2%. The occurrence of nuclear fusion, to produce a diploid, following cell fusion is dependent on the species involved. However, a progression may be seen with the filamentous fungi producing heterokaryons, the dimorphic yeasts unstable diploids, and the true yeasts stable diploids (see Fig. 7.4).

Interspecific fusions also occur at reasonable frequency, especially with fairly closely related species of *Aspergillus* or *Penicillium*. If nuclear fusion occurs it is generally accompanied by chromosome loss. Frequently, spores containing haploid nuclei of only one of the two parental types are recovered following growth under non-selective conditions.

Fusions of yeasts belonging to different genera have been attempted, e.g. *Candida* × *Saccharomycopsis*, *Yarrowia* × *Pichia*, *Yarrowia* × *Kluyveromyces*, *Saccharomyces* × *Schizosaccharomyces*. Fusion frequencies were very low, 10^{-5}–10^{-4}, and the products unstable. In the last cross, multinucleate cells were common. There has been one report of the successful fusion of a filamentous fungus (the cellulytic organism, *Trichoderma reesii*) and a yeast (*Saccharomyces cerevisiae*), but this has yet to be confirmed.

Many fusions lead to the production of unstable heterokaryons, in which the nucleus of one parental type or the other eventually prevails. However, this is not necessarily a barrier to genetic exchange. There have been convincing demonstrations of internuclear genetic exchange in heterokaryons of *Schizophylum commune* and *Saccharomyces cerevisiae*. Protoplast fusion may be a useful method for establishing the heterokaryotic state which permits this more limited form of gene transfer.

Heterokaryons
Geotrichum candidum
Phycomyces blakesleeanus
Mucor racemosus

Heterokaryons with frequent transient diploids
Cephalosporium acremonium

Heterokaryons with occasional stable diploids
Aspergillus
Penicillium

Heterokaryons with frequent stable diploids
Candida tropicalis

Unstable diploids
Yarrowia lipolytica

Stable diploids
Saccharomyces cerevisiae
Kluyveromyces lactis

Fig. 7.4. Intraspecific protoplast fusion.

Formation of plant hybrids by protoplast fusion

Protoplasts of plant cells can be obtained by either the mechanical or enzymic removal of their cell walls. Protoplasts are an extremely useful tool for the plant geneticist. They rapidly and efficiently regenerate their cell walls and can be induced to divide to generate callus tissue from which whole plants that are able to flower and set fertile seeds may be grown. The availability of viable protoplasts has permitted the plant geneticist to carry out mutagenesis and selection regimes at the level of the single cell in a manner comparable to that used by the microbiologist. For instance, tobacco plants resistant to wild-fire disease have been obtained by subjecting protoplasts to ethyl methane sulphonate mutagenesis and selecting mutants resistant to methionine sulphoxime. The toxin produced by *Pseudomonas tabaci*, the organism which causes wild-fire disease, is also a methionine analogue.

Plant protoplasts may be induced to fuse by a variety of agents, including sodium nitrate, alkaline calcium, and PEG. The highest frequency of protoplast fusion is obtained with PEG but hybrids produced with calcium can most readily be regenerated into complete plants. Hybrids between members of the same species, between different species of the same genus and even between different genera have been produced in this way. The plants involved include such important species as potato, tobacco and asparagus.

The fusion process can result in the coalescence of cells from the same parental line as well as producing hybrids between the two parents. As with most mating procedures, the problem is to find some means of selecting the hybrids and the paucity of genetic markers available in many plants is a severe handicap here. In most of the cases where hybrids have been successfully obtained by protoplast fusion, the cross had already been made by the conventional sexual route. The phenotype of the hybrid was known, therefore, and could serve as a model for the selection system. Great strides should be made with this novel system of plant breeding once a wide range of selectable markers (such as drug and toxin resistance) or auxotrophic mutations are available in plants.

7.3.3   *Cell fusion*

The production of inter- or intraspecific hybrids of mammalian cells by cell fusion is technically simpler than with microorganisms and plants as mammalian cells are not covered with a cell wall which must be removed before fusion can be induced. The first method of deliberate fusion of mammalian cells employed inactivated Sendai virus as a fusing agent. Sendai is a paramyxovirus which is covered with a lipid-containing envelope and can enter cells by coalescence with the host's membrane; it may mediate fusion by simultaneously coalescing with the membranes of two individual cells. Chemical agents such as calcium ions, lysolecithin and PEG can also be used to fuse mammalian cells. PEG is now the most commonly used agent.

A selection technique is desirable in order to isolate hybrid cells from a mixture of fusion products, although laborious screening procedures have been used. It is possible to select for the complementation of auxotrophic or temperature-sensitive mutations but the most widely used techniques employ selection for drug resistance markers. The method was introduced by Littlefield who fused a bromodeoxy-uridine-resistant (BUdR^r) line of mouse L cells with another which was resistant to azaguanine (AG^r). The BUdR^r cells are deficient in the enzyme thymidine kinase (TK) while the AG^r line lacks inosine pyrophosphorylase (IPP) activity. Neither the TK$^-$ IPP$^+$ or the TK$^+$ IPP$^-$ line will grow in HAT medium, which contains hypoxanthine, aminopterin and thymidine. The aminopterin inhibits nucleotide biosynthesis and pyrimidine and purine triphosphates can only be synthesized in cells able to utilize both thymidine and hypoxanthine. Such cells require both the TK and IPP enzymes to be functional. Thus in a fusion mixture containing TK$^+$ IPP$^-$, TK$^-$ IPP$^+$ and TK$^+$ IPP$^+$ cells, only the hybrid cells which have both activities are able to grow.

The use of such selective devices has permitted the production of hybrid cells not only by the fusion of different lines of cells from the same species but also the formation of intergeneric hybrids between human cells and those of rat or mouse. These rodent/human lines are unstable and human chromosomes appear to be preferentially lost so that, after thirty divisions in culture, the cells usually contain only seven of the twenty-four human chromosomes originally present in the hybrid. This phenomenon of chromosome elimination has been put to good use in mapping the human genome since it permits the rapid assignment of a gene to a particular chromosome.

Monoclonal antibodies

The mammalian cell fusion technique which has brought the most immediate biotechnological pay-off is the formation of hybrid-myeloma or 'hybridoma' cell lines which enables the production of monoclonal antibodies (McAb). This technique, which was developed by Milstein and his collaborators in Cambridge, involves immortalizing antibody (Ab)-producing cells by fusion with myeloma cells.

Most antigens provoke the production of a range of antibodies by the body but each individual lymphocyte secretes only a single Ab. Myelomas are tumours of the immune system in which a single lymphocyte line proliferates in an uncontrolled manner, synthesizing large amounts of a single Ab protein. Myelomas are, therefore, natural sources of McAb. Milstein's technique involves fusing normal (non-tumourous) lymphocytes with myeloma cells. The hybrids initially secrete a mixture of Abs—the two parental types and hybrid molecules formed by the association of heavy and light chains from the two parents. Eventually, however, chromosome exclusion leads to the generation of cells which secrete only a single Ab. This may be either that encoded by the lymphocyte genome or that of the myeloma cell. Biochemical and immunochemical techniques are used to screen the clones in order to discover the one producing the desired Ab. The technique is outlined in Fig. 7.5.

The potential applications of McAb are widespread. They may be used for diagnostic purposes, e.g. in rapid and accurate typing of tissues and organs for transplantation. They may also be used for the purification of proteins and other bioproducts by immunoabsorption. Secher & Burke have used McAb for the large-scale purification of interferon (Chapter 8) from human lymphocytes. The technique can obviously also be applied to purifying interferon produced by recombinant bacteria. The potential for using McAb for passive immunization is restricted, at least as far as human patients are concerned, because it is not yet possible to produce stable clones with human cells.

7.4   *In vitro* genetic manipulation

7.4.1   *Technology*

Bacteria and the yeast *Saccharomyces cerevisiae* have been used to produce a variety of eukaryotic proteins and proteins encoded by animal viruses. Several different procedures have been used but the following methods are generally employed.

First of all, it is necessary to isolate the gene. If an animal gene is to be expressed in a bacterium or a yeast, the usual method for obtaining the gene is to isolate its messenger RNA (mRNA). It is often impossible to express an animal gene directly in bacteria because of introns within the gene. Introns are sequences within genes which do not code for parts of the protein specified by the gene. The protein-encoding parts of genes are exons. Mature mRNA molecules in animal cells do not contain sequences complementary to introns as these have been removed by an enzymic process known as 'splicing'. These molecules can, therefore, be directly transcribed into DNA molecules using the enzymes reverse transcriptase and DNA polymerase. Some animal genes, such as those

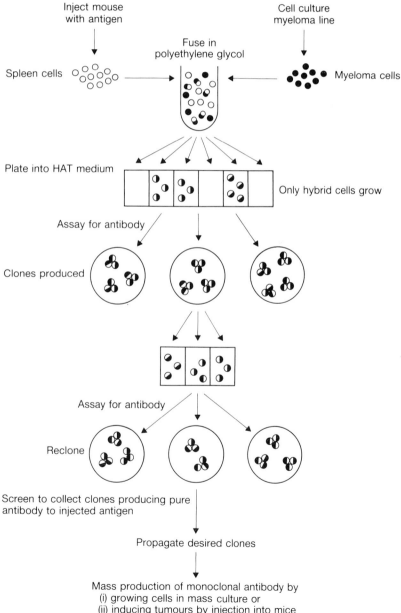

Fig. 7.5. Milstein's technique.

encoding some α-leucocyte interferons do not have introns and in these cases the chromosomal genes can be used directly as the material to be expressed in bacteria or yeast. If mRNA is to be used as the source of the gene, it is important to identify the tissue which produces the largest amounts of the gene product, so that this can be used as the source of the mRNA. Cancerous tissues or cell lines producing the mRNA in question can sometimes be used.

Once the complementary DNA (cDNA) from mRNA has been made, the next stage is to identify which of the many cDNA molecules made from a particular tissue has the gene to be isolated and expressed. At this stage the cDNA molecules are usually attached to plasmids or bacteriophage genomes which are used as vectors. The recombinant plasmids which have the cDNA molecules as inserts are put into a suitable host bacterium.

Identification of particular clones can be time consuming. However, if a gene has already been cloned, identification of related genes from a cDNA bank is relatively straightforward using a technique developed by Grunstein & Hogness. In this procedure, colonies of bacteria containing recombinant plasmids or phages are grown on nitrocellulose filters placed on the surface of nutrient agar plates. The colonies are broken open on the filters and the DNA which is released is denatured and remains stuck to the filter. A radioactive probe, comprising denatured DNA or RNA, is then added to the filter and binds to homologous, or partly homologous, DNA molecules derived from the bacterial colonies. Radioactive probe which is not specifically attached to any homologous DNA is removed by washing so that the positions of lysed colonies which have bound the probe can be identified by autoradiography. The bacteria which react positively in the test are recovered from a replica set of colonies.

If a specific probe is not available, a DNA probe can be synthesized chemically, if all or part of the protein sequence of the gene product is known. Of course, the precise DNA sequence cannot be deduced directly from the protein sequence because each amino acid may be encoded by one of up to four different triplet codons. Nevertheless, a series of probes can be constructed based on the protein sequence to identify plasmids carrying a partially homologous sequence.

Other methods for identifying recombinant plasmids depend on expression of the inserted gene so that the protein can be identified by immunological methods or from its biological activity. The technique devised by Broome & Gilbert is often used to screen bacterial colonies for antigens which specifically react with an antibody. A plastic disc is coated with specific antibody and is placed on top of the colonies which have been previously lysed. Antigens from the lysed bacteria are adsorbed to the antibodies on the disc. They are labelled by treating the disc with radioactive antibodies against the same antigen. The positions of antigens which have adsorbed to the disc are revealed by autoradiography.

Insertion of DNA into a plasmid or phage vector

The choice of vector depends largely on the host strain which is to be used to express the cloned DNA. If *E. coli* is the host the plasmid vector will probably be derived from pBR322. Plasmids isolated from various *Bacillus* spp. or from *Staphylococcus* are used as vectors for *Bacillus* hosts, and the vectors for *Saccharomyces cerevisiae* are usually derived from the 2 μm plasmid or from parts of yeast chromosomes which can replicate as plasmids. Often, shuttle vectors are employed which can replicate in one or more of these hosts. These are useful as it is often simpler to use *E. coli* as a host for producing large quantities of plasmids and for transforming DNA. The structure of pBR322 is shown in Fig. 7.3. This plasmid codes for resistance to tetracycline and ampicillin. Restriction endonucleases cleave the molecule at the positions indicated.

The restriction endonucleases used in cloning experiments are derived from a wide variety of prokaryotes. More than one hundred and fifty different enzymes have been isolated. They differ in the nucleotide sequences which they recognize and cleave. The sites recognized by most of the enzymes used in genetic engineering are palindromes of either 4, 5 or 6 bases. These may be cleaved to leave blunt ends, as in the case of *Alu*I, or sticky (or cohesive) ends, as in the case of *Bam*HI (Fig. 7.6). The unpaired bases of the sticky end terminate with a 3′ hydroxyl or a 5′-monophosphate group, depending on the enzyme.

A common method for cloning DNA molecules is to cleave both plasmid and insert DNA with the same restriction endonuclease. The sticky ends generated by the enzyme can then be annealed to create circular recombinant molecules which may be sealed with phage T4 DNA ligase. Insertions are often made within antibiotic resistance genes of pBR322, since the recombinant plasmids can be easily recognized by their failure to confer resistance.

T4 DNA ligase can also be used to join blunt-ended DNA molecules together, although the reaction is less efficient than with sticky ends. Sticky ends can be converted into blunt ones by digesting away the single-stranded portions with single-strand-specific nucleases. Alternatively, DNA polymerase, which adds nucleotides complementary to a 5′

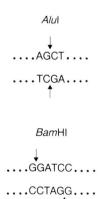

*Alu*I

↓

....AGCT....

....TCGA....

↑

*Bam*HI

↓

....GGATCC....

....CCTAGG....

↑

Fig. 7.6. DNA sequences recognized and cleaved by restriction endonucleases *Alu*I and *Bam*HI.

single-strand extension, can be used. The ligase enzyme is also employed
to add synthetic DNA molecules to plasmids so that a suitable sequence
is obtained at the junction of insert and vector DNA. Complementary
single-strand ends on vector and insert can be made using terminal
deoxynucleotidyl transferase, which adds bases to the 3′ ends of DNA
molecules. Thus, a tail of several deoxyguanosine residues (oligo(dG))
can be added to the 3′ ends of the vector and complementary oligo(dC)
tails to the ends of the insert. This has the advantage of preventing the
re-ligation of the vector molecule.

Transformation

Once recombinant plasmids have been made *in vitro*, the next step is to
transfer them into a suitable host organism. The exact method of
transformation depends on the host. *E. coli* is made competent (capable
of taking up purified DNA) by treating it with cold $CaCl_2$. *Bacillus
subtilis* becomes competent at a particular point in its growth cycle in
batch culture, i.e. when it becomes starved of nutrients. Many other
bacteria, including *Streptomyces* spp., can only be transformed if
converted to protoplasts by removing their cell walls. Yeasts are often
converted to protoplasts in order to facilitate the uptake of DNA
following treatment with $CaCl_2$ and PEG. Intact yeast cells become
transformable when treated with alkali metal ions such as Li^+.

7.4.2   *Expression of cloned genes*

The purpose of many cloning experiments is to obtain large amounts of
some eukaryotic protein and, to this end, the eukaryotic gene is inserted
into a bacterial plasmid. To achieve a high level of gene expression, the
eukaryotic DNA must be inserted close to an efficient promoter for
transcription and the mRNA synthesized must be efficiently translated.
The promoters used for high levels of expression are usually not those
which naturally occur on plasmids but, instead, are derived from
chromosomal or phage genes which have been inserted into plasmid
vectors. A typical example is the P_L promoter of bacteriophage λ. This is
a promoter which is transcribed to give high concentrations of mRNA.
It can be controlled by the repressor protein encoded by the C_I gene of λ.
The C_{I857} gene encodes a temperature-sensitive repressor protein which
is active at 30 °C but inactive at 42 °C. Inclusion of this gene, either on a
plasmid or a phage, in the same cell as the P_L promoter means that
expression of cloned genes under P_L control can be regulated by
changing the temperature.

In the yeast *Saccharomyces cerevisiae*, promoters from genes which encode major cellular proteins, such as alcohol dehydrogenase (*ADC*1) or phosphoglycerate kinase (*PGK*), have been used in the construction of high expression vectors. The *ADC*1 promoter has been used to direct the transcription of a human insulin gene within yeast, and high levels of expression of interferon genes have been obtained with PGK.

Several proteins specified by animal genes or the genes of animal viruses have been produced in high yields in bacteria. For example, *E. coli* strains have been made in which about 20% of total protein is the core antigen of hepatitis B virus, human growth hormone or the major capsid antigen of foot-and-mouth disease virus. A strain of *B. subtilis* has been made which produces about 1% of its total protein as the major antigen of foot-and-mouth disease virus. However, it seems that some of the proteins encoded by animal genes or animal viruses cannot be readily expressed in bacterial cells, even when they are coupled to initiation signals for transcription and translation which normally lead to high levels of expression of prokaryotic genes. The reasons for poor expression are not always clear, but in some cases it is known that the animal or viral proteins are rapidly degraded by bacterial proteases. In such cases, it may be possible to increase the yield by employing protease-deficient mutants. For the production of proinsulin, the precursor of insulin, some protection against proteases can also be obtained by ensuring that the polypeptide is secreted into the periplasmic space of the *E. coli* cell envelope. Preproinsulin has a sequence of hydrophobic amino acids at its N-terminus which transports it, with a concomitant removal of the hydrophobic sequence, across the membrane and into the periplasmic space. The half-life of proinsulin is about 2 min inside the cell and about 20 min in the periplasm.

Another way to synthesize protease-sensitive proteins is to obtain them as excretion products of *Bacillus*. *Bacillus* spp., unlike *E. coli*, excrete several proteins into their growth media. Strains of *Bacillus* that excrete proinsulin have been made. An excreted product is usually easier to purify than an intracellular product, and *Bacillus* has the additional advantage over *E. coli* in not producing endotoxins. However, probably only certain kinds of polypeptide can be efficiently excreted by *Bacillus*. In yeast, the problem of secretion has been tackled by constructing vectors which contain the signal sequences of the genes encoding the α-factor mating hormone or the killer toxin.

7.4.3 *Applications of genetic engineering*

It is evident that the techniques of genetic engineering will play a major role in the development of biotechnology and will be applied in many ways. Bacteria or yeasts are now used to produce eukaryotic or viral proteins of medical or veterinary importance in large amounts. Table 7.1 lists some of the proteins which can be synthesized in this way. The production of considerable amounts of a protein which is of no use to it places a severe selective disadvantage on a microbial cell and serious strain stability problems can be envisaged. These may be solved, at least in the case of very valuable products, by the imposition of strong selection for the presence of the plasmid vector if not for the gene cloned into it. The use of inducible promoters, such as λP_L, ensures that the product is not synthesized continually, but only when it is induced at the appropriate time in the growth cycle.

Stability problems should not arise when the cloned gene confers some selective advantage on its new host. Genetic engineering can be used to extend the substrate range of a microorganism. The brewing industry, for example, is interested in constructing yeast strains able to secrete starch-degrading and proteolytic enzymes. The introduction of some new enzymic capability into an organism may increase its metabolic efficiency. For example, ICI has increased the efficiency of nitrogen assimilation by its single cell protein (SCP) organism, *Methylophilus methylotrophus*, by providing it with the *E. coli* gene for the enzyme, glutamate dehydrogenase (see Fig. 7.7). This has increased the yield of SCP for the organism growing on methanol and ammonium by some 7%. In *Saccharomyces cerevisiae* it is now possible to construct artificial linear chromosomes; these should permit the stable introduction of whole novel metabolic pathways into the organism.

The introduction of some foreign gene into a process organism is not the only way that genetic engineering can be used in strain development. It may be advantageous to clone some of the organism's own genes. For instance, if a particular enzymic step is known to be rate limiting for a critical metabolic pathway then the presence of many copies of the gene for that enzyme on recombinant plasmids may relieve the restraint via the production of more enzyme molecules. A major

Human insulin α-, β-, and γ-interferons Hepatitis B viral core antigen Hepatitis B viral surface antigen Human growth hormone Foot-and-mouth disease viral capsid protein Calf rennin	**Table 7.1.** Eukaryotic and viral proteins produced from cloned genes within microorganisms

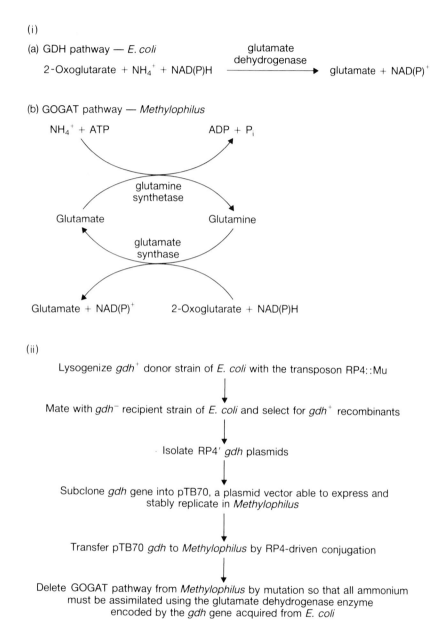

(i)

(a) GDH pathway — *E. coli*

$$\text{2-Oxoglutarate} + \text{NH}_4^+ + \text{NAD(P)H} \xrightarrow{\text{glutamate dehydrogenase}} \text{glutamate} + \text{NAD(P)}^+$$

(b) GOGAT pathway — *Methylophilus*

NH$_4^+$ + ATP → ADP + P$_i$

glutamine synthetase

Glutamate   Glutamine

glutamate synthase

Glutamate + NAD(P)$^+$ 2-Oxoglutarate + NAD(P)H

(ii)

Lysogenize *gdh*$^+$ donor strain of *E. coli* with the transposon RP4::Mu

↓

Mate with *gdh*$^-$ recipient strain of *E. coli* and select for *gdh*$^+$ recombinants

↓

Isolate RP4′ *gdh* plasmids

↓

Subclone *gdh* gene into pTB70, a plasmid vector able to express and stably replicate in *Methylophilus*

↓

Transfer pTB70 *gdh* to *Methylophilus* by RP4-driven conjugation

↓

Delete GOGAT pathway from *Methylophilus* by mutation so that all ammonium must be assimilated using the glutamate dehydrogenase enzyme encoded by the *gdh* gene acquired from *E. coli*

Fig. 7.7. Improving the efficiency of nitrogen assimilation in the SCP organism *Methylophilus methylotrophus*. (i) Pathways of ammonium assimilation in bacteria. Route (a) is more efficient than route (b) since it does not consume an ATP. (ii) Cloning of the *E. coli* gene *gdh* which specifies glutamate dehydrogenase and its introduction into *Methylophilus*.

application of 'self-cloning' is likely to be directed mutagenesis. The conventional route to strain improvement by mutagenesis and selection involves the exposure of the entire genome of the process organism to the mutagen. There is no guarantee that favourable mutations will be obtained in the genes critical to the process. Mutations will also be induced in other genes and some of these may be detrimental to the general fitness of the process organism. If the critical genes can be cloned, then they may be mutagenized *in vitro* and returned to the organism. This ensures that the desired mutations will be obtained and that harmful mutations in other genes will be avoided.

Screening, both of microorganisms and synthetic chemicals, for biological activity has been a mainstay of the pharmaceutical industry. The advent of improved techniques for the chemical synthesis of genes will probably mean that random nucleotide sequences will be constructed, cloned, and their peptide or protein products screened for biological activity. While this may be compared with setting a group of monkeys the task of typing a Shakespearean sonnet, the occasional big winner will ensure that this strategy is not neglected!

7.5   Summary and prospects

The recent technological and conceptual advances in genetics mean that there is now a tremendous armoury of methods which may be applied to strain development and process innovation. The tried and tested approach of mutagenesis and selection will continue to be used, but *in vitro* mutagenesis of cloned genes will increase the specificity of this technique. Improvements in the instrumentation and control of fermenters will greatly facilitate the use of continuous selection to bring about quantitative improvements in process organisms.

Cell fusion techniques should play a dominant role in the development of the genetics of plant and mammalian cells. The production of monoclonal antibodies by hybridoma cell lines is already yielding important diagnostic tools and may also bring therapeutic advances. Bacterial protoplast fusion has become a routine mating procedure for *Streptomyces* spp., but in fungi it is not yet a technique of major importance. The discovery of internuclear gene transfer in the fungi may allow the exploitation of protoplast fusion techniques on a more refined level.

There is every reason to believe that genetic engineering will bring about significant improvements in medicine and agriculture, not least because recombinant DNA technology should greatly advance our understanding of the basic molecular biology of animal and plant cells. This technology is already increasing our knowledge of the processes of such refractory diseases as malaria and Chagas' disease.

Genetic engineering should yield entirely new products which have not previously been used because they are too difficult and expensive to isolate from animal tissues. The interferons are products of this type and interferons produced by bacteria are currently being used in large-scale clinical trials for the treatment of viral diseases and cancers. Existing medical products may become significantly cheaper to manufacture; human insulin produced by yeasts or bacteria may eventually replace that obtained from pigs (Chapter 8).

Many organisms currently used in industrial processes are poorly characterized genetically. There is often a good reason for this, such as the lack of a natural sexual system. The new techniques of genetics should allow the development of these strains to proceed more rapidly and on a more rational basis.

7.6   Recommended reading

BEGGS J.D. (1981) Gene cloning in yeast. In *Genetic Engineering*, Vol. 2, (ed. Williamson R.), pp. 175–203. Academic Press, London.

BOLIVAR F., RODRIGUEZ R.L., GREENE P.J., BETLACH M.C., HEYNECKER H.L. & BOYER H.W. (1977) Construction of new cloning vehicles II. A multipurpose cloning system. *Gene*, 2, 95–113.

BRODA P. (1979) *Plasmids*. W.H. Freeman & Co., Oxford.

BROOME S. & GILBERT W. (1978) Immunological screening method to detect specific translation products. *Proc. natn. Acad. Sci. U.S.A.* **75**, 2746–2749.

BROWN S.W. & OLIVER S.G. (1982) Isolation of ethanol-tolerant mutants of yeast by continuous selection. *Eur. J. appl. Microbiol. Biotechnol.* **16**, 119–122.

CATEN C.E. (1981) Parasexual processes in fungi. In *The Fungal Nucleus*, (eds. Gull K. & Oliver S.G.), pp. 191–214. Cambridge University Press, Cambridge.

FERENCZY L. (1981) Microbial protoplast fusion. In *Genetics as a Tool in Microbiology*, (21st Symposium of the Society for General Microbiology; eds. Glover S.W. & Hopwood D.A.), pp. 187–218. Cambridge University Press, Cambridge.

FINCHAM J.R.S., DAY P.R. & RADFORD A. (1979) *Fungal Genetics*, 4th edn., Blackwell Scientific Publications, Oxford.

GRUNSTEIN M. & HOGNESS D.S. (1975) Colony hybridisation: A method for the isolation of cloned DNAs that contain a specific gene. *Proc. natn. Acad. Sci. U.S.A.* **72**, 3961–3965.

HARDY K.G. (1981) *Bacterial Plasmids*. Van Nostrand Reinhold, London.

HOLLOWAY B.W. (1979) Plasmids that mobilise the bacterial chromosome. *Plasmid*, **2**, 1–19.

HOPWOOD D.A. (1981) Genetic studies of antibiotics and other secondary metabolites. In *Genetics as a Tool In Microbiology*, (21st Symposium of the Society for General Microbiology; eds. Glover S.W. & Hopwood D.A.), pp. 187–218. Cambridge University Press, Cambridge.

KLECKNER N. (1981) Transposable elements in prokaryotes. *Ann. Rev. Genet.* **15**, 341–404.

KÖHLER F. & MILSTEIN C. (1975) Continuous culture of fused cells secreting antibody of predefined specificity. *Nature*, **256**, 495–497.

KUBITSCHEK H.E. (1970) *Introduction to Research with Continuous Culture*. Prentice-Hall, Englewood Cliffs, New Jersey.

MAXAM A. & GILBERT W. (1980) Sequencing end-labelled DNA with base-specific chemical cleavages. *Meth. Enzym.* **65**, 499–560.

MILSTEIN C. (1980) Monoclonal antibodies. *Scient. Am.* **243**, no. 4, 56–64.

OLD R.W. & PRIMROSE S.B. (1985) *Principles of Gene Manipulation: An Introduction to Genetic Engineering*. Blackwell Scientific Publications, Oxford.

SANGER F., COULSON A.R., BARRELL B.G., SMITH A.J.H. & ROE B.A. (1980) Cloning in single-stranded bacteriophage as an aid to rapid DNA sequencing. *J. Mol. Biol.* **143**, 161–178.

SHORTLE D., DIMAIO D. & NATHANS D. (1981) Directed mutagenesis. *Ann. Rev. Genet.* **15**, 265–294.

STRUHL K. (1983) The new yeast genetics. *Nature*, **305**, 391–397.

VASIL I.K., AHUJA M.R. & VASIL V. (1978) Plant tissue cultures in genetics and plant breeding. *Adv. Genet.* **20**, 127–215.

WINDASS J.D., WORSEY M.J., PIOLI E.M., PIOLI D., BARTH P.T., ATHERTON K.T., DART E.C., BYROM D., POWELL K. & SENIOR P.J. (1980) Improved conversion of methanol to single-cell protein by *Methylophilus methylotrophus*. *Nature*, **287**, 396–401.

WILLIAMS B.G. & BLATTNER F.R. (1980) Bacteriophage lambda vectors for DNA cloning. In *Genetic Engineering. Principles and Methods*, Vol. 2, (eds. Setlow J.K. & Hollaender A.). Plenum Press, New York.

8 Medicine and Biotechnology

J. C. PICKUP

8.1 Introduction

The recent advances in biotechnology are especially exciting for practising doctors, not least because the progression of knowledge and techniques from the laboratory to industrial production and thence to patient care has taken place with quite astonishing speed. The expression of the human insulin gene in *E. coli*, for example, was first reported in 1979. Human insulin, of recombinant DNA origin, was tested in non-diabetic volunteers in 1980 and clinical trials in diabetic patients were well under way during 1981.

Important discoveries in research usually come from work on important problems and this chapter expresses the belief of most doctors that the current applications of biotechnology in medicine are in extremely important areas and are already revolutionizing the diagnosis, treatment and understanding of the pathology of many serious diseases. For the benefit of those without medical training, the importance in clinical practice of a number of these conditions and some commonly used diagnostic techniques are summarized. It has only been possible to give examples; the reader will readily think of countless other therapeutically useful proteins that could be synthesized by 'genetically engineered' microorganisms, applications for monoclonal antibodies, enzymes and so on. Details of the technologies involved (which are well reviewed in other chapters) have been omitted, although some basic methodology is described in the case of human insulin manufacture. This is largely because insulin has pride of place as the first protein produced by recombinant DNA technology to be tested in man and probably the first, or one of the first, to be used therapeutically.

8.2 Unmodified or mutant cells and their products (see also Chapter 7)

8.2.1 *Antibiotics*

Clinical biotechnology might be said to have started with the therapeutic application and industrial production of penicillin in the 1940s. The prototype penicillin has probably had more impact on the reduction of disease-induced morbidity and mortality than any other drug but itself led to a number of secondary problems, which subsequent biotechnological efforts have been able to solve. Firstly, the very success of penicillin created an enormous demand for the drug and, to maintain supplies, a high yield from the fermentation processes was necessary. Secondly, the original penicillin G (benzylpenicillin) is active mainly on Gram-positive organisms, such as the *Streptococci* and *Staphylococci*, and antibiotics with a broader spectrum of activity and/or effectiveness against Gram-negative bacteria, such as *E. coli* and *Pseudomonas*, were needed. Thirdly, allergic reactions to antibiotics, often minor manifestations such as skin rashes, but occasionally severe life-threatening anaphylaxis, require a range of antibacterials so that an equally effective agent can be chosen to which the particular patient is not allergic. Fourthly, penicillin G is unstable under the acid conditions of the stomach and cannot be given orally. Finally, many bacteria develop resistance against antibiotics, the classic example of this being the elaboration of an enzyme by *Staphylococci*, penicillinase (or more correctly β-lactamase), which hydrolyses the amide bond in the β-lactam ring to produce the pharmacologically inactive penicilloic acid (Fig. 8.1).

The yield of penicillin production was vastly improved by many mutations of the strain *Pencicillum chrysogenum* (by such measures as UV and X-ray irradiation, nitrogen mustard and spontaneous mutation), coupled with fermentation efficiencies. New antibiotics were isolated which were active against Gram-negative bacteria: e.g. streptomycin from filamentous bacteria (*Actinomycetes*) of the genus *Streptomycetes* and the cephalosporins from the mould, *Cephalosporium*. Although cephalosporins are β-lactam antibiotics containing the β-lactam core common to penicillins, they are sufficiently different in structure to be safely used in patients who are allergic to penicillin (Fig. 8.1).

Semisynthesis is a technique for manufacturing new antibiotics by substituting a side chain on the native β-lactam molecule by a group which thereby alters the overall properties of the molecule: e.g. spectrum of activity, sensitivity to penicillinase and gastric contents, etc. Originally, chemical means were used to remove the benzyl group from penicillin G, giving the 6-aminopenicillanic acid nucleus, but the addition of bacteria secreting amidases to the fermentation broth accomplishes this step biologically. Ampicillin (Fig. 8.1), for example, is a semisynthetic derivative which differs from benzylpenicillin only in that there is an additional amino group on the side chain; it is, nevertheless, orally active, with a wide antibacterial range, including the Gram-negative organisms commonly causing respiratory (*Haemophilus influenzae*), intestinal (*Shigella* and *Salmonella*) and urinary (*E. coli, Proteus*) infections. Cloxacillin (Fig. 8.1) is also acid resistant but is not destroyed by β-lactamases; it is often given in combination with ampicillin when penicillinase-producing *Staphylococci* are known to be

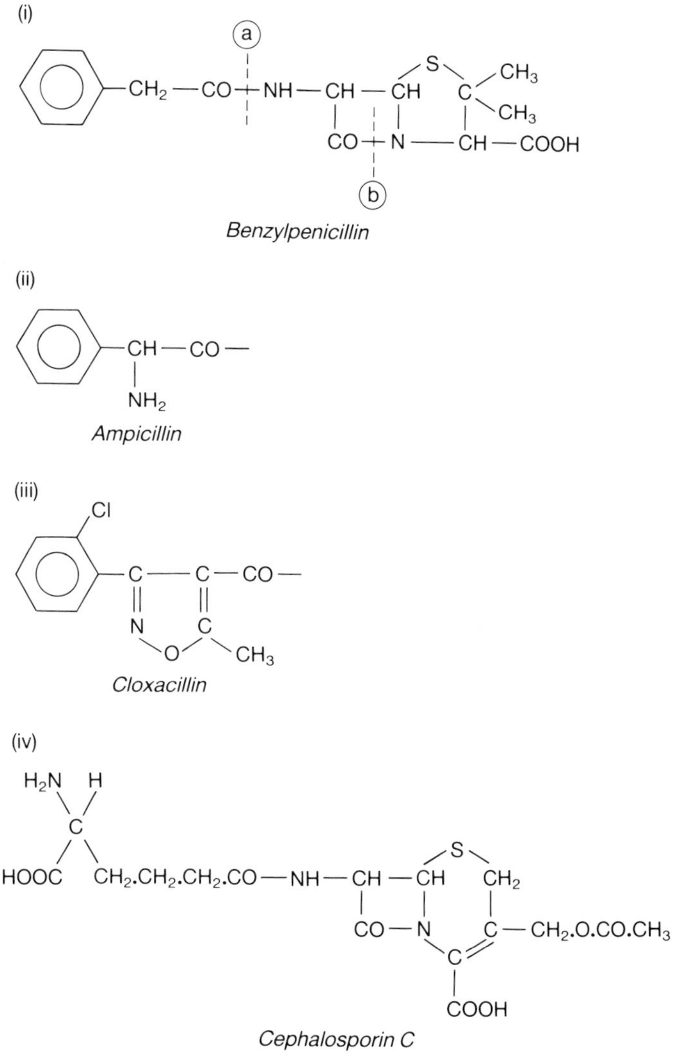

Fig. 8.1. Chemical structure of natural benzylpenicillin and its relationship to the semisynthetic penicillins, ampicillin and cloxacillin, and to another natural β-lactam antibiotic, cephalosporin. (i) Benzylpenicillin: (a) the site of cleavage of the side chain from the 6-aminopenicillanic acid residue and (b) the site of cleavage of the β-lactam ring by penicillinases; (ii) side chain of ampicillin; (iii) side chain of cloxacillin; (iv) cephalosporin C.

present or are highly likely to be involved, e.g. infections originating in hospital.

8.2.2   *Other cell cultures*

Other examples of the manufacture of pharmaceuticals using intact cells with unmodified genome are the Wellcome Foundation process for interferon synthesis by lymphoblastoid cells in culture (cf. section 8.3.2) and the production of viral antigens for vaccines by growth of cell cultures on inert microcarriers.

8.2.3 *Bioconversions* (see Chapter 4)

Microorganisms can also be used to carry out some of the individual steps in the synthesis of drugs which were formerly made by multiple and costly chemical reactions. A strain of the bread mould, *Rhizopus arrhizus*, for example, is able to hydroxylate progesterone at position 11, an early stage in the synthesis of the steroid cortisol. This strategy of bioconversions plus traditional chemical steps has enabled a variety of steroids to be manufactured much more simply and economically from sterol raw materials of plant origin. Steroids such as prednisone, dexamethasone, testosterone and oestradiol can now be in widespread everyday clinical use because of this ease of production. Some idea of the tremendous therapeutic importance of these substances can be seen from Table 8.1, which lists major applications for some of these steroids.

Table 8.1. Examples of the wide range of clinical uses for common steroid drugs

Steroid	Disease

Glucocorticoids
    Cortisone
    Hydrocortisone
    Prednisone
    Prednisolone
    Dexamethasone
	Blood diseases: haemolytic anaemia, acute
	        leukaemia, idiopathic thrombocytopaenic
	        purpura, bone marrow depression
	Replacement: Addison's disease
	Allergies: asthma, eczema (topical),
	        anaphylaxis
	Immunosuppression: post-organ transplant
	Autoimmune disease: systemic lupus
	erythematosis, polyarteritis nodosa,
	rheumatoid arthritis, temporal arteritis
	Miscellaneous: ulcerative colitis, sarcoid, nephrotic syndrome

Sex steroids
    Testosterone
    Oestradiol
    Progesterone
	Replacement: hypogonadism
	Gynaecological: menopausal symptoms,
	        dysmenorrhoea, endometriosis, excessive menstrual
	        bleeding, oral contraceptives
	Miscellaneous: breast cancer, osteoporosis

8.3 Modified cells and their products

8.3.1 *Monoclonal antibodies*

Antibodies had already had a profound impact in clinical medicine before the development of hybridoma technology and the availability of homogeneous antibodies. However, the supply of suitable antibodies was limited by the unpredictability of the immune response, both the titre and the cross-reactivity varying from animal to animal and bleed to bleed. The latter difficulty is chiefly because of the mixture of antibodies which is produced against a given antigen. The pioneering work of Köhler & Milstein in the early 1970s established a procedure for making almost unlimited quantities of homogenous antibody with widespread applications, some examples of which are given below.

Immunoassay

Many areas of clinical medicine and research have been revolutionized by Yalow & Berson's development of radioimmunoassay (RIA), whereby measurement of extremely small concentrations of substance is made by displacement of the radioactively labelled antigen from its specific antibody by the addition of increasing quantities of unlabelled test or standard antigen. This is particularly true of endocrinology, where circulating hormone levels are often very low and bioassay is protracted, tedious or virtually impossible. Samples are sometimes labile, even without further processing, often need to be concentrated, and contain interfering substances which may, for example, have biological activity resembling the hormone being measured (e.g. the insulin-like growth factors, somatomedins, are detectable by bioassay but not suppressable by anti-insulin antiserum). Just one example of a hormone which has achieved a new significance since the advent of radioimmunoassay is prolactin. Its rapid and relatively simple measurement by RIA in the plasma of large numbers of patients and normal volunteers has revealed that the most common anterior pituitary gland tumour is associated with elevated plasma prolactin concentrations (prolactinoma) and often with the prolactin-induced symptoms of galactorrhoea and amenorrhoea or other menstrual disturbance. In fact, prolactin measurement has become an essential part of the modern investigation of infertility and amenorrhoea; about 20% of women with amenorrhoea have raised plasma prolactin concentrations, whether associated with a pituitary tumour, ingestion of drugs which stimulate prolactin secretion or other causes.

Monoclonal antibodies are already being used in commercial radioimmunoassay kits and the ready availability allows alternative assay procedures using excess labelled antibody (immunoradiometric assay), which may have the advantage of increased sensitivity.

Immunohistochemistry

Labelled antibodies may be used to identify the light microscopic or ultrastructural (i.e. electron microscopic) distribution of antigens in tissue sections. In the commonly employed sandwich technique, unlabelled, specific antibody is applied to the section, excess washed off and then (usually fluorescence labelled or enzyme-linked) antibody against immunoglobulin is layered on, indirectly marking the antigen under investigation. This technique has been enormously productive in medical and biological research: just one example (using ordinary polyclonal antisera) is the topographical mapping of different cell types in tissues such as the islets of Langerhans. Immunohistochemistry with different antisera showed that the insulin-containing B cells were located in a central mass within the islet, whilst glucagon-secreting A cells were at the periphery. At the junction of the two layers were somatostatin D cells. This arrangement may have functional significance.

An example of the current use of monoclonal antibodies in the histopathological diagnosis of human disease is the examination of lymph node biopsies as an aid to the classification of the particular type of lymphoid tumour (e.g. Hodgkin's disease, various lymphomas).

Immunohistochemistry can also be of use in diagnosis by, for example, identifying the cell of origin of a tumour section by the antigen it contains.

Tissue typing for transplantation

The humoral and cellular response which causes rejection of tissues and organs transplanted across species or from one unrelated person into another is chiefly directed against so-called histocompatibility antigens on the surface of the cells. These proteins were first recognized as human leucocyte antigens (HLA), the gene complex which codes for them being located on chromosome 6. These surface markers are essentially how the body recognizes a cell as 'self' or 'non-self' and matching between donor and recipient HLA haplotypes has been found to improve immensely clinical organ transplantation results. Four of the five main loci (A, B, C and Dr) at the HLA complex (Fig. 8.2) code for antigens which are serologically defined, i.e. are typed by techniques

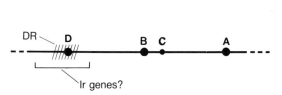

Fig. 8.2. The major HLA loci (A, B, C, D and Dr) on chromosome 6. Dr, the hypothetical immune response gene, is thought to be closely related to the D locus. The link between HLA and disease susceptibility may possibly be mediated through this locus.

involving antibodies. In contrast, the D locus determines the proliferative response in culture when lymphocytes are mixed, although it may be the same as the Dr locus. The genes are extremely polymorphic at each locus—there are up to thirty alleles for a given antigen. HLA typing of patients will be considerably improved by the reliability and standardization offered by monoclonal antibodies directed against the various antigens of the HLA system.

HLA antigens are also associated with certain diseases in the sense that studies of the frequency of these antigens in populations suggest that some conditions occur much more often in groups with certain HLA haplotypes. For example, coeliac disease is associated with a relative risk of about 73 for HLA D3 and rheumatoid arthritis has a relative risk of about 4 associated with D4, i.e. the diseases occur 73 and 4 times more frequently in those possessing the antigen than in those without the antigen. HLA typing, therefore, identifies populations at risk for a given disease, aiding in detection programmes and preventative measures, where these exist.

Diagnosis and monitoring of malignancy

There are several tumour-derived markers which are helpful in the diagnosis, prognosis and detection of distant tumour spread (i.e. metastases). Some of these are detected in the circulation (usually by radioimmunoassay) and some are studied in tumour specimens. α-Fetoprotein (AFP), for example, is the main serum protein of the fetus, but levels fall during the first year of life. Radioimmunoassay of plasma AFP has shown elevated concentrations in many patients with hepatoma (liver cancer) and a testicular cancer (teratoma). Plasma levels of carcinoembryonic antigen (CEA) are raised in many patients with colorectal cancer and rising concentrations suggest failure of the tumour to respond to chemotherapy or radiotherapy.

As well as providing reagents for these standard tests, monoclonal antibodies may be able to identify new, more specific tumour markers. Ritz *et al.* (1980) were able to develop a monoclonal antibody to an antigen associated with human acute lymphoblastic leukaemia cells. Homogeneous antibodies have also been produced against human malignant melanoma cells (a skin cancer), which do not cross-react with normal skin cells. Labelling of tumour-specific antibodies with radioactive or fluorescent moieties may lead to easier detection of metastases and primary tumour response to treatment, both *in vivo* and in biopsy specimens. Targeting of drugs linked to tumour antibodies may also produce new advances in treatment (see below).

Drug targeting

Monoclonal antibodies could be used to direct drugs and toxins to a particular site in the body (say a tumour) either by being directly coupled to the active agent or by being linked to the outside of liposome-encapsulated drugs.

Various non-specific anti-cancer drugs have been coupled to antibodies against tumour surface antigens, with limited success to date. Perhaps a more promising approach is the use of certain powerful bacterial and plant toxins, where it is thought that a single molecule may kill a cell. Diphtheria and ricin toxins consist of two polypeptide chains, joined by disulphide bonds. Chain B binds to the cell surface, but chain A is an enzymatically active species which enters the cell and destroys the protein-synthesizing machinery. Attempts are, therefore, being made to replace the B chain toxin with specific antibodies, preferably homogenous. A monoclonal antibody to a colorectal-associated antigen linked to a toxin A chain, which is selectively toxic to cancer cells in culture, has recently been described.

Liposomes (phospholipid membrane microspheres) containing methotrexate (an inhibitor of dihydrofolate reductase and an anti-cancer drug) have been coupled to monoclonal antibodies against mouse histocompatability antigens and shown to have specific effect on mouse spleen cells. Much more work with the targeting of liposomally contained drugs is expected in the future.

Drug overdose

Antibodies against drugs such as digoxin may be helpful in the treatment of drug overdose, although it is not yet clear if the antibodies should be used extra-corporeally, bound to a solid support, through which blood circulates, or injected into the circulation.

Preparation of medically important products

An example of the use of monoclonal antibodies for the purification and large-scale preparation of a medically important substance is the work of Secher & Burke, who covalently coupled mouse monoclonal antibody against human leucocyte interferon to a solid support of Sepharose. This specific immunoabsorbent allows a purification of lymphoblastoid-derived interferon of up to 5000-fold in a single step. Particular advantages are that only small quantities of partially pure antigen are necessary for antibody production, since the hybrid of interest can be selected and cloned.

8.3.2   *Recombinant DNA technology* (see also Chapter 7)

The development of techniques for the genetic manipulation of cells so that foreign gene products can be inserted, cloned, expressed and harvested has often been called a biological revolution and is certainly an advance with major medical implications. Proteins and peptides which have been previously available in only very small quantities are now promised in sufficient amounts to treat many patients. A few examples follow.

Insulin

About 1–2% of the population in the UK, Europe and USA suffer from diabetes mellitus and some 20% of these are dependent on insulin injections. Since the first clinical use of insulin for the treatment of diabetes, in 1922, the hormone has been extracted and purified from animal pancreases, either beef or pork. There are several reasons for developing methods for the large-scale production and marketing of human insulin, involving commercial interest and emotional reactions as well as scientific advance or possible therapeutic advantages.

Both bovine and porcine insulins differ slightly in amino acid sequence from human insulin. Porcine is very similar to human insulin in that the C-terminal threonine on the B chain is replaced by alanine, whilst bovine is different at three positions. Associated with the structural difference is the greater immunogenicity of bovine insulin, compared to porcine. Practically all patients being treated by traditional beef insulin injections develop circulating anti-insulin antibodies, although this antigenicity is also compounded by the lack of purity of these preparations. Insulin antibodies are almost certainly responsible for fairly minor problems such as subcutaneous fat atrophy at the site of repeated injection of bovine insulin and this usually resolves on changing to highly purified porcine insulin. Insulin antibodies also neutralize insulin in the circulation, so that the patient needs a greater dose than he otherwise would, and they probably modify the duration of biological action of injected insulin. There is no good evidence that antibodies cause more serious long-term damage, although one of the hopes for human insulin is that it will be even less immunogenic than porcine insulin (since, of course, it is not a foreign protein).

It is possible that the supply of animal pancreases will be a limiting factor in the availability of insulin in the future, as the number of diabetics in the world and, therefore, the demand for insulin increase. It is also undoubtedly true that there is an emotional appeal to using human sequence insulin, rather than that from animal sources.

Several insulin manufacturers, led by Eli Lilly & Co., have used recombinant DNA technology as the basis for human insulin manufacture. The process developed by Lilly, in collaboration with Genentech Inc. (Fig. 8.3), consisted of initially designing a DNA sequence from the known amino acid sequence of insulin and then chemically synthesizing separate artificial insulin A chain genes and B chain genes. Each contained a methionine codon at the 5′ end (which, of course, becomes the amino terminal of the translated protein) and stop sequences at the 3′ ends. Each gene was inserted into the β-galactosidase gene of plasmids which were themselves inserted into *E. coli*. Because the

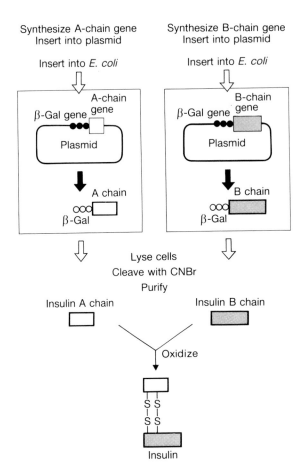

Fig. 8.3. Principle of human insulin production by recombinant DNA technology (after Miller & Baxter, 1980).

bacteria were grown in a medium containing galactose, but not glucose, β-galactosidase was induced and with it the insulin A or B chain, attached by a methionine residue. After lysis of the bacteria, cyanogen bromide treatment, which cleaves proteins only at methionine, allowed the separation of the chains from β-galactosidase (insulin contains no methionine). The chains were purified and recombined (in low yield) to produce the native two-chain insulin. This product was shown to be free from *E. coli* proteins, endotoxins and pyrogens, to be chemically and physically equivalent to human insulin of pancreatic origin and had full biological activity in animal hypoglycaemia assays.

More recently, an alternative synthetic approach has been adopted, where the gene for the precursor molecule, proinsulin, is constructed and inserted into *E. coli*. After purification of the proinsulin, native insulin is derived by trypsin and carboxypeptidase β digestion of the proinsulin.

Human insulin produced from *E. coli* was the first 'genetically engineered' protein to be tested in man. Studies in normal volunteers showed that it was safe, at least in the short term (no allergic or other adverse reactions), and had almost identical blood glucose lowering activity to purified porcine insulin (Fig. 8.4) when injected subcutaneously or infused intravenously. There are now many diabetics throughout the world being treated by this human insulin, initially as part of clinical trials assessing its metabolic and immunological effects, and now as part of routine clinical practice.

Interferon

Interferons are a family of proteins, originally discovered as the material elaborated by virally infected cells and having the property of inducing an antiviral state in other cells, both locally and systemically. Interferons have two other important biological effects: inhibition of cellular proliferation (thus, potentially an anti-cancer drug), and modulation of the immune system. The interferons have recently been reclassified into subtypes: α (formerly leucocyte interferon), β (formerly fibroblast interferon) and γ (formerly immune interferon), with several proteins in each category.

Interferons have been previously available in very small quantities, mostly obtained in rather impure condition from human white blood cells. Because of the marked species specificity it has been considered necessary to produce human interferon for clinical evaluation and this has not been tested widely in experimental animals. Properly conducted large-scale trials of interferon in patients with viral diseases and cancer

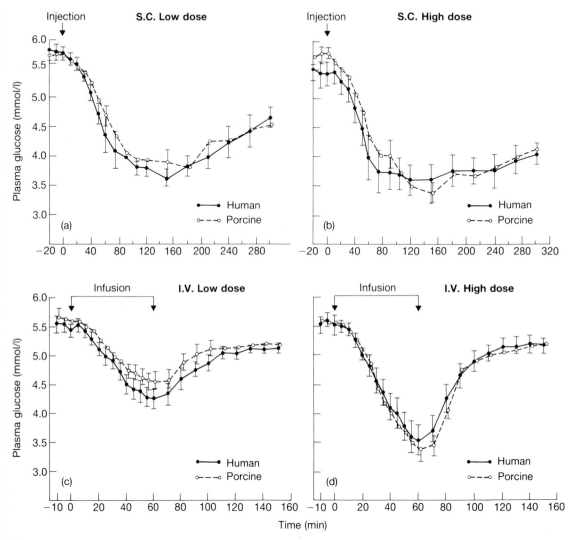

Fig. 8.4. Comparison in normal volunteers of the plasma glucose lowering by high and low doses of human insulin (produced by recombinant DNA technology) and porcine insulin (of pancreatic origin). Upper panels: biological responses after subcutaneous injection; lower panels: responses after intravenous infusion for 1 h (from Keen *et al.*, 1980).

are, therefore, severely limited but there are already indications of its potential. For example, leucocyte interferon significantly decreased the spread, duration of vesical formation and persisting pain in patients with herpes zoster infection (shingles). Hepatitis B infection (see below) has also shown response to interferon treatment. In malignancies there have been trials, for example, with metastatic breast cancer (twelve out of forty-three patients experienced at least a 50% reduction in tumour diameter), non-Hodgkin's lymphoma, osteosarcoma and malignant melanoma. However, much work needs to be done to find the optimal dosage, frequency and route of administration of interferon and the value of combining therapy with existing anti-cancer drugs. Side effects of fever, malaise and weight loss are often reported.

It is obvious that research, clinical testing and the possible routine therapeutic use of interferon will be advanced tremendously by the ready availability of the pure product which recombinant DNA technology allows. A human leucocyte interferon gene 514 base pairs long has been synthesized, incorporated into a plasmid and subsequently cloned into *E. coli*. Also, a human interferon gene has been expressed in yeast and the fibroblast interferon gene has been inserted into *E. coli* and expression demonstrated with a product possessing antiviral activity. This biological potency is interesting as natural interferon is a glycosylated protein, whereas that produced by recombinant DNA technology is not. At least two pharmaceutical companies are now producing interferon of recombinant DNA origin and clinical trials of this material have begun.

Growth hormone

Human growth hormone is a 191 amino acid protein (molecular weight 22 000), synthesized and secreted by the anterior pituitary gland. It is needed for longitudinal growth of the skeleton and it is estimated that seven to ten people per million of the population suffer from growth hormone deficiency, resulting in dwarfism. Although the defect is usually congenital, the growth retardation is not noticed until later in childhood because the hormone is not necessary for intra-uterine growth. As growth hormone is species specific only the human hormone can be used for replacement therapy. This has, until now, been extracted from cadaver pituitaries, but the supply is limited. The pharmaceutical company Kabi Vitrum have collaborated with Genentech Inc. to produce growth hormone from *E. coli*, using recombinant DNA techniques, and thus there is promise of an inexaustible source for the future. The purified bacterial growth hormone has comparable biological activity to the pituitary hormone and clinical trials with growth hormone deficient children are under way.

Vaccines

An example of disease where there is difficulty in obtaining a suitable antigenic source for immunization is hepatitis B. This viral infection of the liver can have a poor prognosis, particularly in aged and debilitated patients. Transmission occurs by transfusion of blood and plasma products, inadequately sterilized syringes, needles and medical instruments and by sexual contact (particularly in homosexual men). High risk groups that would benefit from vaccination include patients with natural or acquired immune deficiency states, those with malignancy, those requiring multiple blood transfusions, staff of haemodialysis, transplantation and cancer units, drug addicts, the mentally handicapped and their support staff, prostitutes and homosexuals.

The hepatitis B virus cannot be grown in tissue culture and vaccines have been prepared from the heat-treated serum of asymptomatic carriers. More recently, vaccine has been developed from the inactivated small spherical surface antigen particle (HBsAg) found in carrier serum. Although these vaccines are safe and effective, large quantities of pooled serum are required and expensive containment and production facilities are needed.

An alternative source of antigen is expression of hepatitis B proteins in prokaryotic cells and this has been achieved by Edman *et al.* (1981). Another approach is the inoculation of synthetic surface antigen peptides. The amino acid sequence of HBsAg has been deduced from the nucleotide sequence of the viral genome and various authors have used computer programmes to predict antigenic sequences of the HBsAg. Dreesman *et al.*, for example, synthesized two cyclic peptides which raised an antibody response in mice after a single injection.

Enzymes

Enzymes form the basis for many tests used in clinical medicine and are increasingly a part of the automated analyses and biochemical screens of body fluids, which are undertaken in modern clinical chemistry departments. Examples are: glucose oxidase, hexokinase, cholesterol oxidase and esterase, urease, uricase, lactate dehydrogenase and alcohol dehydrogenase. Occasionally, enzymes are used therapeutically, e.g. streptokinase and urokinase, which are effective fibrinolytic agents for serious vascular thrombosis.

In the future, enzymes, probably immobilized, may be used much more extensively for *in vitro* and *in vivo* monitoring purposes as part of enzyme electrodes. Many of these devices have already been constructed but are not yet used clinically. A glucose oxidase enzyme electrode, for example, has been constructed by many workers and that of Updike & Hicks is essentially glucose oxidase layered over a conventional platinum oxygen electrode (Fig. 8.5); the more oxygen consumed by the reaction (glucose + oxygen → gluconic acid + hydrogen peroxide), the smaller the amount of oxygen detected by the underlying electrode. Unfortunately, there are still many difficulties with this type

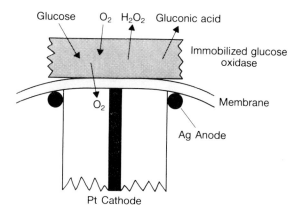

Fig. 8.5. Diagrammatic representation of a glucose oxidase enzyme electrode (after Updike & Hicks, 1967).

of device, which have prevented its use as an implantable glucose monitor: competition between glucose and oxygen in the body fluids, inactivation of the enzyme *in vivo* and calibration and electrode drift problems. Research in this area is very active and it is hoped that some modification of the enzyme electrode principle will eventually form the glucose sensing limb of a closed-loop, completely automatic, miniature artificial pancreas for the treatment of diabetics. A recent development of the enzyme electrode principle is of considerable interest in this context. Workers at Cranfield Institute of Technology, Oxford University, and Guy's Hospital, London, are evaluating a glucose-sensing electrode which uses an organic mediator (ferrocene, for example) to transfer electrons from the prosthetic group (FAD) of glucose oxidase to a graphite electrode, without the intervention of oxygen, the usual final electron acceptor. The electrode is, therefore, oxygen independent and promises to be particularly suitable for testing as an implantable, *in vivo* device in diabetic patients.

It is likely that there will be major efforts in the field of biosensor technology in the next few years. Enzymes (and other ligands such as antibodies; see section 8.5) are potentially useful for monitoring many substances of clinical importance, e.g. intermediary metabolites, drugs and hormones. Biotechnology will play a part by providing both common and unusual microbial enzymes by large-scale fermentation processes and perhaps recombinant DNA techniques.

Concerning the therapeutic possibilities of enzymes, administration of encapsulated enzymes within liposomes has been suggested as a treatment for certain lysosomal storage diseases and this approach has been tested in a few patients with Gaucher's disease (inborn deficiency of glucocerebroside: β-glucosidase). Results were, however, equivocal and further trials are necessary.

8.4 Application of the techniques of molecular genetics and recombinant DNA technology in diagnosis and pathology of human disease

8.4.1 *Prenatal diagnosis of inherited disease*

The most immediate clinical application for many of the techniques of molecular genetics is in the field of prenatal diagnosis of inherited disorders, such as the haemoglobinopathies. For example, in 1978, Kan & Dozy developed a method for diagnosis of sickle cell anaemia by analysis of the DNA of amniotic fluid cells. This is much safer than obtaining samples of fetal blood for analysis, where there is up to 7% risk of abortion. Sickle cell disease is the commonest congenital disorder of haemoglobin synthesis and is due to the substitution of valine for glutamic acid at position 6 of the β chain of haemoglobin. Red blood cells containing the abnormal HbS (two α and two abnormal β chains) assume a crescent shape (sickle-like) on deoxygenation. Non-pliable sickle cells obstruct the capillary circulation and are rapidly destroyed. Amongst the clinical manifestations are, therefore, anaemia, failure to thrive, and acute painful episodes (crises) or acute and chronic organ damage due to vaso-occlusion (e.g. stroke).

In the Kan & Dozy study (see Fig. 8.6) DNA was isolated from the white blood cells of two parents, from their affected child with sickle cell anaemia and from a normal control. DNA was also obtained from amniotic fluid cells, aspirated during the third pregnancy of the mother. The now established technique of restriction endonuclease mapping (Chapter 7) allows the DNA to be compared; ^{32}P-labelled cDNA, complimentary to the globin gene, can be prepared by extracting globin mRNA from reticulocytes and using reverse transcriptase in the presence of ^{32}P nucleotides to synthesize cDNA. Cleavage of the subjects' DNA with an endonuclease (in this case HpaI) results in DNA fragments of various lengths, which are separated by gel electrophoresis. The fragments are blotted on to a cellulose nitrate filter and hybridized with the radioactive cDNA probe (Southern blot technique). The nucleotide sequences which bind to the probe locate (by autoradiography) the globin-containing fragments. In a normal person the β globin gene was contained in a 7.6 kilobase pair (kbp) segment. In the affected child the gene was located in a 13.0 kbp fragment and both for parents and the fetus the DNA digestion map showed both 7.6 and 13.0 kbp segments (i.e. they were heterozygous and had sickle cell trait). The polymorphism is due to variations in the recognition site of the

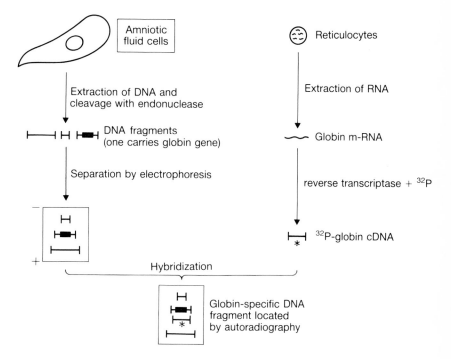

Fig. 8.6. Diagram to illustrate the use of gene probes and restriction fragment length polymorphism in the prenatal diagnosis of haemoglobinopathies, such as sickle cell anaemia (from Emery, 1981).

endonuclease, some five thousand nucleotides from the gene on the 3′ side.

A similar technique has also been used for the antenatal diagnosis of β-thalassaemia, a haemoglobinopathy characterized by low or absent synthesis of the β chain of haemoglobin. Because the detection of restriction fragment length polymorphism (RFLP) requires no detailed knowledge of the molecular pathology of a disease and is simply a finger-print which picks out abnormalities linked to the clinical picture, the procedure can be applied to a wide variety of genetic disorders.

8.4.2   *Genetic influences in disease pathology*

Diabetes is a disease where there is evidence for genetic influences but the mode of inheritance and molecular operation is unclear. Type 1 or juvenile-onset diabetic patients have suffered complete or near-complete destruction of the β cells of the islets of Langerhans and consequently have absolute insulin deficiency. This type of diabetes is strongly associated with the Dr3 and Dr4 HLA haplotypes. Type 2 or maturity-onset diabetes patients have normal or increased blood insulin levels but with abnormalities including receptor insensitivity to insulin, i.e. a relative lack. There is no association with HLA types in type 2 diabetics. Rotwein *et al.* (1981) used RFLP to examine the DNA in thirty-five non-diabetics, seventeen type 1 and thirty-five type 2 diabetics. Insertions of 1.5–3.4 kbp were found at the 5′ flanking region on the human insulin gene in 26% of non-diabetics, 35% of type 1 and 66% of type 2 diabetics (Fig. 8.7). The polymorphic region near the human insulin gene has been sequenced and the variation seems to be due to the number and arrangement of a family of repeating nucleotides. A number of studies have indicated that sequences in the 5′ flanking region may

Fig. 8.7. Restriction fragment length polymorphism in the 5′ flanking region of the human insulin gene. The position of 1.5 and 3.4 kilobase pair (kbp) inserts is shown. Pre: nucleotide sequence coding for the amino acids which precede the N terminus of proinsulin and which are thought to facilitate secretion into the endoplasmic reticulum; B, C, A: sequences coding for B chain, connecting peptide and A chain of insulin; hatched area: non-coding intron (see Chapter 7).

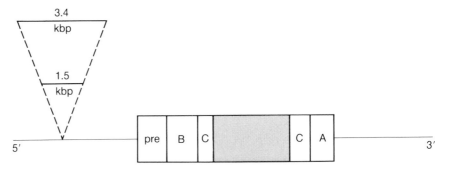

modify expression of the gene and so the significantly greater polymorphism associated with type 2 diabetes is of great interest (although one should note that some workers have been unable to confirm the association). Unfortunately, it cannot be said that this is sufficient or necessary for the development of the disease and the implications are still uncertain.

Two further inherited diseases where the exact genetic defect is unknown but where linkage has recently been established between a particular DNA polymorphism and the presence of the disease gene are Duchenne muscular dystrophy and Huntington's chorea. The latter is an incurable disease, inherited in an autosomal dominant manner, and causing progressive dementia and paralysis in the third or fourth decade of life. Unfortunately, there has been no way of distinguishing those who have inherited the gene and such close linkages with DNA markers may offer an aid to genetic counselling and perhaps eventually to identification of the Huntington's chorea gene and gene product itself.

8.5   Future prospects

In the future, new opportunities for biosensing may stem from advances in microelectronics in combination with biotechnological products such as enzymes and antibodies. Field effect transistors (FETs), for example, have already been made ion selective by placing a selectively permeable membrane on the insulator layer of the transistor. It may be possible to coat enzymes or antibodies onto the transistor surface so that the binding and/or possible conformational change in the protein and/or substrate reaction are sensed by the FET. As a preliminary step towards this, Caras & Janata (1980) reported an FET sensitive to penicillin, which was constructed with the enzyme penicillinase.

Can human diseases be cured by transplantation of normal or genetically manipulated cultured cells or by cellular engineering of the body's own tissue? For example, interest in treatment of inherited disorders of metabolism by transplantation has recently been rekindled by studies on human amnionitic membrane epithelial cells. These cells, readily obtainable from placentae, can easily be grown in tissue culture, secrete enzymes which are absent in lysosomal storage diseases, do not seem to express HLA antigens on their cell surface and would not, therefore, be rejected by the patient.

However, treatment of diseases caused by deficiency of products normally synthesized by a specific cell type, such as the hormones from endocrine tissue and globin chains from bone marrow cells, present formidable problems. Human cells that produce the substance and are of non-syngeneic origin (cadaver, fetal, non-identical donor) will be rejected unless the surface HLA antigens or the body's immunological response to them can be modified or suppressed in some way. An alternative would be to engineer the genome of one of the patients own cell lines, such as fibroblasts, probably by inserting genes (by DNA transformation or recombinant viral vectors) but perhaps by derepression, or cell fusion techniques. Problems that immediately spring to mind are the fears about reimplantation in humans of genetically altered, virally infected cells, with the risks of neoproliferation, and the fact that other cellular functions would presumably need to be transferred as well as the synthetic gene. To cure diabetes, for example, the cell must not only produce insulin, but package insulin in secretory granules and regulate its synthesis and exocytosis according to the ambient blood glucose (i.e. it must have a glucose sensor). The genetic control of these processes is likely to be complex and, to date, completely unknown.

It has been argued that the logical first application of cytogenetic manipulation in the cure of human disease should be a potentially lethal disease such as sickle cell anaemia or thalassaemia. Here, the disorder results from an abnormality of a single gene or gene complex, where expression takes place in a single tissue. Pluripotential blood-forming stem cells from the bone marrow continue to replicate throughout life and, if genetically transformed and reinserted into the marrow, could become a significant part of the haemoglobin-synthesizing tissue. Whilst workers such as Mercola & Cline raise the question of whether the time has now come to start such gene transplantation experiments in humans, there are many who would demand, before proceeding, a much sounder base of animal data on the evaluation of techniques, outcome of studies, risks, etc. It is certain, then, that both the scientific and medical potential of clinical biotechnology and its associated ethical dilemmas will continue to challenge us for many years to come.

8.6   Recommended reading

Antibiotics and bioconversions

AHARONOWITZ Y. & COHEN G. (1981) The microbiological production of pharmaceuticals. *Scient. Am.* **245**, 106–118.
FINTER N.B. & FANTES K.H. (1980) The purity and safety of interferons prepared for clinical use: the case for lymphoblastoid interferon. In *Interferon 2* (ed. Gresser I.), pp. 65–80. Academic Press, London, New York.
VAN WEZEL A.L., VAN STEENIS G., HANNICK C.A. & COHEN H. (1978) New approach to the production of concentrated and purified inactivated polio and rabies tissue culture vaccines. *Devl Biol. Standard* **41**, 159–168.

Monoclonal antibodies and their potential applications

BELCHETZ P.E., BRAIDMAN I.P., CRALWY J.C.W. & GREGORIADIS G. (1977) Treatment of Gaucher's disease with liposome-entrapped glucocerebroside: β-glucosidase. *Lancet*, **ii**, 116–117.
BUCKMAN R. (1982) Tumour markers in clinical practice. *Br. J. Hosp. Med.* **27**, 9–20.
COONS A.H. & KAPLAN M.H. (1950) Localisation of antigen in tissue cells. Improvements in a method for the detection of antigen by means of fluorescent antibody. *J. exp. Med.* **91**, 1–13.
DAUSSET J. & SVEJGAARD A. (1977) *HLA and Disease*, pp. 316. Munksgaard, Copenhagen.
DIAMOND B.A., YELTON D.E. & SCHARFF M.D. (1981) Monoclonal antibodies. A new technology for producing serological reagents. *New Engl. J. Med.* **304**, 1344–1349.
GILLILAND D.G., STEPLEWSKI Z., COLLIER R.J., MITCHELL K.F., CHANG T.H. & KOPROWSKI H. (1980) Antibody directed cytotoxic agents: use of monoclonal antibody to direct the action of toxin A chains to colorectal carcinoma cells. *Proc. natn. Acad. Sci. U.S.A.* **77**, 4539.
GREGORIADIS G. (1981) Targeting of drugs: implications in medicine. *Lancet*, **ii**, 241–246.
KÖHLER G. & MILSTEIN C. (1975) Continuous cultures of fused cells secreting antibody of predefined specificity. *Nature*, **256**, 495–497.
KOPROWSKI H., STEPLEWSKI Z., HERLYN D. & HERLYN M. (1978) Study of antibodies against human melanoma produced by somatic cell hybrids. *Proc. natn. Acad. Sci. U.S.A.* **75**, 3405–3409.
LESERMAN L.D., MACHY P. & BARBET J. (1981) Cell-specific drug transfer from liposomes bearing monoclonal antibodies. *Nature*, **293**, 226–228.
NAIRN R.C. (1976) *Fluorescent Protein Tracing*, 4th edn., pp. 648. Livingstone, Edinburgh.
OLSNES S. (1981) Directing toxins to cancer cells. *Nature*, **290**, 84.
RITZ J., PESANDO J.M., NOTIS-McCONARTY J., LAZARUS H. & SCHLOSSMAN S.F. (1980) A monoclonal antibody to human acute lymphoblastic leukaemia antigen. *Nature*, **283**, 583–585.

SECHER D.S. & BURKE D.C. (1980) A monoclonal antibody for large-scale purification of human leukocyte interferon. *Nature*, **285**, 446–450.

YALOW R.S. & BERSON S. (1960) Immunoassay of endogenous insulin in man. *J. clin. Invest.* **39**, 1157–1175.

YEH M.Y., HELLSTRÖM I., BROWN J.P., WARNER G.A., HANSEN J.A. & HELLSTRÖM, K.E. (1979) Cell surface antigens of human melanoma identified by monoclonal antibody *Proc. natn. Acad. Sci. U.S.A.* **76**, 2927–2931.

Medically important products made by recombinant DNA technology

BLOOM B.R. (1980) Interferons and the immune system. *Nature*, **284**, 593–595.

CREA R., KRASZEWSKI A., HIROSE T. & ITAKURA K. (1978) Chemical synthesis of genes for human insulin. *Proc. natn. Acad. Sci. U.S.A.* **75**, 5764–5769.

DERYNCK R., REMANT E., SAMAN E., STANSSENS P., DE CLERQ E., CONTENT J. & FIERS W. (1980) Expression of human fibroblast interferon gene in *Escherichia coli*. *Nature*, **287**, 193–197.

DREESMAN G.R., SANCHEQ Y., IONESCU-MATIA I., SPARROW J.T., SIX H.R., PETERSON D.L., HOLLINGER F.B. & MELNICK J.L. (1982) Antibody to hepatitis B surface antigen after a single inoculation of uncoupled synthetic HBsAg peptides. *Nature*, **295**, 158–160.

EDGE M.D., GREENE A.R., HEATHCLIFFE G.R., MEACOCK P.A., SCHUCH W., SCANLON D.B., ATKINSON T.C., NEWTON C.R. & MARKHAM A.F. (1981) Total synthesis of a human leukocyte interferon gene. *Nature*, **292**, 756–762.

EDMAN J.C., HALLEWELL R.A., VALENZUELA P., GOODMAN H.M. & RUTTER W.J. (1981) Synthesis of hepatitis B surface and core antigens in *E. coli*. *Nature*, **291**, 503–506.

GOEDDEL D.V., KLEID D.G., BOLIVAR F., HEYNEKER H.L., YANSURA D.G., CREA R., HIROSE T., KRASZEWSKI A., ITAKUA K., RIGGS A.D. (1979) Expression in *Escherichia coli* of chemically synthesized genes for human insulin. *Proc. natn. Acad. Sci. U.S.A.* **76**, 106–110.

GOEDDEL D.V. *et al.* (1979) Direct expression in *Escherichia coli* of a DNA sequence coding for human growth hormone. *Nature*, **281**, 544–548.

HITZEMAN R.A., HAGIE F.E., LEVINE H.L., GOEDDEL D.V., AMMERER G. & HALL B.D. (1981) Expression of a human gene for interferon in yeast. *Nature*, **293**, 717–722.

KEEN H., GLYNNE A., PICKUP J.C., VIBERTI G.C., BILOUS R.W., JARRETT R.J. & MARSDEN R. (1980) Human insulin produced by recombinant DNA technology: safety and hypoglycaemic potency in healthy men. *Lancet*, **ii**, 398–401.

OLSON K., FENNO J., LIN N., HARKINS R.N., SNIDER C., KOHR W.H., ROSS M.J., FODGE D., PRENDER G. & STEBBING N. (1981) Purified human growth hormone from *E. coli* is biologically active. *Nature*, **293**, 408–411.

MILLER W.L. & BAXTER J.D. (1980) Recombinant DNA—A new source of insulin. *Diabetologia*, **18**, 431–436.

SOUTHERN E.M. (1975) Detection of specific sequences among DNA fragments separated by gel electrophoresis. *J. Mol. Biol.* **98**, 503–517.

STIEHM R., KRONENBERG L.H., ROSENBLATT H.M., BRYSON Y. & MERIGAN T.C. (1982) Interferon: immunobiology and clinical significance. *Ann. Intern. Med.* **96**, 80–93.

TURNER A.P.F. & PICKUP J.C. (1985) Diabetes mellitus: biosensors for research and management. *Biosensors*, **1**, 85–115.

UPDIKE S.J. & HICKS G.P. (1967) The enzyme electrode. *Nature*, **214**, 986–988.

ZUCKERMAN A.J. (1982) Developing synthetic vaccines. *Nature*, **295**, 98–99.

Molecular genetics of human disease

ANONYMOUS (1984) Molecular genetics for the clinician. *Lancet*, **i**, 257–259.

BELL G.I., SELBY M.J. & RUTTER W.J. (1982) The highly polymorphic region near the human insulin gene is composed of simple randomly repeating sequences. *Nature*, **295**, 31–35.

DAVIES K.E., PEARSON P.L., HARPER P.S., MURRAY J.M., O'BRIEN T., SARFARAZI M. & WILLIAMSON R. (1983) Linkage analysis of two cloned DNA flanking the Duchenne muscular dystrophy locus on the short arm of the human X chromosome. *Nucleic Acids Res.* **11**, 2303–2312.

EMERY A.E.H. (1981) Recombinant DNA technology. *Lancet*, **ii**, 1406–1409.

GUSELLA J.F., WEXLER N.S., CONNEALLY P.M. *et al.* (1983) A polymorphic DNA marker genetically linked to Huntington's disease. *Nature*, **306**, 234–238.

KAN Y.W. & DOZY A.M. (1978) Antenatal diagnosis of sickle-cell anaemia by DNA analysis of amniotic cells. *Lancet*, **ii**, 910–912.

LITTLE P.F.R., ANNISON G., DARLING S., WILLIAMSON R., CAMBA L. & MODELL B. (1980) Model for antenatal diagnosis of β-thalassaemia and other monogenic disorders by molecular analysis of linked DNA polymorphisms. *Nature*, **285**, 144–147.

ROTWEIN P., CHYN R., CHIRGWIN J., CORDELL B. & GOODMAN H.M. (1981) Polymorphism in the 5′ flanking region of the human insulin gene and its possible relation to type 2 diabetes. *Science*, **213**, 1117–1120.

WEATHERALL D.J. (1979) Mapping haemoglobin genes. *Br. med. J.* **2**, 352–354.

Future prospects: biosensors, transplantation and gene therapy

ADINOLPHI M., ACKLE C.A., McCOLL I., FENSON A.H., TANSLEY L., CONNOLLY P., HSI B.L., FAULK W.P., TRAVES P. & BODMER W.F. (1982) Expression of HLA antigens, β_2-microglobulin and enzymes by human amniotic epithelial cells. *Nature*, **295**, 325–327.

CARAS S. & JANATA J. (1980) Field effect transistor sensitive to penicillin. *Analyt. Chem.* **52**, 1935–1937.

MERCOLA K.E. & CLINE M.J. (1980) The potentials of inserting new genetic information. *New Engl. J. Med.* **303**, 1297–1300.

TURNER A.P.F. & PICKUP J.C. (1985) Diabetes mellitus: biosensors for research and treatment. *Biosensors*, **1**, 85–115.

9   Agriculture and Biotechnology

F. A. SKINNER

9.1   Introduction

The agricultural scene is so familiar that its artificiality tends to pass unnoticed. Yet the landscape that man inhabits bears the impress of his efforts to cultivate crop plants and domesticate animals from time immemorial. The farmer may be seen as the forerunner of the modern biotechnologist for he has manipulated living creatures for his own advantage, although, because his activities were largely empirical, he should not be termed a biotechnologist: biotechnology refers properly to the intentional and science-based application of biological processes to the production, modification and use of materials. In practice, many such processes are mediated by microorganisms.

A farm is essentially a centre for the production of food from the land, but in earlier times it was more than this because food manufacture and processing were also conducted there. The traditional crafts of baking, brewing, cheese making, meat curing and the preparation of fermented products were necessary extensions to the production of crops, meat and milk and had to be practised on the spot if the produce was not to be wasted by microbial spoilage. Today, many of these secondary activities in which the growth of beneficial microorganisms plays an essential role have, in the developed countries at least, been transferred from farm to factory, where the processes have been developed and refined, and subjected to increasing control of production efficiency and quality.

Modification of food products in the factory, often by fermentation processes, is carried a stage further when bulk supplies of agricultural produce form feedstocks for industry. The manufacture of industrial ethanol from surplus, low-grade wine is familiar; less well known is the large-scale fermentation of purpose-grown crops to yield the same product (see Chapter 2). Industrial biotechnology also helps the farmer directly by providing vaccines and antibiotics for veterinary use, and additives for animal feedstuffs. Thus, biotechnology and agriculture interact in different ways and it is clearly impossible to survey the entire field in a single article. A seemingly satisfactory procedure is to consider the ways in which biotechnology can assist the arable farmer and horticulturist by improving the crop and improving the plant.

There is a need to increase agricultural production and to improve the quality, in terms of nutrition, palatability and market appeal, of the produce. There is also the need to effect these improvements in ways that are economic and environmentally acceptable.

Developed countries can still afford to use chemical fertilizers in abundance, but many countries are less fortunate and must secure adequate amounts of plant nutrients in other ways. Nitrogen is a case in point. There is no scarcity of this element but it must be fixed to make it available to plants. Fortunately, symbiotic biological nitrogen fixation has generated an established, effective technology that will be considered in some detail. The related topic of non-symbiotic nitrogen fixation is under active investigation but is, as yet, of limited practical application.

Biological means for improving phosphate nutrition and for controlling pests and diseases are attracting much attention and giving indication of passing from a stage of research and development to one of practical utility, and controlling the environment to favour growth of high-value crops is a new and promising practice. Perhaps the greatest benefit of biotechnology to farmers throughout the world will derive from improvements in the crop plants themselves, improvements likely to come about in a greatly accelerated way by genetic manipulation and plant protoplast technology.

9.2   The nutrient film technique

9.2.1   *Principle of the technique*

Plant roots need both water and oxygen for satisfactory plant growth, but these essentials are rarely present together in adequate amounts when a plant is growing in soil or other solid rooting medium. Abundant water fills soil spaces and deprives roots of oxygen; conversely, good aeration is frequently associated with lack of water. The nutrient film technique (NFT) avoids this disadvantage of conventional cultivation.

In NFT plants are allowed to develop their root systems in a very shallow stream of recirculating solution containing all the required plant nutrients. The depth of solution is carefully controlled so that the lower part of the developing root mat grows wholly in the solution and the upper part projects just above the surface, but remains covered by a liquid film. It is essential that this film of solution be maintained for if all the roots become submerged, oxygen becomes limiting.

A most important feature of NFT is that the root mat of each plant develops so extensively and interlaces so intimately with those of its neighbours that the plants become self-supporting (Fig. 9.1).

NFT thus differs essentially from water-culture (hydroponics) which provides abundant water and nutrients, but usually inadequate aeration of root systems. The difficulty in supporting hydroponic plants is not encountered with aggregate culture where the plant roots develop, as in soil, in a mass of inert particles (e.g. sand, gravel, pumice)

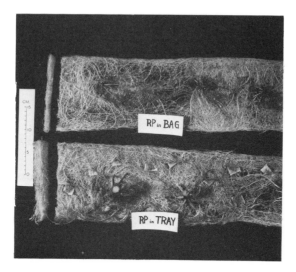

Fig. 9.1. Mats of maize roots grown by the nutrient film technique (NFT) with rock phosphate placed in the tray (channel) or in a muslin bag situated below the outlet pipe, viewed from above. Sectional view of root mats on left. Note how the root systems have grown together and made the plants self-supporting (from Elmes & Mosse, 1984).

moistened with nutrient solution by capillarity: aggregate culture corresponds more closely to soil culture with its characteristic disadvantages.

Details of construction, operation and uses of NFT systems are given in the comprehensive volume of Cooper (1979) from which much of the following information has been obtained.

9.2.2 *Basic requirements for a NFT installation*

A smooth rectangle of land to be cropped should have a two-way fall and a catchment trench along the full length of the lowest side. The trench should be lined with waterproof plastic film, and covered with a lid to exclude light and prevent evaporation; both trench and lid must be insulated if the nutrient solution is to be heated. A sump and circulating pump are sited at the lowest corner of the area. Rigid channels in which the plants are to grow can be laid directly on the smoothed soil from the top of the area to the catchment trench into which they discharge. An alternative system is to cover the whole area with concrete or construct a strip of concrete along the line of each channel, and then to lay flexible, flat-based channels on the smooth, accurate gradient.

A polythene supply pipe delivers nutrient solution from the pump to the highest side of the area and from it a small-bore polythene tube delivers solution to each channel. The solution then flows by gravity as a very shallow stream down the channels to the catchment trench and thence back to the pump. A direct return pipe fitted with an adjustable valve is always inserted into the supply pipe near the pump to allow a proportion of the solution being pumped to return directly to the trench without flowing down the channels.

With this system:

(1) the rate of flow of solution in the channels can be readily controlled by adjusting the valve;

(2) the returning solution can be made to fall through air to the trench to aid mixing and aeration;

(3) the trench can be emptied, for maintenance purposes, without interrupting the circulation of solution in the channels simply by allowing the direct return flow to run to waste. Sufficient solution will remain in the sump to maintain circulation of the plants while the trench is being emptied.

A standby pump should be available to maintain circulation of the solution in the event of mechanical breakdown and an emergency generator provided to guard against failure of the power supply.

An NFT system can range in size from a very large commercial installation for the bulk production of, say, tomatoes or lettuces, to a small apparatus for research purposes. An experimental NFT system, illustrating all the essential features, is shown in Fig. 9.2.

9.2.3   *Phytotoxicity*

Clearly, the materials in contact with the nutrient solution must not liberate toxic substances that might impair growth or even kill the

Fig. 9.2. Small-scale, experimental NFT apparatus with bean plants (*Phaseolus vulgaris*). Pump delivers nutrient solution along plastic pipe to the far end of the channel. Solution drains from near end of channel through wide-bore pipe to plastic dustbin acting as reservoir (foreground).

plants. Polythene is safe to use and so, apparently, is polypropylene and acrylonitrile butadiene styrene (ABS). Rigid PVC seems to be non-phytotoxic but flexible PVC is best avoided as it has, on occasion, been phytotoxic. Metals which are also trace elements should not be used in the system because toxic concentrations can build up in the nutrient solution; galvanized piping, for example, should not be used because the zinc gradually dissolves. The continual circulation of solution in an NFT system extracts toxic materials very efficiently. Constructional materials not known to be safe must be tested by keeping samples in contact with aerated nutrient solutions in which seedlings are being grown.

9.2.4   *NFT channel design*

NFT channels usually range in width from 10 to 23 cm depending on the size of plant to be grown, and vary greatly in constructional detail. The guiding principle in design is that the channel should be flat and rigid. Where irregularities of surface are likely, the channel itself must be rigid, but if the base is firm and flat, then cheaper, flexible channelling can be used. A black polythene channel, supported if necessary within a rigid guide channel, should have sides deep enough to form flaps that curve inwards against each other and the young plants, thereby minimizing evaporation and excluding light from the roots while allowing the plant stems to grow freely upwards. An alternative system is to make channels in a concrete base when it is laid, but this is only satisfactory for growing crops such as lettuce with spreading leaves that enclose the channel and prevent excessive evaporation loss. Many variations of design are possible.

9.2.5   *Supporting young plants*

Young plants transferred to the NFT system must be supported in the channels so that the leaves are in the light while the roots are in contact with the nutrient solution. It is customary to prick off seedlings into an absorbent cube such as a peat block, or into a rooting medium contained in a pot pervious to root growth. Capillarity ensures that the roots receive sufficient water and nutrients from the shallow solution flow.

9.2.6   *Disease*

The design of an NFT system seems ideal for the rapid spread of any plant pathogen gaining access to it yet, in practice, disease spread is less of a problem than was at first envisaged. It has been suggested that the resistance of NFT-grown plants to pathogens circulating in the nutrient solution is the lack of mechanical or insect-caused damage to the root mat, damage that is common with soil-grown plants and which provides the pathogens with points of entry.

9.2.7 *Prospects*

NFT is already in commercial use in the UK and abroad for both outdoor and glasshouse crops. In addition, it has great potential as a research tool consequent upon the close control that can be exerted on growth conditions. Temperature of the nutrient solution can be controlled thermostatically and pH value can be monitored and adjusted manually or automatically. Nutrients can be maintained at optimal levels, and supplements such as fungicides, pesticides or antibiotics can be added as required to combat disease or prevent undesirable microbial development in the system. Small-scale NFT installations can be constructed in the research glasshouse and preliminary attempts have already been made to devise sterile systems. Many interesting uses, and potential uses, of NFT, such as indoor gardening and production of turf, fodder and rubber, are discussed by Cooper (1979). The scope for development is considerable.

9.3.   Leguminous crops and symbiotic nitrogen fixation

9.3.1 *Historical introduction*

The Leguminoseae, with some 625 genera and 18 000 species, is one of the largest families of flowering plants and also one of the most important economically. Many species yield seeds that are rich in protein and constitute valuable sources of food and fodder. Examples are the peas and beans of temperate climates and the wide variety of crops, including soybean, cowpea, lentil and groundnut, that are grown so extensively in tropical and subtropical regions. Other legumes are valuable as pasture plants (e.g. clovers, *Stylosanthes* spp.), green manures (e.g. lucerne) or providers of timber, fruits, vegetables, gums, fibre, drugs and spices. The importance of legumes is enhanced by their ability to fix atmospheric nitrogen in root nodules formed with soil bacteria of the genus *Rhizobium* and to use the nitrogen compounds for growth.

Although the role of nodule bacteria in fixing nitrogen symbiotically has been known for less than a century, the benefits that result from the cultivation of legumes have been recognized for at least two thousand years. A crop that produces food rich in protein, while needing little, if any, nitrogenous fertilizer, has obvious potential, so it is no surprise that nodulated legumes became major components of farming systems of Europe, North America and elsewhere.

Of the legumes important in agriculture (mostly of the Papilionoideae) about 98% of species are estimated to be able to form root nodules and wild species are mostly nodulated by specific strains of rhizobia in their natural habitats. Continual natural growth, or cultivation, of a legume in one area usually leads to a gradual accumulation of suitable rhizobia in the soil. Satisfactory crop yields often depend, especially when nitrogen compounds are deficient, on the establishment of effective (i.e. nitrogen-fixing) associations of the host plants with an appropriate strain of rhizobia, but the right strain does not always occur naturally where the crop is to be grown, in which case the desired bacteria should be introduced.

As soon as the importance of the legume–*Rhizobium* symbiosis was recognized, methods were devised to introduce rhizobia into soil to benefit particular crops: the practice of soil and seed inoculation had begun. The capital cost of introducing inoculation technology is low, transportation costs are negligible and the techniques are simple enough for agricultural systems in the developing countries where present-day high fertilizer costs bear heavily on the farmers. Legume cultivation, aided by seed inoculation, can often improve an environment by promoting vegetation to combat desertification, assist the control of soil erosion, reduce wind movement of soil and help to reclaim degraded land. Most nodulated legumes are fully capable of meeting all their nitrogen requirements, provided, of course, that other conditions, including availability of water and non-nitrogenous nutrients, are satisfactory.

9.3.2 *Inoculation with rhizobia*

An obvious method of inoculation was to use soil itself, taken from an area where the legume to be cultivated normally grew well. This method works, and it was used during the later years of the nineteenth century.

A disadvantage is that large amounts of soil are needed, if applied as a general dressing, because the rhizobia form a minor part of the microflora and only a minute fraction of the soil mass. It became a common practice in America to broadcast 100–1000 lb/acre of soil taken from a nearby field where a good crop of the desired legume had been grown. A further disadvantage is that soil-borne plant diseases can be spread.

A much smaller amount of soil could be placed by seed drill, thus placing the inoculum near the seed where it was needed. From this simple method developed the technique of applying an inoculum directly to the seed: at first powdered soil was used, say only 0·5 lb of soil to 1 lb of seed but, later, methods based on the use of pure cultures of bacteria took their place.

Nobbe & Hiltner patented the first type of commercial inoculant culture in 1896, marketed under the name 'Nitragin'; seventeen types were produced for use with different leguminous crops. Many commercial inoculants were available in the USA by the 1920s. Some inoculants were pure culture mixed with soil, sand, peat, manure and powdered rock; some were agar, and others liquid, cultures. Inoculants were not always of good quality, nor were they always applied correctly, so successful crops were not always obtained.

9.3.3 Seed inoculation

The principle of the method is to place a large number of effective rhizobia, specific for that particular host, on a seed to increase the chance that the seedling will quickly become nodulated by the selected bacteria. Sufficient bacteria must be present and they must survive well until soil conditions allow the root hairs to be invaded.

Agar cultures

Cultures of rhizobia on slopes of agar medium, in test-tubes or bottles, can be used to inoculate seeds directly. A suitable medium for such cultures is yeast extract–mannitol agar (YMA) used for the routine cultivation of rhizobia, consisting of: K_2HPO_4, 0·5 g; $MgSO_4.7H_2O$, 0·2 g; NaCl, 0·1 g; mannitol, 10 g; yeast extract (e.g. Difco, Oxoid), 0·4 g; agar, 15 g; distilled water to one litre. The medium is adjusted to pH 6.8–7.0 and autoclaved at 121 °C for 15 min. A 10 ml slope of such a medium can yield c. 10^{10} rhizobium cells and can be stored for several weeks when refrigerated.

Tube cultures of *Rhizobium meliloti* for inoculating lucerne (alfalfa: *Medicago sativa*) were produced commercially in the UK from the 1920s until 1963 and the quality of the inoculum monitored by the Soil Microbiology Department of Rothamsted Experimental Station. To inoculate seeds, the bacteria were scraped from the agar slope culture and suspended in skim-milk to which a little calcium phosphate, supplied with the culture, had been added. The mixture was then poured over the heap of seeds, the whole well mixed and allowed to air-dry in the shade. On drying, the bacteria were held to the seed coat and protected by the dry milk film. Ideally, the seeds were sown soon after. This method is well suited to the small-scale inoculant production required by lucerne growers in the UK, but the cultures are no longer home-produced. Each test-tube culture contained enough rhizobia to inoculate seed for one acre of lucerne. About 40 000 such lucerne cultures were sold annually in the mid-1940s, but declined to 9–10 000 in the late 1950s, and to only 1–2000 by 1964. This decline reflected, at least partly, the decreasing acreage sown with lucerne. There are no commercial inoculants produced today in the UK but good quality peat-based inoculants are imported from Australia and the USA.

9.3.4   *Modern solid-support inoculants*

Rhizobia from agar slope and broth cultures do not survive for long when dried on seed and the cultures themselves have a short shelf-life. These disadvantages have been overcome by peat-based inoculants, developed originally in the USA and now used universally.

A broth (e.g. yeast–mannitol) culture of the required *Rhizobium* strain is grown conveniently in a fermenter of several litres capacity which can be simple, provided it is constructed well enough to prevent contamination. Temperature should be controlled within the range 25–28 °C and enough sterile air to ensure vigorous aeration of the culture should be supplied, but automatic control of pH is unnecessary, especially if the medium contains a little $CaCO_3$. The culture should have a high viable count (5×10^8–4×10^9/ml) at the time of mixing with the peat carrier which may be sterile or unsterile. Slow-growing rhizobia (e.g. *R. lupini*) with a mean generation time (m.g.t.) of *c.* 10 h) are often more difficult to build up into a really large population and even fast growers (m.g.t. = 3 h) can easily be overtaken by contaminants. It is advisable to use a large inoculum to start the liquid culture growing quickly.

Preparation of peat cultures

Peat is dried, either at ambient temperature or by careful heating, to *c.* 10% moisture (on a dry weight basis), ground by hammer mill to pass a 200-mesh sieve and brought to pH 6.5–7.0 by adding finely divided $CaCO_3$. This material can be used as sterile or unsterile carrier. The suitability of a batch of peat for making *Rhizobium* inoculants cannot be inferred from its general properties; each batch must be tested with the strains of rhizobia that are to be used with it. Many factors influence the growth and survival of rhizobia in peat.

Unsterile peat inoculants

Sufficient broth culture is added to the prepared ground peat to provide c. 40% of the total weight; it is thoroughly mixed and allowed to mature in shallow trays. After a few days, the material is again mixed and packaged, usually in sealed polyethylene film (0.0375 mm) bags. An increase (two- to fivefold) in the number of viable rhizobia usually occurs in the first few weeks after mixing. Such packs have a long shelf-life, especially when stored at low temperature. The time taken for the viable count to decrease by 90% (decimal reduction time) is used as a guide to indicate the safe storage time. Decimal reduction times can vary from ninety weeks (at 5 °C) to eight weeks (at 25 °C). Storage for one year is possible in favourable circumstances (decimal reduction time > sixteen weeks) but can be much less than this when stored at ambient temperatures. Time and conditions for satisfactory storage cannot be predicted accurately but must be determined for each production batch.

Sterile peat inoculants

Sealed polyethylene or polyamide bags containing prepared, ground moist peat are sterilized by autoclaving (e.g. at 115 °C for 4 h) or by γ-irradiation (e.g. 4.5×10^6 rad). When sterile, a suitable volume of broth culture (e.g. 10 ml) is introduced by hypodermic syringe, the puncture sealed and the bag incubated. In this process, considerable growth of rhizobia occurs during the incubation period.

Carriers of rhizobia

A solid inoculant consists of the bacteria plus a carrier, whose function is to maintain the viability of the cells, partly by protecting the bacteria against desiccation, to dilute them so that they will be distributed more evenly among the seeds being treated, and to help the bacteria to adhere to the seeds. Although good results were often obtained with bacterial suspensions, the superiority of peat as a carrier for the cells is generally recognized. Liquid cultures or suspensions seem to lack the protective effect afforded by peat to the rhizobia on seed following inoculation. Peat has been the most commonly used carrier for commercial inoculants since inoculation became an established practice, but it is not found everywhere and even where it does occur, its value as a carrier is unpredictable. The search for alternative carriers with similar protective properties to peat has engaged the attention of many workers.

A wide variety of materials has been used with varying degrees of success: various peat and soil mixtures supplemented with lucerne meal, ground straw, yeast or sugar; Nile silt with added nutrients; soil mixed with coir dust or wood charcoal. Vermiculite, decomposed sawdust, bentonite, kieselguhr and ground coal have also been used. A satisfactory carrier was made in Africa by composting maize cobs for thirty weeks with ground limestone, superphosphate and ammonium nitrate, the resulting product being air-dried, hammer-milled, sifted, filled in 100 g portions into high density polythene bags, and autoclaved.

The conclusion is that although is is relatively easy to devise an inoculant carrier from a variety of materials that support satisfactory growth and survival of rhizobia, peat is still the best carrier. Cheap and locally available alternatives to peat may have considerable value in certain areas, such as the developing countries, and should not be judged solely as a potential replacement for peat in general.

9.3.5   *Methods of inoculation*

The easiest and probably least effective method is simply to mix the dry inoculant with the seed during drilling; a little will adhere to the seed but much will separate out and be dispersed. Good adhesion of inoculant to seed is essential to supply sufficient rhizobia where they are needed. Inoculant is better applied as a water slurry. Peat inoculant adheres well, especially if water-soluble gums are added (e.g. carboxymethyl-cellulose, gum arabic); the latter also improve survival of rhizobia during drying of the seed.

Seed pelleting

In this technique relatively insoluble inorganic particulate substances such as calcium phosphate and limestone are added to legume seeds, together with the rhizobia to improve nodulation. Lime pelleting was very successful in establishing subterranean clover (*Trifolium subterraneum*) on acid soils in Australia. At first, seeds were coated with fine limestone and then inoculated with a moist peat-based inoculum just before sowing. A later method facilitates application of a much larger inoculum and improves survival of rhizobia on the seed. A solution of gum arabic (c. 40%) is added to peat inoculant and mixed mechanically (e.g. in a small, clean concrete mixer) with the seeds until they are evenly coated. The required amount of calcium carbonate (*not* calcium hydroxide, which is too alkaline) is then added all at once and mixing continued until the inoculated seeds are coated evenly with the limestone. The quantities and timing must be carefully controlled to produce firm pellets with a high content of viable rhizobia. The seed should be sown as soon as possible after pelleting, to take advantage of the improvement in survival of the rhizobia on the seed.

Modern soil inoculants

Although good seed inoculants are now available, it is not always possible to use them satisfactorily. Fungicide and pesticide seed dressings must often be used, and some of these preparations are lethal or inhibitory to the rhizobia. Moreover, some legume seed coats contain natural antibiotic substances. A further difficulty arises when inoculated seeds are sown in hot, dry soils where the rhizobia may die off before there is sufficient rain to allow germination. There is thus a reason to place the rhizobia near enough to the seeds to be available to the emerging roots, while sufficiently far removed to be scarcely affected by seed-borne toxins. A return to soil, as distinct from seed, inoculation is coming about but with vastly improved soil inoculant.

An inoculant consisting of porous granules is convenient to use and retains adequate numbers of rhizobia during several months of storage. Granules can be made from moistened plaster of Paris containing 0.2% of sodium carboxymethylcellulose to delay setting. Rhizobia are grown in broth culture in the presence of the formed granules which are finally air-dried and packed for storage.

Alternatively, a granular peat inoculum can be prepared in which the rhizobia are enabled to proliferate. These, like the plaster granules, can be sown with the seed or placed in a layer below it when necessary to protect the rhizobia from hot, arid surface conditions.

9.3.6 *Improving the legume–Rhizobium symbiosis*

The nitrogen-fixing symbioses are at present the most efficient sources of biologically produced ammonia for agricultural crops. Future developments in the use of biological nitrogen fixation for food production will, therefore, be governed by our ability to manipulate these symbioses. An increase in our knowledge of the genetics of *Rhizobium* will be useful in suggesting ways in which man can increase the range and efficiency of nitrogen-fixing symbioses so that, in future, we need not be dependent upon the existing (naturally occurring) symbioses but can produce them at will for desirable food plants.

Genetic studies of nitrogen-fixing organisms

In the last decade, the techniques of microbial genetics and molecular biology have been applied in studies on biological nitrogen fixation. Since it was demonstrated by Streicher *et al.* (1971) that coliphage P_1 could be propagated in *Klebsiella pneumoniae* M5 al, a free-living, nitrogen-fixing bacterium, and could transduce the *nif* (nitrogen-fixing) genes, it has been hoped to enhance symbiotic nitrogen fixation by genetic means for more efficient crop production.

Until recent times, there was some doubt as to which partner in the *Rhizobium*–legume symbiosis carried the *nif* genes governing nitrogen fixation but, in 1975, several research groups announced that some slow-growing *Rhizobium* spp. displayed nitrogenase activity in pure culture, thereby indicating that the bacterial genome, rather than the host plant, contains the *nif* genes. Support for this came from the observation that a plasmid mediated the transfer of *nif* genes from *R. trifolii* to a non-nitrogen fixing strain of *Klebsiella aerogenes*. It is now realized that some of the plasmids carrying genes involved in nodulation and nitrogen fixation can be transferred relatively easily between *Rhizobium* strains by conjugation.

9.4   Vesicular-arbuscular mycorrhiza

9.4.1   *Significance of vesicular-arbuscular mycorrhiza*

Sterile plant roots do not occur in nature. Roots carry a general, rhizoplane microflora on their surfaces and many are invaded by soil-borne, microbial pathogens. A different phenomenon is caused by certain non-pathogenic fungi that proliferate to form a sheath around the fine roots (ectomycorrhiza), or by others that invade root cortical tissues (endomycorrhiza) and establish symbiotic associations beneficial to the hosts. Ectomycorrhiza, caused by basidiomycete and ascomycete fungi, occurs on some trees and shrubs, especially northern temperate species, and two of the three known types of endomycorrhiza are restricted to the *Ericaceae* and *Orchidaceae*, respectively. The remaining type of endomycorrhiza, known as vesicular-arbuscular (VA) mycorrhiza because of the characteristic structures formed within cortical cells, has a special agricultural importance (Figs. 9.3, 9.4, 9.5). VA mycorrhiza, formed by phycomycete fungi of the family *Endogon-*

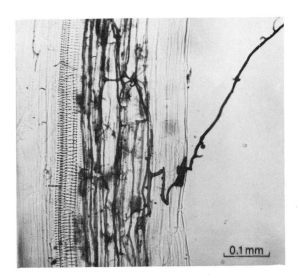

Fig. 9.3. Intact roots of onion infected with vesicular-arbuscular mycorrhizal fungus, cleared and stained with KOH–Trypan blue treatment. Roots inoculated with the yellow vacuolate spore type of *Endogone*. Root has external hyphae and several entry points but little internal infection. Bar = 0·1 mm.

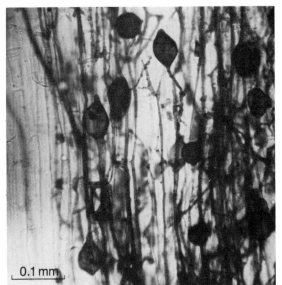

Fig. 9.4. Onion root as in Fig. 9.3. Vesicles and longitudinally running hyphae in root cortex. Bar = 0·1 mm.

aceae, is extremely widespread, occurring in most angiosperms, many gymnosperms and some ferns and liverworts, and in most soil of all climatic regions. Most plants of agricultural importance have VA mycorrhiza; it is common in the important crop families *Gramineae* and *Leguminoseae,* but absent from the *Cruciferae* and from sugar beet.

The agricultural significance of VA mycorrhiza lies principally in its ability to assist the plant to absorb phosphate from the surrounding soil. Phosphate ions are not very mobile in soil and a phosphate depletion zone often develops around roots when phosphate is scarce. Mycor-

0.05 mm

Fig. 9.5. Onion root as in Fig. 9.3. Arbuscles branching off along the longitudinally running hyphae. Bar = 0·05 mm. Figs. 9.3, 9.4 and 9.5 are from Phillips & Hayman (1970).

rhizal hyphae are able to extend from the mycelium within the root to beyond the depletion zone and translocate phosphate directly to the host. Plants with restricted root systems, having short stubby roots with few root hairs (e.g. onion, citrus) are generally not very efficient in extracting soil phosphate and tend to benefit considerably from the establishment of VA mycorrhiza.

Establishment of a mycorrhizal association often causes a dramatic improvement in plant growth in phosphate-deficient soil. Absorption of trace elements such as zinc and copper can also be enhanced and there is evidence that mycorrhizal plants can contain higher concentrations of growth hormones than their non-mycorrhizal equivalents. Among the benefits found is the improved nodulation and nitrogen fixation in legumes resulting from mycorrhizal inoculation in the presence of appropriate rhizobia.

The need to use phosphatic fertilizers economically has stimulated great interest in ways to exploit the advantages of VA mycorrhiza in agricultural crops. Although mycorrhiza is widespread, suitable types may be absent from land in which a new crop is to be grown, and will certainly be absent from sterilized glasshouse soils and potting media. Partial sterilization treatments, which are so necessary to destroy plant pathogens and pests, also kill the useful endophyte fungi. Introduction or augmentation of the endophyte population in soil requires inoculation, and here a particular difficulty arises because the fungus cannot be grown in pure culture away from a host plant. The large resting spores (c. 0.15 mm diameter) of the endophyte, which became detached from roots and can be harvested from soil by wet sieving, will germinate on agar media but will form only limited mycelial growth unless a host plant is available to be invaded. Hence the only way to produce large quantities of the endophyte is to grow it on suitable stock plants and to use as inoculum the resulting mixture of root tissue, mycelium and spores. The beneficial effect of inoculating plants with VA mycorrhiza is illustrated in Fig. 9.6.

Fig. 9.6. White clover seedlings growing in unsterile peat from a hill grassland site in mid-Wales. Both sets of plants were infected by the same indigenous mycorrhizal fungi, but they were ineffective (L); only those plants inoculated with effective mycorrhizal fungi indigenous to a nearby hill grassland site (R) grew wel!. All pots had received superphosphate at 40 kg P/ha.

9.4.2   *Inoculation with the endophyte*

Isolation of endophyte spores from soil

A portion of soil containing fine roots is soaked in water to disperse the soil particles and washed gently through a sieve with holes of 500–600 μm diameter. The sieved slurry is resuspended by shaking, allowed to stand for the coarser mineral particles to settle, passed through another sieve with holes 70 or 100 μm diameter and the residue remaining on this sieve examined microscopically. This sieving technique is satisfactory because the resting spores of the endophyte, with a diameter of c. 100–200 μm, are by far the largest fungal spores known. The method works best with light sandy soils because excessive organic matter makes it difficult to identify the spores. Spores have characteristic morphological features upon which the taxonomy of these fungi is based.

An alternative method is to isolate spores from an enrichment culture prepared by inoculating potted seedlings of a stock plant with pieces of mycorrhizal roots. The resulting infections yield a much higher density of spores in the pot (e.g. several thousand spores per pot) than in field soil (e.g. a few hundred per 100 g soil), and consequently it is easier to isolate them from such enriched soil.

Preparation of single cultures

Ten to twenty identical spores can be used to inoculate the roots of seedlings in sterilized soil to establish single cultures on stock plants. Satisfactory infection is indicated by the appearance of a uniform spore population in the pot some three to four months later. These spores can be harvested as already described and used to inoculate further batches of stock plants.

Large-scale inoculum production

The scheme proposed by Menge et al., in 1977, is a development of the small-scale pot culture method described above. Root material of a VA mycorrhizal plant, or infected soil, is used to inoculate a host plant that has no root diseases in common with the host plant for which the inoculum is ultimately intended. For example, an inoculum for citrus can be produced on Sudan grass but should not be produced on citrus.

Spores harvested from this culture can be surface-sterilized and used to inoculate other stock plants of the same kind. When the pot culture is considered to be free from diseases and pests likely to attack the selected host, young plants of the latter can be inoculated with the pot contents. Suitable pesticides and fungicides that are ineffective against VA mycorrhiza should be used throughout the production process. Methods such as this should yield enough inoculum for the commercial production of container-grown plants with well-established VA mycorrhiza. Nevertheless, production by pot culture is tedious and it is difficult to produce enough inoculum for field-scale trials, so it is not surprising that other methods are being tried. Also, there is always the hope that research on the physiology of the endophyte will lead to a method for growing it in pure culture, apart from any host plant, in the laboratory; such an advance would greatly simplify inoculum production, and yield a more uniform, plant-pathogen free product. For the time being, the infected plant remains the inoculum production unit.

Production of clean infected roots

The use of the NFT (see section 9.2) to produce clean roots infected with the VA endophyte is currently being investigated and some preliminary results have been published. Bean (*Phaseolus vulgaris*) and maize (*Zea mays*) were used to produce inoculum by NFT. Spores or mycorrhizal roots placed on small pieces of capillary matting and inserted between roots in the nutrient flow channels produced much cleaner infected roots than those grown in soil but spread of the endophyte within the root mass was slow. When used as inoculum these NFT-grown mycorrhizal roots produced normal VA infections in lettuce, onion, bean, clover and maize seedlings. NFT-grown inocula of three different endophytes (*Glomus* spp.) yielded infections characteristic of each endophyte in bean and lettuce seedlings. Each inoculum was as good as the corresponding pot-grown inoculum of similar weight and its infectivity was decreased, but not destroyed, by air-drying (Elmes & Mosse, 1980).

A great advantage of NFT is the close control of growth conditions available to the operator; these can be regulated to ensure good mycorrhizal development without regard to the complicating effects of soil colloids. For example, in the work reported above, the maize required 5.6 mg/l of iron as NaFe EDTA for satisfactory growth of the plants and development of mycorrhiza. Also, infection by the endophyte was inhibited by 11 mg/l of N (95% as NO_3^-; 5% as NH_4^+) in the circulating solution but developed well when the nitrogen was maintained at 1 mg/l of solution. Thus, NFT has clearly great potential for both research on mycorrhizal infections and the large-scale production of inoculum for field work and commercial practice.

9.4.3 *Inoculation techniques*

It is relatively easy to inoculate young perennial plants such as tree seedlings or cuttings with VA endophytes, especially when grown in containers. A few grams of crude inoculum from stock plant pot cultures can be incorporated with the potting medium or placed in contact with the young roots, to produce a heavy mycorrhizal infection with a selected endophyte before transplanting. This is proving to be a valuable technique for the forester and citrus grower.

Inoculation of field crops with VA endophytes is not yet an agricultural practice, although it is employed experimentally. The technique, which is satisfactory for container-grown perennials, is not, however, practicable for inoculation on a field scale because the inoculum is so bulky. About 2–3 t of crude inoculum per hectare would be required, and such large amounts cannot be produced by stock plant pot culture. Even when working with small experimental plots, careful placement of the scarce inoculum is necessary to achieve the best results, as the following example indicates.

Four methods of inoculation using a crude stock plant soil inoculum of two selected endophytes were tested on red clover (*Trifolium pratense*) at Rothamsted in 1979. The inoculum was applied: (1) below the seeds in furrows; (2) broadcast and raked in with the seeds; and (3) as multi-seeded pellets each *c.* 1 cm^3. In a fourth method, the crude inoculum was concentrated by wet sieving, suspended in 4% methylcellulose solution together with germinated seeds and applied as a slurry to the furrows, a technique known as fluid drilling. After nine weeks the broadcast method was found to be unsatisfactory; plants were infected with VA mycorrhiza to only 5–10%, about the same as the degree of infection of uninoculated control plants by indigenous endophytes. Plants from pelleted seeds were well infected (*c.* 25%), but the best results were obtained by placing inoculum in the furrows or by fluid drilling: both methods gave heavily infected (65–70%) plants.

Fluid drilling may be a suitable technique for field inoculation because the quantity of inoculum needed is smaller than that for other techniques, and other inocula, such as legume rhizobia, can be incorporated in the methylcellulose gel. Pelleting also has advantages in that large quantities of sparingly soluble material, such as lime, can be added with seed and inoculum (Hayman *et al.*, 1981).

9.5   Biological control

9.5.1   *Principles of biological control*

The fact that certain microorganisms can inhibit the growth of others has been known since the earliest days of microbiology and has stimulated much research, of which the discovery and development of antibiotics for clinical use is probably the most important. The possibility that a population of one microorganism might, by antagonistic or competitive mechanisms, be used to control a different microbial population, a plant pathogen, for example, has also incited great interest but, regrettably, has led to few developments of agricultural significance. Nevertheless, the possibility of such biological control on a wide scale continues to engage attention.

Biological control occurs naturally and helps to keep plant diseases in check but it is rarely possible to explain how such control operates or how it might be manipulated to agricultural advantage. Progress in this applied research area is slow, undoubtedly because it must wait upon the accumulation of much fundamental knowledge on the behaviour of mixed populations in soil and on plant surfaces, work that may not attract adequate notice or funding. There are, however, some examples of biological control systems that have been developed to the point where they may reasonably be regarded as biotechnological.

9.5.2   *Examples of biological control*

Antagonistic action of Trichoderma

The antagonistic properties of the fungus *Trichoderma* have been known for a long time. Heavy inoculation of *Trichoderma lignorum* into wet soils could suppress damping-off of seedlings, primarily by the action of a toxin that could be isolated from culture filtrates of the fungus. Some direct parasitism of the fungi causing damping-off was also demonstrated. Other species of *Trichoderma* are now known to be antagonistic to or parasitic upon a wide range of fungi and can effectively reduce diseases caused by several soil-borne plant pathogens. Certain species of *Trichoderma* show distinct promise as biological control organisms. Wells *et al.*, in 1972, reported that they could control stem and root diseases caused by *Sclerotium rolfsii* under field conditions by applying *Trichoderma harzianum* inoculum at a high ratio of inoculum to soil (1:10 v/v). Backman & Rodriguez-Kabana (1975) developed a diatomaceous earth granule carrier impregnated with a 10% solution of molasses to support growth and placement of *Trichoderma harzianum*. This inoculum gave significant control of *Sclerotium rolfsii* diseases in the field with an inoculum of 140 kg/ha of the granules, an economic level of treatment that compares favourably with chemical pesticide treatments.

Trichoderma harzianum controlled damping-off of bean and radish seedlings in glasshouse conditions but its value as a control organism in the field was not established. *Trichoderma hamatum* inhibits *Rhizoctonia solani* and, when inoculated on to the seeds, protected pea or radish seedlings from damping-off caused by *Rhizoctonia solani* or *Pythium* spp.

Annosus rot

The basidiomycete *Fomes annosus* is the most important cause in Europe of heart rot of conifers, especially Norway Spruce (*Picea abies*); it also attacks frondose trees and structural timbers. The infection of cut pine tree stumps with *Fomes annosus* is undesirable because the disease can progress through the stumps to the roots and thence to the roots of nearby healthy trees. The colonization of stumps by this fungus can fortunately be prevented by inoculating them with spores of another basidiomycete fungus, *Peniophora gigantea*. The *Peniophora gigantea* then colonizes the stumps and, by competition, prevents the development of heart rot. This method is used in the UK and in the USA, where commercial inoculant is available, to control *Fomes annosus* rot.

It is most probable that any biological control organisms developed in the future will act adversely on pathogens, either by producing inhibitory substances or by competing for limiting nutrients, or both. Most of the reports on this subject are of academic interest and the potential practical value of the organisms discussed is impossible to evaluate at present.

9.5.3   *Crown gall disease and its control*

Nature of the disease

Certain strains of the soil bacterium *Agrobacterium tumefaciens* can enter the tissues of dicotyledonous plants through wounds and give rise to tumours, the condition being known as crown gall disease. Although the relationship between plant and bacterium was first demonstrated in 1907, it is only recently that the mechanism of pathogenicity has been elucidated. Pathogenic strains of this bacterium carry a large plasmid, the Ti plasmid, which has a molecular weight ranging from 90×10^6 to 150×10^6, depending on which functions, other than tumorogenicity, that it encodes. The evidence indicates that the transformation of normal plant cells to tumour cells is caused by the transfer of Ti plasmid DNA from the bacterium and its integration into the chromosomal DNA of the host.

Plant tumour cells can grow as an undifferentiated mass in tissue culture but, sometimes, teratomas arise and can be cultivated as apparently normal plants. These teratomas still contain Ti plasmid

DNA and synthesize opines but do not usually exhibit tumours, although these can arise spontaneously or following damage.

It has now been shown that the Ti plasmid can be modified by genetic engineering techniques and can then convey non-Ti plasmid DNA into host plants where it is maintained. Thus, it is possible to transfer intact genes via Ti plasmids to a range of host plants. The existence of viable teratomas indicates that stable transformed plants might be produced by this kind of plasmid-mediated DNA transfer.

A curious fact is that, although teratomas can flower and set seed, the plants grown from such seed do not carry Ti plasmid DNA, presumably because it is lost in some way at meiosis. At the present time, plants containing Ti plasmid DNA can only be propagated vegetatively.

The fact that *Agrobacterium tumefaciens* cannot infect monocotyledons is a disadvantage in that the Ti plasmid vector system cannot be used in genetic work with cereal crop plants.

Biological control

In Australia, almond, peach and rose are severely affected by crown gall disease but, since 1973, commercial growers have protected their crops by dipping the planting stock (seedlings, transplants or cuttings) in a suspension of bacteria that can inhibit growth of the pathogen. This treatment achieves nearly complete control of the disease. It is the first commercial use of a bacterium to control a plant disease.

Not all agrobacteria are pathogenic. Many strains isolated from soil in a stone-fruit nursery were non-pathogenic but could not be distinguished from the pathogen by any test other than pathogenicity. In the soil around diseased plants the ratio of numbers of pathogenic to non-pathogenic strains was high, but in soil around healthy plants it was low.

In a laboratory study of interaction between the two types, using strain 84 as the non-pathogen, it was found that when the ratio of pathogenic to non-pathogenic cells was 1 or < 1 on tomato or peach seedlings, no galls developed. The control objective was to ensure that the ratio of pathogen to non-pathogen at the root surface never exceeded unity, so attempts were made to establish high numbers of the non-pathogen around the seeds or roots of young trees when transplanted. In experiments with peach, *c.* 95% control was achieved by root inoculation of transplants after growth for one year in pathogen-free soil, and 99% control by combined seed and root inoculation.

Treatment procedure

The treatment is simple. Seeds, cuttings or roots of young plants are dipped in the bacterial suspension and sown or planted immediately. About 10^7–10^8 bacteria/ml of suspension gives satisfactory results, giving almost complete control of the disease in practice. Commercial inoculants are available in Australia, New Zealand and the USA. Some inoculants are peat-based, similar to those prepared with rhizobia, whilst others are agar cultures.

Mechanism of disease control

Strain 84 produces a highly specific antibiotic that inhibits most pathogenic agrobacteria: it is a nucleotide bacteriocin known as agrocin 84. This strain carries a small plasmid (mol. wt. 30×10^6) that codes for agrocin production, and a large plasmid (mol. wt. 124×10^6) that codes for utilization of nopaline, an unusual amino acid of the opine type found only in crown gall tumour tissue. This strain does not carry the large Ti plasmid that determines pathogenicity and, also, sensitivity to agrocin. When a pathogenic strain is exposed to agrocin 84, some resistant colonies arise but most of these are non-pathogenic; either the Ti plasmid is lost or a deletion mutation has occurred that results not only in resistance to agrocin 84 but also in non-pathogenicity.

Disease control problems

Unfortunately, control of crown gall by strain 84 cannot be extended to all vulnerable crops because some, such as grapevine, are infected by strains resistant to agrocin 84. Moreover, pathogenic, agrocin 84 producing strains have been isolated from galls in some field experiments. All known agrocin 84 producing strains are resistant to this bacteriocin, so resistant pathogens cannot be controlled by strain 84.

Laboratory experiments have shown that such resistant pathogens can arise by conjugation between strain 84 and pathogenic strains. During conjugation, the two plasmids from strain 84 transfer independently while the Ti plasmid in the pathogen may or may not appear in the recipient cells. Of the six possible transconjugants, three contain the Ti plasmid, two of them also containing the small plasmid coding for agrocin 84 production. In the laboratory, such conjugation only occurs in the presence of nopaline and is likely to occur in nature only within or near crown gall tissue, which contains it. It is to be expected that pathogenic, agrocin-producing strains will eventually appear. The best answer that the plant pathologist can give would seem to be the discovery, or construction by genetic manipulation, of a mutant of strain 84 which would either not conjugate with the pathogen or would prevent transfer of the agrocin-84 plasmid.

9.6   The aerobic treatment of agricultural wastes (see also Chapter 6)

9.6.1   *The problem of waste storage and disposal*

Traditional methods of agriculture produce animal wastes in small quantities, which are easily returned to the land as fertilizer. Present-day, intensive rearing of animals, however, produces large volumes of liquid and solid wastes that cannot always be disposed of on nearby land areas that may be too small for the purpose, and that are troublesome to handle and store. Moreover, the disposal of all farm wastes is now scrutinized by environmentalists and public health bodies concerned by the drainage of pollutants to water supplies and the possible spread of pathogens. The real problem is how to make the fullest and most economic use of the fertilizer value of these wastes while avoiding pollutional and nuisance disadvantages caused by their presence in large quantity. This problem has stimulated the development and establishment of various treatment systems for the wastes.

The essential feature of aerobic biological treatment, practised so successfully with domestic sewage, is the provision of an abundant air supply to the aerobic microorganisms responsible for converting waste material into a relatively stable end-product. Controlled aerobic treatment of agricultural wastes has been shown to be feasible and several centres in Europe and North America have engaged in research and development programmes on the various treatment systems available. The essential features of a treatment system for farm use are that it should be robust, simple to operate and maintain and, preferably, automatic. Several small-scale systems are available.

9.6.2   *Aerobic treatment systems*

The handling of solid wastes is time consuming and expensive so it has become commonplace for intensive animal units to remove wastes by water, the resulting slurry being pumpable to storage tanks or treatment systems.

The oxidation pond

This is the simplest type of plant, consisting of a static tank or lagoon of liquid waste that relies upon its dimensions to provide adequate aeration; there should be a large surface area of liquid to volume and a good growth of photosynthetic algae on the surface. The rate of waste loading must also be low. Depth should not exceed 5 ft and the area should be as large as possible because oxygen is not very soluble in water. Oxygen released by algae improve the efficiency of the system. The practical value of such a plant is limited for the following reasons:

(1) a long period is required for the waste treatment, which means that an extensive lagoon area is needed if the amount of waste to be treated is large;

(2) solids are deposited and decompose anaerobically;

(3) it provides a breeding-ground for insects.

The advantages are twofold:

(1) there is no mechanical equipment;

(2) it is maintenance free.

The aerated lagoon is similar to an oxidation pond but incorporates a mechanical device to facilitate aeration, mixing and the maintenance of solids in suspension. It can be smaller and deeper than an oxidation pond with the same loading rate. Treatment rate and effluent quality are more predictable.

Lagoons and ponds are essentially storage treatment systems favoured when liquid cannot be discharged to a watercourse, or for the conservation of wastewater for irrigation.

The barrier ditch

The barrier ditch is also a simple, non-mechanical system used in England but, unlike the oxidation pond, the waste is fed to it continuously and the residence time is short. The system comprises a primary settling tank, preferably with baffles to control the flow, in which settling of gross solids occurs. This is followed by a series of shallow basins separated by barriers or weirs over which the liquid falls and becomes aerated on its travel from basin to basin. If the retention time is adequate, it gives a degree of treatment similar to that in an oxidation pond. As in the non-aerated oxidation pond, there is little mixing of the slurry and activity of the microorganisms may be inhibited by inadequate oxygen or the slow diffusion of end-products of metabolism from them.

The Pasveer ditch

This oxidation ditch, developed as a modification of the activated sludge tank, is a continuous, elongated ditch, often situated beneath the animal house that provides the waste, in which the slurry is oxidized. The liquid (0.3–0.6 m deep) is aerated, mixed and circulated by a rotor situated at one point on the channel, and provision is made to add fresh waste and remove treated effluent. This, and other systems relying on forced aeration, are similar to a continuous culture apparatus—which contains a developed 'treatment' microflora (analogous to activated sludge floc) able to metabolize the substrate—in that additions of waste (substrate) are small compared with the total volume of material being treated. Many investigations have been made with this type of oxidation ditch to determine its ability to provide controlled treatment of animal wastes.

A major problem with these systems is the inability to relate oxygen demand of the waste to oxygen supply. Better understanding of the rate at which different substrates are degraded and of the microorganisms responsible for it may lead to improved efficiency.

9.7   Anaerobic digestion of agricultural wastes (see also Chapter 6)

9.7.1   *Essential features of the process*

The anaerobic, bacterial fermentation of organic wastes yields combustible gas containing c. 60% methane, a residue containing all or most of the nitrogen, and all other plant nutrients present in the original material. In nature, the process occurs wherever plant or animal residues accumulate in oxygen-deficient situations, such as marshes and lake sediments, and in the rumen of herbivores; it can also be made to take place in a closed container supplied with suitable organic matter and from which air is excluded. The methanogenic bacteria and others that supply substrates for them develop as a stable microbial symbiosis that operates indefinitely as long as fresh waste material is fed into the system at a suitable rate.

A temperature range of 30–35 °C is optimal and the heat to maintain this must be supplied. Since the process was first studied systematically in the time of Pasteur, it has been the subject of much research and development, especially for wastewaster treatment; it is a highly successful example of biotechnology in action, provided it can be made to work on a sufficiently large scale to justify capital and running costs of the installation.

Methanogenic sludge digestion is practised at some large sewage treatment plants in the UK and elsewhere and makes a significant contribution to reducing treatment costs. An example is the Mogden

Sewage Treatment Works in West London (Thames Water Authority), where the sludge digestion plant comprises twenty tanks each of 21.3 m diameter and 10.4–16.0 m depth, with a total capacity of 79 000 m³. This installation is still functioning after continuous operation for about fifty years and provides enough methane for dual-fuel engines continuously supplying compressed air for the activated sludge tanks and electric power for the works generally.

9.7.2 *Approach to agricultural waste treatment*

The possibility of using manure to produce fuel gas on the farm while retaining fertilizer value in the residue was realized early in this century and a few attempts were made to use the process. Interest in this topic was subdued, however, until revived by wartime fuel scarcity.

Many types of methane plant have been designed but all contain two essential parts: (1) a closed tank or silo in which the fermentation takes place; (2) a container for the gas, usually a conventional floating bell gas-holder of about the same capacity as the digester.

Simple plants of this kind with digesters of *c.* 10 m³ capacity were developed by the French in North Africa in the 1940s and many were in use there and in France in the early 1950s. At that time in the UK several manufacturers offered such small-scale plants for sale to farms but it is unlikely that many were actually installed.

The simple batch-fed plant suffers from several serious defects, the most important being the difficulty of filling and emptying the digesters with fairly solid waste material, a task that usually has to be done by hand. Other drawbacks were the need to install several digesters with contents at different stages of maturity to minimize fluctuations in the gas production flow, and to provide some means of warming the digesters to 30–35 °C. Such difficulties undoubtedly prevented widespread adoption of farm methane plants and they tended to disappear altogether when abundant supplies of bottled propane and butane appeared in the 1960s.

A different approach, directed towards development of large-scale, highly mechanized plant, was adopted in Germany in the post-war period. The classical example of this type is the 'Bihugas' plant developed by F. Schmidt of Allerhop. This was on a farm of 220 acres which carried, in 1950, 55 head of cattle, 180 pigs and 650 chickens. It was claimed that the manure from these animals plus straw could produce $1\cdot5 \times 10^6$ cubic feet of gas per annum. The plant consisted of one or more cylindrical, heat-insulated fermentation tanks above ground, fed with manure and chopped straw as a pumpable slurry. A high rate of gas production was achieved by taking advantage of brisk gas evolution during a short fermentation period of 5–21 d; no attempt was made to obtain maximum gas yields by prolonged, uneconomic treatment. There is no doubt that this type of plant was very efficient but its size and cost rendered it suitable only for the largest farms with abundant livestock. Despite their efficiency it seems that these large-scale installations also fell into disuse under the competition from abundant and cheap energy sources.

The enormous increase in oil prices in the 1970s again focused attention on the need for economic farm operation and it is no surprise to find renewed interest in methane generation and efficient manure handling. Indeed, an oil crisis is the signal for letters and articles to appear in the popular press urging the adoption of methane generation for the nation, farm, home or even car: such suggestions are often bizarre and show imperfect understanding of the issues involved. This interest continues and considerable progress has been made with all aspects of the methanogenic fermentation.

9.7.3   *Microbiology of the process*

The methanogenic bacteria are such strict anaerobes that the study of their characteristics has always posed problems for the research worker. Isolation has always been difficult, although it is now made easier by the development of satisfactory anaerobic cabinets. A number of them have now been studied and much learned of their metabolic activities.

The formation of volatile fatty acids (acetic, butyric) from plant and faecal residues is the first stage of the fermentation process, one in which clostridia can play an important part. Acids other than acetic then act as substrates for a group of acetogenic bacteria: thus, the combined activities of these two groups yield acetic acid, hydrogen and carbon dioxide, all substrates for the methanogenic bacteria proper. Methanogenic bacteria are of the acetoclastic type, e.g. *Methanosarcina barkeri*, which converts acetic acid to methane and CO_2, and of the hydrogen-utilizing type, e.g. *Methanospirillum hungatei*, which synthesizes methane from hydrogen and CO_2. The interplay of these groups and the manner in which metabolism of the digestion process is regulated are discussed by Mosey, 1982, (see 'Recommended reading').

9.7.4   *Modern developments*

The main problem on a farm carrying much livestock is to store and distribute the manure in the most profitable way. If methane could be a by-product without increasing costs of the manure storage installation, then it would, doubtless, find favour with farmers. There seems little chance, however, that this can be done because efficient, mechanized systems suitable for advanced agriculture are expensive. Commercial farm waste digesters of improved design are available in the UK but their cost may be difficult to recover from the value of the methane produced. Several interesting new types of anaerobic digester, operating on the fluidized-bed principle, have been developed for use with dilute industrial wastes, but here the primary interest is in purification of effluent rather than production of fuel gas.

Many thousands of simple digesters have been built in India, Africa, the Far East and China, where there is ample labour to operate them. It is, however, very difficult to ascertain how useful they really are and how long they remain in service.

9.8   Plant breeding and protoplast techniques

9.8.1   *Vegetative propagation*

In higher plants variation generally results from the mixing of genes by sexual reproduction. Although such variation is important in an evolutionary sense it is also inconvenient to those engaged in maintaining stocks of plants with desirable characteristics. Fortunately, not all flowering plants depend entirely on sexual reproduction for their propagation; many develop specialized structures for vegetative reproduction, the stolons of strawberry and the tubers of potato being familiar examples of modified stems. The ability of many plants to regenerate completely from small stem pieces (cuttings), is a valuable property and it is a universal practice to make use of it for the propagation of plants with useful varietal characteristics. It is the only practicable method to propagate plants that are sexually infertile. Some plants also have the property of being able to regenerate completely from whole leaves (many succulents) or leaf fragments (e.g. *Begonia*).

An entire plant can also arise from callus tissue grown *in vitro* from meristem, a technique that has made possible the production of virus-free potato stocks. Tissue culture techniques have now been used for some time to produce disease-free and disease-resistant plants, for the multiplication of genotypes and for storing or maintaining stocks of valuable plants.

It is now known that the ability to regenerate an entire plant resides in many individual cells of a parent plant, given suitable conditions of

cultivation. A single cell that has this potential to grow and form a
tissue from which an entire plant can be regenerated is described as
totipotent. Totipotency was established for many plant species by about
1965.

Wall-free protoplasts of totipotent cells can be produced easily and
methods have been devised for their cultivation to callus tissue and
thence to plantlets which can then be propagated by conventional
means.

Protoplasts have been isolated from tissues of most plant organs,
including roots, leaves, petals, fruits, coleoptiles, storage organs,
microspore tetrads (from anthers), root nodules of legumes and callus
tissue. Leaf mesophyll tissue is often used because yields of protoplasts
from this source can be high and uniform. Protoplasts often yield callus
cultures from which whole plants can be regenerated. Unfortunately,
little success has so far been achieved in trying to recover whole plants
from isolated protoplasts of legumes or cereals, both groups of important crop plants. Recently, the ability of *Medicago sativa* (lucerne,
alfalfa) protoplasts to regenerate into plants at high frequency under
relatively simple growth conditions has been established. This is an
important indication that work with legumes will progress satisfactorily when suitable growth conditions are devised. So far, we still lack the
ability to regenerate plants from protoplasts of cereals and grain
legumes. Wernicke & Brettell, in 1980, obtained callus formation and
plant regeneration from cereal leaves so useful progress with cereal
protoplasts is to be expected.

The particular usefulness of protoplasts in plant breeding stems
from several of their properties. First, many protoplasts can be
produced and screened for useful variation. Although the protoplasts
when formed are genetically homogeneous, the calluses derived from
them yield plantlets that can show considerable variation in characteristics. Secondly, the absence of cell wall facilitates fusion between
protoplasts and the consequent initiation of hybrids. The fact that
fusion can occur between somatic cells that are at least diploid provides
the plant breeder with a powerful technique. Thirdly, the absence of cell
wall also facilitates the uptake of DNA, as fragments or plasmids of
bacterial origin, to produce plants with entirely new features.

Variation

Although plants derived by vegetative means (clones) are usually like
the parent plant, it does not follow that all the clones are genetically
similar. Clones that differ significantly from the parent can appear; they

are known as somatic variants or 'sports' and they result from genetic changes in the meristematic cells that generate all or part of the new plant. Occasionally such sports become established as important new varieties, e.g. the navel orange and nectarine. Several genetic mechanisms can lead to somatic variation: changes in number of nuclear chromosomes, mutation of single genes, and modifications of extranuclear genes carried by organelles such as chloroplasts and mitochondria. Although there is little variation between potato plants regenerated from protoplasts, considerable variation has been found among protoclones of the tetraploid variety Russet Burbank. A study of variation amongst regenerated potato plants (cv. Maris Bard) indicated that this variation arose during growth of the callus from genetically homogeneous protoplasts and was not a consequence of the protoplast technique itself.

The occurrence of protoclonal variation, even without artificial mutagenesis, is likely to be important for agriculture for it provides the plant breeder with a wealth of new material. Potato variants have been regenerated with several potentially useful characteristics including compact growth habit, variable maturity date and altered tuber characteristics.

9.8.2   *Regeneration of plants from protoplasts*

Shephard & Totten (1977), working with potato which does not regenerate easily from tissue culture, devised a detailed procedure which enabled them to raise plants reliably from protoplasts. Although their method was unusual in that growth of the source plants was carefully controlled before protoplast isolation and a different medium was used at each stage of development of the new plants, it serves to illustrate the general procedure.

Cultivation of source plants

Potato plants of the American variety Russet Burbank were raised from virus X free tubers under high illumination with a 12 h daily light period at 24 °C and 70–75% relative humidity. Before use the plants were held for 4–10 d under the same conditions of temperature and humidity but with reduced illumination (7000 lx) and only a 6 h daily light period. Subjecting the plants to such short periods of dim light ensured consistently high yields (e.g. 2–3×10^6/g of tissue) of viable protoplasts from the leaf tissue.

Media for isolation and propagation

The five media (A–E) used throughout the isolation and propagation procedures were variants of a basal medium described by Lam (1975). The media were prepared from a range of major plant nutrients, trace elements, amino acids, growth factors, phytohormones and osmotic pressure regulators, the amounts of each being varied to give optimal growth at each stage of the regenerating plants. All media contained agar at 1.5–2.0% except the semi-solid medium A with only 0.5%. Media D and E, for the later stages of growth, each contained NH_4NO_3 at 1650 mg/l.

Isolation of protoplasts

Four grams of tissue cut from surface-sterilized leaflets were conditioned in 200 ml of modified medium A (without sucrose or agar) in the dark at 4 °C for 16–24 h. The medium was then replaced with 100 ml of a buffered (pH 5.6) mixed enzyme solution to separate the mesophyll cells and digest away the cell walls. Digestion was complete after 4 h at 28 °C in shaken culture. Viable protoplasts were recovered by centrifuging the strained mixture, washing with medium A, collecting and storing in liquid medium A at 6×10^5 cells/ml.

Growth of callus

Incubation of a protoplast on solid medium B or on a modified solid medium system promoted cell wall regeneration and limited division to form a microcallus. Transfer of these to medium C enabled growth of cultures to progress until they could be transferred to medium D, which induced shoot morphogenesis. Finally, calluses with shoots *c.* 1 mm in length were transferred to medium E for the final stages of shoot development and root initiation. Rooted plantlets were transferred to small pots of vermiculite for further growth.

This description indicates the general procedures that must be followed to regenerate entire plants from single protoplasts, and illustrates how the experimenters may need to vary the growth conditions at different stages of development to ensure a successful outcome. The appearance of freshly isolated protoplasts, young calluses, and plantlets regenerating from callus tissue is shown in Figs. 9.7, 9.8, 9.9 and 9.10.

Providing precisely the right conditions is critical, although one

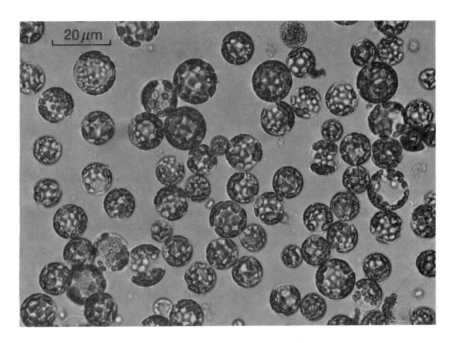

Fig. 9.7. Freshly isolated protoplasts of *Solanum brevidens*. Bar = 20 μm.

may confidently expect that the details will need to be varied according to the particular plant being studied.

Thomas (1981) established shoot cultures of the tetraploid potato cultivar Maris Bard for use as source material for protoplast production. Axillary buds were surface sterilized with sodium hypochlorite solution and cultivated on the medium of Murashige & Skoog (1962) containing, in addition, sucrose (30 g/l), agar (6 g/l) and 6-benzylaminopurine (6-BAP: 0.5 mg/l), at pH 5.8. Incubation at 26 °C under illumination of *c.* 400 lx (16 h day/8 h night) yielded shoots *c.* 3 cm long in three to five weeks. These shoots could be propagated further from short (3–8 mm) stem segments with a leaf and axillary bud on the same medium without 6-BAP for three weeks, and then propagated again or used to prepare protoplasts. Complete shoots, but not their leaves alone, reliably yielded viable protoplasts capable of sustained division in culture.

9.8.3   *Protoplast fusion techniques*

Although the fusion of plant protoplasts had been known to occur sporadically, Power *et al.* (1970) were first to devise a method for controlled, reproducible fusion, and so achieve the first step towards somatic hybridization in plants. Suspension of protoplasts in 0.25 mol/l sodium nitrate solution induced rapid fusion. A solution of 10.2%

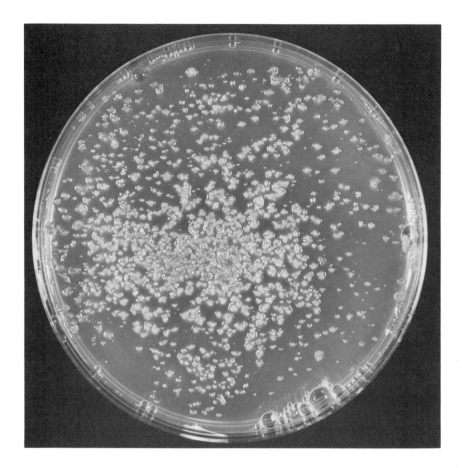

Fig. 9.8. Developing callus tissue from protoplasts of *Solanum brevidens* on agar medium, *c.* one month after isolation. Growth in 9 cm diameter Petri dish.

sucrose and 5.5% sodium nitrate was used later to induce fusion of protoplasts of *Parthenocissus tricuspidata* with those of *Petunia hybrida*; calcium chloride was also used.

Kao & Michayluk (1974) discovered that high molecular weight polyethylene glycol (PEG) also induced non-specific adhesion, then fusion, between plant protoplasts. Very few fusions occurred during incubation with PEG, but up to 10% of hybrid fusions were formed as the PEG was diluted. PEG could also induce the uptake of chloroplasts derived from the alga *Vaucheria dichotoma* by protoplasts of carrot. Many details of protoplast fusion techniques are given in the laboratory manual of Power & Davey (1979).

The production of somatic hybrids

The following stages are essential for the generation of somatic hybrids by the protoplast fusion technique:

(1) isolation of protoplasts;

(2) fusion;

(3) regeneration of cell wall;

(4) fusion of nuclei to give a true hybrid nucleus;

(5) growth of the hybrid cell in culture; and, finally,

(6) regeneration of the whole plant.

In general, the fusion of plant protoplasts is not, now, difficult to achieve, although it is not necessarily easy to grow hybrid cells satisfactorily or to regenerate a hybrid plant. It is clearly necessary to separate hybrid protoplasts, which will be present as a minority in the cell mixture, and to encourage their development by selection procedures. An example of this is afforded by experiments on somatic hybridization between *Petunia hybrida* and *Petunia parodii*, in which a selection procedure made use of the difference in growth potential between leaf protoplasts of the two species. Thus, the protoplasts of

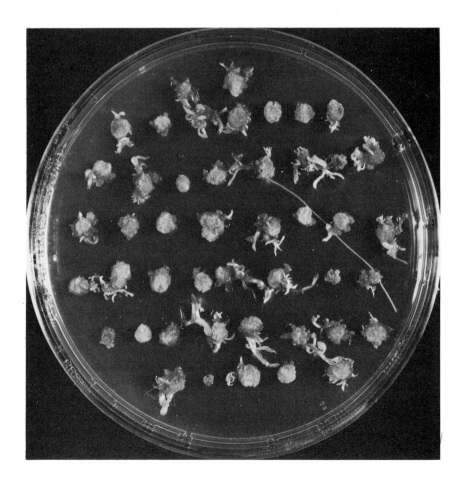

Fig. 9.9. Regeneration of plantlets of *Solanum tuberosum* from callus tissue derived from protoplasts, *c.* 100 d after planting out on agar medium favouring regeneration.

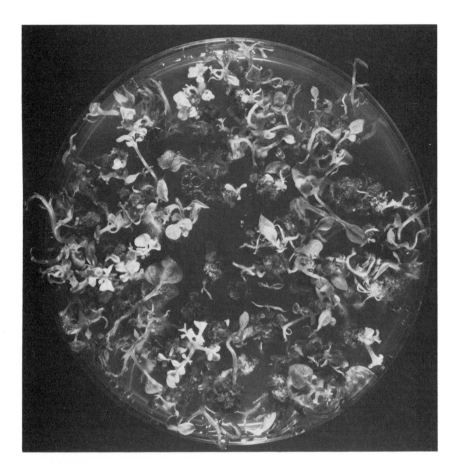

Fig. 9.10. Regeneration of plantlets of *Solanum brevidens* from callus tissue derived from protoplasts, *c*. 100 d after planting out on agar medium favouring regeneration.

Petunia parodii never formed more than a microcallus of about fifty cells on the medium used, whereas those of *Petunia hybrida* made continued callus growth. Advantage was also taken of the fact that *Petunia hybrida* protoplasts were more sensitive to actinomycin D than were those of *Petunia parodii*.

A heterokaryon resulting from the fusion of two dissimilar protoplasts can develop, by fusion of the nuclei, into a hybrid cell. By this means all the gene-carrying, self-replicating organelles of both protoplasts are combined whereas, in normal sexual crosses, one nucleus bearing the chromosomal genes (karyome) derives from each parent but, usually, the plastid-inherited genes (plastidome) and the mitochondria-inherited genes (chondriome) derive only from the maternal parent. Thus, protoplast fusion techniques offer the opportunity to produce combinations of two complete parental genomes. Fusions between protoplasts of sexually compatible species can often result in the formation of stable amphiploid hybrid cells from which somatic hybrid plants may be regenerated. The protoplast fusion technique is not, however, limited solely to closely related species or to hybrid cells.

Thomas *et al.* (1979) list numbers of references to: plants regenerated from intraspecific (7) and interspecific (14) fusion experiments; cell lines established from interspecific fusion experiments (4); uptake of cell organelles into plant protoplasts (4); uptake and expression of DNA in plant protoplasts (2), and mutant plants from protoclones (3). The possible ways in which plant protoplast fusion can be exploited are manifold.

An interesting aspect of somatic hybrid formation is the opportunity it affords to vary the extra-nuclear gene component of the hybrid cells, e.g. by the use of albino protoplasts that carry no chloroplast-borne genes, or by protoplast fusion coupled with procedures to inactivate the nuclear genome of one parent. Enucleate subprotoplasts or microplasts may provide cytoplasm-borne genes for insertion, by fusion, into whole protoplasts.

Uptake of microorganisms and DNA by protoplasts

Davey & Power (1975) found that PEG could induce the uptake of yeast cells, yeast protoplasts and blue-green algal cells into protoplasts of *Parthenocissus tricuspidata*, the microorganisms becoming localized in membrane-bounded vesicles in the protoplast cytoplasm. Considerable interest attaches to the possibility of introducing entire microorganisms with useful properties, such as the ability to fix nitrogen, or genetic material from them into higher plants in such a way that the useful property is expressed. Protoplast fusion techniques offer opportunities to do this although, so far, little success has been achieved (see Thomas *et al.*, 1979).

New genetic information, introduced into protoplasts in the form of purified DNA, is not necessarily maintained and expressed in the host cells. Use of the *Agrobacterium tumefaciens* Ti plasmid (see section 9.5.3) as a vector system is particularly useful in investigations of this kind because it enters directly into the DNA of the host and finds ready expression in tumorigenicity. Ti plasmids, with or without foreign DNA incorporated, can combine with plant protoplasts and transform them. A considerable extension of such work can be envisaged.

9.9   **Recommended reading**

ALLEN M.F., MOORE T.S. JUN. & CHRISTENSEN M. (1980) Phytohormone changes in *Bouteloua gracilis* infected by vesicular-arbuscular mycorrhizae. I. Cytokinin increases in the host plant. *Canad. J. Bot.* **58**, 371–374.

Aviv D., Fluhr R., Edelman M. & Galun E. (1980) Progeny analysis of the interspecific hybrids: *Nicotiana tabacum* (CMS) + *Nicotiana sylvestris* with respect to nuclear and chloroplast markers. *Theor. appl. Genet.* **56**, 145–150.

Backman P.A. & Rodriguez-Kabana R. (1975) A system for the growth and delivery of biological control agents to the soil. *Phytopathology*, **65**, 819–821.

Beringer J.E. (1982) Microbial genetics and biological nitrogen fixation. In *Advances in Agricultural Microbiology*, (ed. Subba Rao N.S.), pp. 2–23. Oxford & IBM Publishing Co., New Delhi.

Bilkey P.C. & Cocking E.C. (1980) Isolation and properties of plant microplasts: newly identified subcellular units capable of wall synthesis and division into separate microcells. *Eur. J. Cell Biol.* **22**, 502.

Bonnett H.T. & Eriksson T. (1974) Transfer of algal chloroplasts into protoplasts of higher plants. *Planta*, **120**, 71–79.

Brown M.E. (1982) Nitrogen fixation by free-living bacteria associated with plants—fact or fiction? In *Bacteria and Plants*, (eds. Rhodes-Roberts M.E. & Skinner F.A.), pp. 25–41. Academic Press, London.

Burton J.C. (1980) New developments in inoculating legumes. In *Recent Advances in Biological Nitrogen Fixation*, (ed. Subba Rao N.S.), pp. 380–405. Edward Arnold, London.

Burton J.C. & Curley R.L. (1965) Comparative efficiency of liquid and peat-base inoculants on field-grown soybeans (*Glycine max*). *Agron. J.* **57**, 379–381.

Carr J.G. (1982) The production of foods and beverages from plant materials by micro-organisms. In *Bacteria and Plants*, (eds. Rhodes-Roberts M.E. & Skinner F.A.), pp. 155–167. Academic Press, London.

Cocking E.C. (1981) Opportunities from the use of protoplasts. *Phil. Trans. Roy. Soc. Lond.* **B292**, 557–568.

Coombs J. (1981) Biogas and power alcohol. *Chemy Ind.* **4 April**, 223–229.

Cooper A. (1979) *The ABC of NFT*. Grower Books, London.

Corby H.D.L. (1975) A method of making a pure-culture, peat-type, legume inoculant, using a substitute for peat. In *Symbiotic Nitrogen Fixation in Plants*, (ed. Nutman P.S.), pp. 169–173. Cambridge University Press, Cambridge.

Corby H.D.L. (1980) The systematic value of leguminous root nodules. In *Advances in Legume Systematics*, Part 2, (eds. Polhill R.M. & Raven P.H.), pp. 657–669. Royal Botanic Gardens, Kew.

Davey M.R. & Power J.B.(1975) Polyethylene glycol-induced uptake of micro-organisms into higher plant protoplasts. An ultrastructural study. *Plant Sci. Lett.* **5**, 269–274.

Deacon J.W. (1983) *Microbial Control of Plant Pests and Diseases*. Aspects of Microbiology 7. Van Nostrand Reinhold (UK) Co. Ltd., Nottingham, England.

Dunican L.K., O'Gara F. & Tierney A.B. (1976) Plasmid control of effectiveness in *Rhizobium*. Transfer of nitrogen fixing genes on a plasmid from *Rhizobium trifolii* to *Klebsiella aerogenes*. In *Symbiotic Nitrogen Fixation in Plants*, (ed. Nutman P.S.), pp. 77–90. Cambridge University Press, Cambridge.

Dye M. (1979) The Rothamsted *Rhizobium* culture collection and inoculant use in the UK. *Rothamsted Report for 1978, Part 2*, 119–130.

Elmes R.P. & Mosse B. (1980) Nutrient film technique. *Rothamsted Report for 1979, Part 1*, 188.

Elmes R.P. & Mosse B. (1984) Vesicular-arbuscular endomycorrhizal inoculum production. II. Experiments with maize (*Zea mays*) and other hosts in nutrient flow culture. *Canad. J. Bot.* **62**, 1531–1536.

Fraser M.E. (1966) Pre-inoculation of lucerne seed. *J. appl. Bacteriol.* **29**, 587–595.

Fraser M.E. (1975) A method of culturing *Rhizobium meliloti* on porous granules to form a pre-inoculant for lucerne seed. *J. appl. Bacteriol.* **39**, 345–351.

Fred E.B., Baldwin I.L. & McCoy E. (1932) *Root Nodule Bacteria and Leguminous Plants*. University of Wisconsin Press, Madison, Wisconsin.

HADAR Y., CHET I. & HENIS Y. (1979) Biological control of *Rhizoctonia solani* damping-off with wheat bran culture of *Trichoderma harzianum*. *Phytopathology*, **69**, 64–68.

HAYMAN D.S. (1977) Mycorrhizal effects on white clover in relation to hill land improvement. *ARC Res. Rev.* **3**, 82–85.

HAYMAN D.S. (1980) Mycorrhiza and crop production, *Nature*, **287**, 487–488.

HAYMAN D.S. & MOSSE B. (1979) Improved growth of white clover in hill grasslands by mycorrhizal inoculation. *Ann. appl. Biol.* **93**, 141–148.

HAYMAN D.S., MORRIS E.J. & PAGE R.J. (1981) Methods for inoculating field crops with mycorrhizal fungi. *Ann. appl. Biol.* **99**, 247–253.

HEPBURN A.G. (1982) The biology of the crown gall—a plant tumour induced by *Agrobacterium tumefaciens*. In *Bacteria and Plants*, (eds. Rhodes-Roberts M.E. & Skinner F.A.), pp. 101–113. Academic Press, London.

HOBSON P.N. & FEILDEN N.E.H. (1982) Production and use of biogas in agriculture. *Prog. Energy Combust. Sci.* **8**, 135–158.

KAO K.N. & MICHAYLUK M.R. (1974) A method for high-frequency intergeneric fusion of plant protoplasts. *Planta*, **115**, 355–367.

KERR A. (1972) Biological control of crown gall: seed inoculation. *J. appl. Bacteriol.* **35**, 493–497.

KERR A. (1980) Biological control of crown gall through production of agrocin 84. *Pl. Disease*, **64**, 25–30.

KHIN HTAY & KERR A. (1974) Biological control of crown gall: seed and root inoculation. *J. appl. Bacteriol.* **37**, 525–530.

KUMAR A., WILSON D. & COCKING E.C. (1981) Polypeptide composition of Fraction 1 protein of the somatic hybrid between *Petunia parodii* and *Petunia parviflora*. *Biochem. Genet.* **19**, 255–261.

LAM S. (1975) Shoot formation in potato tuber discs in tissue culture. *Am. Potato J.* **52**, 103–106.

LAWRIE J. (1961) *Natural Gas and Methane Sources*, pp. 82–88. Chapman & Hall, London.

LIU S. & BAKER R. (1980) Mechanism of biological control in soil suppressive to *Rhizoctonia solani*. *Phytopathology*, **70**, 404–412.

MENGE J.A. (1983) Utilization of vesicular-arbuscular mycorrhizal fungi in agriculture. *Canad. J. Bot.* **61**, 1015–1024.

MENGE J.A., LEMBRIGHT H. & JOHNSON E.L.V. (1977) Utilization of mycorrhizal fungi in citrus nurseries. *Proc. Int. Soc. Citriculture*, **1**, 129–132.

MOSEY F.E. (1982) New developments in the anaerobic treatment of industrial wastes. *Wat. Pollut. Contr.* **81**, 540–552.

MUNNS D.N. & MOSSE B. (1980) Mineral nutrition of legume crops. In *Advances in Legume Science*, (eds. Summerfield R.J. & Bunting A.H.), pp. 115–125. Royal Botanic Gardens, Kew.

MURASHIGE T. & SKOOG F. (1962) A revised medium for rapid growth and bioassays with tobacco tissue cultures. *Physiologia Pl.* **15**, 473–497.

NEMEC S. (1980) Effects of 11 fungicides on endomycorrhizal development in sour orange. *Canad. J. Bot.* **58**, 522–526.

NEW P.B. & KERR A. (1972) Biological control of crown gall: field measurements and glasshouse experiments. *J. appl. Bacteriol.* **35**, 279–287.

NOACK W. (1955) *Biogas in der Landwirtschaft*. Otto Elsner Verlagsgesellschaft, Darmstadt.

PASVEER A. (1959) A contribution to the development of the activated sludge process. *J. Proc. Inst. Sewage Purif.* **4**, 436.

PHILLIPS J.M. & HAYMAN D.S. (1970) Improved procedures for clearing roots and staining parasitic and vesicular-arbuscular mycorrhizal fungi for rapid assessment of infection. *Trans. Br. mycol. Soc.* **55**, 158–161.

POLHILL R.M. & RAVEN P.H. (eds.) (1980) *Advances in Legume Systematics*. Royal Botanic Gardens, Kew.

POWER J.B., CUMMINS S.E. & COCKING E.C. (1970) Fusion of isolated plant protoplasts. *Nature,* **225,** 1016–1018.

POWER J.B. & DAVEY M.R. (1979) *Laboratory Manual: Plant Protoplasts* (isolation, fusion, culture, genetic transformation). Department of Botany, University of Nottingham, Nottingham.

POWER J.B., FREARSON E.M., HAYWARD C. & COCKING E.C. (1975) Some consequences of the fusion and selective culture of *Petunia* and *Parthenocissus* protoplasts. *Plant Sci. Lett.* **5,** 197–207.

POWER J.B., FREARSON E.M., HAYWARD C., GEORGE D., EVANS P.K., BERRY S.F. & COCKING E.C. (1976) Somatic hybridisation of *Petunia hybrida* and *P. parodii. Nature,* **263,** 500–502.

RISHBETH J. (1963) Stump protection against *Fomes annosus.* III. Inoculation with *Peniophora gigantea. Ann. appl. Biol.* **52,** 63–77.

ROBERTS W.P., TATE M.E. & KERR A. (1977) Agrocin 84 is a 6-*N*-phosphoramidate of an adenine nucleotide analogue. *Nature,* **265,** 379–381.

ROBINSON K. (1971) Aerobic treatment of agricultural wastes. In *Microbial Aspects of Pollution,* (eds. Sykes G. & Skinner F.A.), pp. 91–102. Academic Press, London.

ROBINSON K. (1974) The use of aerobic processes for the stabilization of animal wastes. *CRC Crit. Rev. Environ. Contr.* **4,** 193–220.

ROBINSON K. (1979) Developments in aerobic stabilization and utilization of animal wastes in Northeast Scotland: a review. *Environ. Protec. Eng.* **5,** 127–143.

ROUGHLEY R.J. (1968) Some factors influencing the growth and survival of root nodule bacteria in peat culture. *J. appl. Bacteriol.* **31,** 259–265.

ROUGHLEY R.J. (1970) The preparation and use of legume seed inoculants. *Pl. Soil,* **32,** 675–701.

ROUGHLEY R.J., DATE R.A. & WALKER M.H. (1966) Inoculating and lime pelleting legume seed. *Agric. Gaz. N.S.W.,* **March,** 142–146.

ROUGHLEY R.J. & VINCENT J.M. (1967) Growth and survival of *Rhizobium* spp. in peat culture. *J. appl. Bacteriol.* **30,** 362–376.

SANTOS A.V.P.DOS, OUTKA D.E., COCKING E.C. & DAVEY M.R. (1980) Organogenesis and somatic embryogenesis in tissues derived from leaf protoplasts and leaf explants of *Medicago sativa. Z. PflPhysiol.* **99,** 261–270.

SCHROTH M.N. & HANCOCK J.G. (1981) Selected topics in biological control. *Ann. Rev. Microbiol.* **33,** 453–476.

SHANMUGAM K.T. & HENNECKE H. (1980) Microbial genetics and nitrogen fixation. In *Recent Advances in Biological Nitrogen Fixation,* (ed. Subba Rao N.S.), pp. 227–256. Edward Arnold, London.

SHEPHARD J.F., BIDNEY D. & SHAHIN E. (1980) Potato protoplasts in crop improvement. *Science,* **208,** 17–24.

SHEPHARD J.F. & TOTTEN R.E. (1977) Mesophyll cell protoplasts of potato. Isolation, proliferation and plant regeneration. *Pl. Physiol.* **60,** 313–316.

STEINBORN J. & ROUGHLEY R.J. (1974) Sodium chloride as a cause of low numbers of *Rhizobium* in legume inoculants. *J. appl. Bacteriol.* **37,** 93–99.

STEWART W.D.P. (ed.) (1975) *Nitrogen Fixation by Free-living Micro-organisms.* IBP 6. Cambridge University Press, Cambridge.

STREICHER S., GURNEY E. & VALENTINE R.C. (1971) Transduction of the nitrogen-fixation genes in *Klebsiella pneumoniae. Proc. natn. Acad. Sci. U.S.A.* **68,** 1174–1177.

STRIJDOM B.W. & DESCHODT C.C. (1975) Carriers of rhizobia and the effects on the survival of rhizobia. In *Symbiotic Nitrogen Fixation in Plants,* (ed. Nutman P.S.), pp. 151–168. IBP 7. Cambirdge University Press, Cambridge.

THOMAS E. (1981) Plant regeneration from shoot culture-derived protoplasts of tetraploid potato (*Solanum tuberosum* cv. Maris Bard). *Plant Sci. Lett.* **23,** 81–88.

THOMAS E., BRIGHT S.W.J., FRANKLIN J., LANCASTER V.A., MIFLIN B.J. & GIBSON R. (1982) Variation amongst protoplast-derived potato plants (*Solanum tuberosum* cv. 'Maris Bard'). *Theor. appl. Genet.* **62,** 65–68.

THOMAS E., KING P.J. & POTRYKUS I. (1979) Improvement of crop plants via single cells *in vitro*—an assessment. *Z. PflZücht.* **82**, 1–30.

WEINDLING R. (1946) Microbial antagonism and disease control. *Soil Sci.* **61**, 23–30.

WELLS H.D., BELL D.K. & JAWORSKI C.A. (1972) Efficacy of *Trichoderma harzianum* as a biocontrol for *Sclerotium rolfsii*. *Phytopathology*, **62**, 442–447.

WENZEL G., SCHIEDER O., PRZEWOZNY T., SOPORY S.K. & MELCHERS G. (1979) Comparison of single cell culture-derived *Solanum tuberosum* L. plants and a model for their application in breeding programs. *Theor. appl. Genet.* **55**, 49–55.

WERNICKE W. & BRETTELL R. (1980) Somatic embryogenesis from *Sorghum bicolor* leaves. *Nature*, **287**, 138–139.

VINCENT J.M. (1970) *A Manual for the Practical Study of the Root-nodule Bacteria*. IBP Handbook No. 15. Blackwell Scientific Publications, Oxford.

ZAENEN I., VAN LAREBEKE N., TEUCHY H., VAN MONTAGU M. & SCHELL J. (1974) Supercoiled circular DNA in crown gall inducing *Agrobacterium* strains. *J. Mol. Biol.* **86**, 109–127.

10   Chemical Engineering and Biotechnology

G. HAMER

10.1   Introduction

The predominant interaction between chemical engineering and biotechnology involves the research, development, design, construction
and operation of microbiological production processes for the manufacture of products of enhanced value relative to the feedstocks employed.
Biotechnology, as is the case for all other technologies, is a composite of
several disciplines, and can be appropriately oriented by the relative
weights given to the composite disciplines. However, one major problem
that still faces biotechnology in its relationship with chemical
engineering is a clear definition of what should constitute the activity.
As far as microbiological processes are concerned, two important
orientations exist, i.e. the approach based on biochemical engineering
and the approach based on microbial physiology. The former orientation, although frequently favoured by chemical engineering edu-
cators, has largely failed, over the past twenty-five years, to provide the
necessary means for overall microbiological process improvement that
can be compared with the successes achieved by the application of
chemical engineering principles to both bulk chemicals manufacture
and petroleum refining. Part of the lack of success must, of course, be
attributed to the conservative traditions of some microbiological
process industries, their divergent and fragmented patterns of development and, for the most part, their relatively small scale of operation. For
biochemical engineering to achieve an independent and justifiable
existence, it must provide much more effective techniques for solving
microbiological process and processing problems, which are, in turn,
reflected by significant economic benefits. The orientation of biotechnology, where the emphasis is on microbial physiology, admits that the
individual factors affecting the overall optimization of microbiological
production processes are best handled by individual specialists, i.e.
process-oriented microbiologists and chemical engineers; the latter
need not necessarily have a specific knowledge of microbiological
processes. This orientation requires that the process microbiologist or
biotechnologist, in addition to having an in-depth knowledge and wide
experience of the physiological problems pertinent to various microbiological processes, must also have the appropriate knowledge and skill to
communicate and interact effectively with chemical engineers. Effective interaction between the biological scientist and the chemical
engineering practitioner will provide the key for both technological
innovation in and commercial success of the microbiological process
industries in future decades. The primary objective of this chapter is to
contribute towards more effective interaction, and thereby enhance the

ultimate economic success of both established and novel biotechnological production processes.

Many of the established biotechnological industries are strictly traditional in nature, having served mankind for many centuries. Amongst such industries are those for the production of alcoholic beverages, of fermented dairy products, of fermented soya-bean products and of fermented fish products. In such industries, as is common in many traditional areas of processing technology, product quality and value are frequently assessed on the subtleties of either tastes or odours rather than on some more easily definable quality criterion. In the future, similar quality criteria will continue to be of importance with respect to many traditional products, but new products are most unlikely to be valued on such a basis, and market success will depend predominantly on product functionality, on overall production costs, and on possible uncontrollable variations in production costs, i.e. fluctuations in feedstock and energy costs.

Biotechnological manufacturing ventures are usually initiated by ill defined statements of need rather than by the discovery that some novel compound is produced by some newly isolated microbial culture. Examples of problem statements that have resulted in major microbiological process research and development programmes are as follows:

(1) from the late 1960s onwards, both Europe and Japan would experience a critical shortage of protein-rich ingredients for animal feeds unless alternative sources of supply could be developed;

(2) either resource depletion or production restrictions would, from the mid-1970s onwards, result in fluid fossil fuel price increases such that the production of fluid fuels from alternative indigenous resources would become essential in most developed and developing countries;

(3) the future recovery of a significant part of the proven reserves of crude oil, in strategically non-vulnerable regions, will depend on the development of novel chemicals for enhanced oil recovery.

The first example cited was clearly responsible for the extensive research and development programmes directed towards single cell protein (SCP) production; the second was responsible for the establishment of extensive programmes aimed at the conversion of renewable, indigenous biomass resources into ethanol and/or methane; and the third, after the evaluation and rejection of synthetic polymers, was responsible for the establishment of product and process research and development programmes directed towards the production of biopolymers with appropriate properties for use in enhanced oil recovery. The commercially successful outcome of all these programmes has either depended on or depends on both the long-term validity of the particular statement of need and on the absence of political decisions that completely alter the basic assumptions supporting the original statement of need.

Three examples of microbiological products based on the concept of novel compound discovery, but divorced from any justifiable statements of need, are the proposals for the production of poly-β-hydroxybutyric acid for plastics production using *Alcaligenes eutrophus*, of emulsifying agents, using numerous hydrocarbon-oxidizing bacteria, and of microbial insecticides, using *Bacillus thuringiensis*.

The essential interactions between the biotechnologist and the chemical engineer are best examined against the background of major products and major potential products from industrial microbiological processes. In chemicals manufacture, products are described as fine chemical products, as intermediate volume products and as bulk chemical products. In general, the approach adopted with respect to the design of production plants for the manufacture of the three categories of product varies markedly. Fine chemicals are usually produced in batch reactors, which are also used for the production of a variety of similar products. Production from single plants usually falls within the 100 kg/a to 100 t/a range, and usually a very significant fraction of the production costs are involved in product purification and testing, so as to meet the demanding product quality specifications usually required of such products. Vaccines, vitamins, 5'-nucleotides, some amino acids, some antibiotics and some enzymes for medical applications are microbiological products in the fine chemicals category of product classification. Intermediate volume chemicals are usually chemicals that are produced in quantities of between 100 t/a and 20 000 t/a. Such products frequently have less rigorous quality specifications than do fine chemical products and are usually manufactured in product-specific plants, but not necessarily using continuous flow processes. Amongst the intermediate volume chemicals that are produced microbiologically are glutamic acid, which is widely used as a flavour enhancer, antibiotics for the protection of agricultural crops, enzymes for industrial use, organic acids, particularly citric, lactic and gluconic acid and, in earlier times, solvents such as acetone and butanol. Many fermented food products and fermented beverages can also be included in the intermediate volume category. Finally, in the bulk product sector one has those products that are produced, usually in continuous flow process plants, where single plant capacity exceeds 20 000 t/a and where the products are marketed on the basis of commodities and overall product performance criteria rather than on the basis of rigid quality specifications. The microbiological products that fall within the bulk product category are SCP, fuels such as biogas (methane) and fuel alcohol (ethanol) and biopolymers for enhanced oil recovery. In addition, one must, of course, consider water from what are the largest continuous flow microbiological processes of all—biological sewage and wastewater treatment plants—as a bulk product, even though the treated product is usually of indirect rather than direct economic value but, nevertheless, of major importance to the welfare and prosperity of communities. The necessity for interaction between biotechnology and chemical engineering increases with the scale of production but, even in the case of fine chemicals production, the interaction should still be significant if effective process design and operation are to be achieved.

The primary objective of chemical engineering practice is the creation of processing systems which economically transform raw materials or feedstocks, by using energy and scientific and technological knowledge, into marketable products. Biotechnological processes comprise an ordered collection of equipment, which effects the transformation of feedstocks into products by biochemical reactions, phase transitions, agglomeration, separation, extraction, drying, etc.

Manufacturing ventures seek to generate profit by risking capital investment. In the situation where a microbial process route is proposed as a possible solution to a problem, it is necessary to evaluate the route relative to alternative process routes and, in some cases, also relative to other sources of production, such as agriculture in the case of proteinaceous ingredients for animal feeds.

The economic evaluation of industrial processes and process routes is an essential activity that is fraught with difficulty. Evaluation procedures tend, for the most part, to be conservative with respect to the claims of proponents of either the process or process route undergoing evaluation. However, excessive stringency with respect to the evaluation of prospective manufacturing ventures can result in the loss of profitable manufacturing ventures to competitors.

Estimates of the total investment for any proposed manufacturing venture can be divided into three parts, according to the degree of financial risk:

(1) the fixed investment in the immediate processing area, i.e. investment inside battery limits;

(2) the investment in auxilliary services, i.e. investment outside battery limits;

(3) the investment in working capital.

Together, they comprise the money risked in the venture. The fixed investment carries the highest degree of risk, the working capital the least.

Manufacturing costs are the costs incurred in operating a process plant and can also be divided into three parts:

(1) costs proportional to the fixed investment, i.e. factors such as labour and material aspects of maintenance, insurance and administration, which are independent of the rate of production and can be expressed as a percentage of the fixed investment;

(2) costs of raw materials, utilities, chemicals, other materials, quality control, maintenance costs related to operation, royalties and licence fees, which depend on the rate of production;

(3) direct costs of maintaining the operation labour force, including, in addition to salaries, overheads and the costs of supervision.

The gross profit obtained as a result of operating a particular plant is the difference between the net income from the annual sales, after distribution, promotional and sales costs have been deducted, and the annual manufacturing costs. The net profit is the expected annual return on the investment after the deduction of depreciation and taxes. Depreciation schedules are frequently complex, government-stipulated and location-dependent and, frequently, tax incentives demand optimization of permitted depreciation schedules.

When evaluating a process with a view to improving its profitability by integration and optimization, it is essential to understand the technological factors that significantly affect the overall economics of manufacture. In the case of microbiological processes, the technological factors can be of either a microbiological or a process engineering nature. However, before discussing these factors, it is first appropriate to clarify what is understood by process integration.

Traditional process design practice considers the optimization of individual unit operations with little effort devoted to their integration as an overall process. For a decade, much of the world has been subjected to a series of steep increases in the price of crude oil, the primary energy source for the process industries, and of feedstocks derived from oil. Prior to these politically mediated price increases, the bulk product processes industries have always based their strategy on the premise that the cost of products will decrease with increasing scale of production. Escalation in both fuel and feedstock prices and the effects of associated inflation on plant construction costs has come close to destroying this principle and the need for saving in both capital investment and expenditure on energy and feedstocks by improved levels of process integration has become increasingly evident. Process integration can involve both the unit operations that comprise those parts of the process within battery limits and interactions between operations undertaken outside battery limits and those conducted within battery limits. Here, emphasis will be placed on the former interactions, but prior to embarking on further discussion of these, there are a few comments, concerning the latter, which are particularly appropriate to industrial microbiological processes and how they find difficulty in fitting into production sites developed primarily for bulk chemicals production by conventional technology.

Industrial microbiological processes are essentially low temperature processes, with reactor operating temperatures usually less than 40 °C, which require large quantities of both low temperature cooling water and high quality process water, significant quantities of clean steam, large quantities of clean air, feedstock and product storage facilities that minimize contamination of the stored materials, hygienic surroundings, specialized wastewater treatment facilities and sophisticated quality control laboratories. Typical production sites for bulk chemicals manufacture, particularly petrochemicals manufacture, are unlikely to be easily able to satisfy such requirements, thereby forcing an unusually large part of the essential service facilities to be constructed essentially inside battery limits with attendant increases in both the capital investment and the investment risk. The overall economics of microbiological process operation are obviously enhanced at locations where several such processes can receive appropriate services from optimized central facilities located outside the battery limits of the individual processes.

In commercial microbiological processes, the microbiological factors with the greatest impact on the process are as follows:

(1) yield coefficients for the product or products;
(2) growth rates and/or product production rates;
(3) affinity of the production culture for carbon energy substrates;
(4) stability and fastidiousness of the production culture.

The process engineering factors with the greatest impact on the process are as follows:

(1) feedstock conversion and conversion coefficients;
(2) productivity;
(3) product concentrations.

Both the microbiological and the process engineering factors are intimately related in technologically and commercially successful processes.

The concepts of and methodology for profitability studies have been discussed with respect to chemicals manufacture by Rudd & Watson (1968). When evaluating any new production process, it is essential to understand the technical factors that significantly affect the overall economics of manufacture.

10.2   Microbial factors affecting process performance and economics

In any microbiological process, the microbial culture that is employed is of key importance. Moulds, yeasts, actinomycetes, bacteria and algae have all found application in industrial microbiological processes, either as mono- or as mixed cultures. Traditional fermentation processes have tended to favour mixed cultures growing under either open or protected conditions, whilst most modern fermentation processes have employed mono-cultures growing under aseptic conditions. Whilst it is possible to make generalizations with respect to ideal process cultures, it must, of course, be understood that the perfect process microorganism does not exist and that the diversity of products that can be produced by microorganisms is a function of the diversity of microorganisms and their substrates. However, in the future, the construction of ideal process microorganisms using genetic engineering techniques might be possible, thereby invalidating the preceding statement. Most process cultures that are in use today have been obtained from natural sources, but have been subjected to improvement by either growth under specific process conditions for biomass and primary metabolic production or conventional mutational techniques for secondary metabolite production.

At this point, it is useful to summarize the equations that describe microbial growth, particularly the growth of unicellular microorganisms, in fermenters and other microbiological reactors. The essential requirements for the growth of any microbial culture are as follows:

(1) a viable inoculum;
(2) an energy source and a carbon source;
(3) essential nutrients for biomass synthesis;
(4) an absence of growth inhibitors;
(5) suitable physicochemical conditions.

Provided that all the above requirements are satisfied, the rate of increase in microbial biomass concentration for unicellular microorganisms undergoing binary fission in a well mixed batch culture system will be proportional to the microbial biomass concentration, i.e.

$$\frac{dx}{dt} = \mu x \qquad (1)$$

where dx/dt is the growth rate, μ is the proportionality constant, usually described as the specific growth rate constant, and x is the biomass concentration on a dry weight basis. Under conditions where μ remains constant, equation 1 can be rearranged and integrated to give

$$\ln x = \ln x_0 + \mu t \tag{2}$$

where x_0 is the microbial biomass concentration at zero time and t is time. Equation 2 may be rewritten as

$$\ln(x/x_0) = \mu t \tag{3}$$

Hence,

$$x = x_0 e^{\mu t} \tag{4}$$

When $\ln x$ is plotted against time a straight line with slope μ results. Such growth is described as either exponential or logarithmic growth, and the qualifying conditions for such growth are that both the microbial biomass composition and the environmental conditions remain constant. The above equations apply to completely mixed culture systems where unicellular microorganisms are both discretely and uniformly dispersed throughout the growth medium.

Biomass doubling time is widely used for the description of microbial growth by binary fission in batch culture systems. The relationship between the specific growth constant μ and the doubling time, t_d, of the biomass can be obtained from equation 3 by substituting $2x_0$ for x and t_d for t so that

$$t_d = \frac{\ln.2}{\mu} \tag{5}$$

10.2.1 *Yield coefficients*

In any quantitative assessment of growth and/or product formation, it is essential to link the formation of microbial biomass and of products with the utilization of substrates and nutrients. The concepts of growth yield and yield coefficients were introduced by Monod (1942) in order to describe the relationship between microbial growth, on a dry weight basis, and the consumption of carbon energy substrate. In mathematical terms the yield coefficient for microbial biomass production, $Y_{x/s}$, can be expressed as

$$Y_{x/s} = -\frac{dx}{ds} \tag{6}$$

where ds is the infinitely small decrease in substrate concentration corresponding to an infinitely small increase in microbial biomass concentration, dx. The negative sign indicates that x and s vary in opposite senses. Under constant growth conditions, the yield coefficient remains constant so that if x_0 and s_0 represent initial conditions and x and s conditions at time t during microbial growth,

$$(x - x_0) = Y_{x/s} (s_0 - s) \tag{7}$$

Generally, the yield coefficient is presented as a constant, but is, in fact, variable and dependent on the environmental conditions in the culture and on variations therein. From the process point of view, yield coefficient variability is extremely important and yield coefficients must, of course, be optimized. The yield coefficient concept can be extended to all the various nutrients utilized by growing microorganisms and to products produced during microbial growth. In addition, the concept of molar growth yields and of yields based on the available electrons in substrates and on the total energy uptake are widely used in the assessment of the relative efficiencies of the biochemical pathways involved in microbial growth and metabolism.

As far as the growth of microorganisms on highly soluble substrates is concerned, the yield coefficient and the conversion coefficient are essentially identical, but in the case of slightly soluble and immiscible substrates, the two coefficients can often be markedly different. The conversion coefficient can be defined as the microbial biomass produced per unit weight of substrate supplied.

When the yield coefficient concept is applied to growth associated product formation, analogous equations to 6 and 7 can be used to describe the yield of product with respect to substrate utilization, i.e.

$$Y_{p/s} = -\frac{dp}{ds} \tag{8}$$

and

$$(p - p_0) = Y_{p/s} (s_0 - s) \tag{9}$$

where $Y_{p/s}$ is the product yield coefficient, dp is an infinitely small increase in product concentration and p_0 and p are the initial product concentration and the product concentration after time t during microbial growth, respectively.

Essentially, the microbial growth process involves the utilization of a carbon substrate and an energy substrate, which in the case of heterotrophic microorganisms is either a single compound or a mixture of carbonaceous compounds that serve the dual needs of the microorganisms. Autotrophic microorganisms utilize separate carbon and energy substrates. For heterotrophic microorganisms both the biomass and product yield coefficients depend on the distribution of the carbon energy substrate utilized between the anabolic requirements and catabolic requirements of the microorganisms. Microbial biomass, product and energy formation are shown in Fig. 10.1. The nature of both the carbon energy substrate and of other essential nutrients and the biochemical capabilities of the microbial cells affect the yield coefficients and, in some cases where the carbon energy substrate inhibits

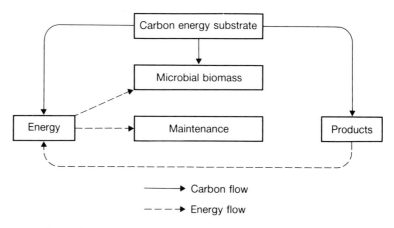

Fig. 10.1. Carbon and energy flows during heterotrophic microbial growth and product formation.

growth, carbon energy substrate concentration also markedly affects the yield coefficients. For carbon energy substrates that contain a high fraction of carbon relative to other elements and, hence, a high energy constant, an imbalance occurs, during aerobic microbial oxidation, between the energy produced and the carbon required for assimilation into microbial biomass, and a limiting maximum biomass yield coefficient results.

Hydrocarbons are carbon energy substrates that fall within this latter category. The excess energy produced from such substrates is dissipated as heat and can present significant problems with respect to the cooling of microbial reactors.

The rate of consumption of particular substrates by growing cultures is frequently expressed as

$$-\frac{ds}{dt} = qx \tag{10}$$

where q is the metabolic quotient. Metabolic quotients, particularly those for oxygen and for carbon energy substrates, are widely used by microbiologists. Substituting for x in equation 10 from equation 1,

$$-\frac{dx}{ds} = \frac{\mu}{q} \tag{11}$$

Therefore, from equation 6,

$$q = \frac{\mu}{Y_{x/s}} \tag{12}$$

10.2.2   *Growth and product formation kinetics*

The exponential growth model described by equation 1 represents only one of several relatively widely accepted microbial growth models, each of which applies to either a specific physical situation or a type of biomass morphology. Amongst these alternative growth models are the linear growth model and the cube-root growth model. The former model applies to either substrate or nutrient diffusion limited growth where the substrate or nutrient is in either a second liquid phase or the gaseous phase in the culture system and the latter model is frequently applied to the growth of mould pellets and bacterial flocs, where it is assumed that growth occurs only in the peripheral zone of either the pellet or the floc. The model is, in fact, a combination of the exponential growth model and mass transfer limitation. The cube-root growth model successfully describes the growth of mycelial pellets in batch culture, but is of little value in describing the continuous culture of bacterial flocs, because it fails to take into account floc instability and disruption as a result of shear forces within bioreactors.

The occurrence of exponential growth in batch cultures shows that the growth rate is virtually unaffected by substrate concentration over a wide range of substrates and substrate concentrations. In fact, the growth process exhibits zero order kinetics. During the growth of microorganisms, it is realistic to expect that substrate consumption will follow enzyme kinetics such that

$$q = \frac{q_m\, s}{(s + K_s)} \tag{13}$$

where q_m is the maximum value of the metabolic quotient, q, when $s \gg K_s$, and K_s is a saturation constant, equivalent to the Michaelis–Menten constant, which is applicable for enzyme reactions. By substituting for q and q_m from equation 12, we obtain the well known Monod relationship:

$$\mu = \frac{\mu_m\, s}{(s + K_s)} \tag{14}$$

The form of the Monod relationship is shown in Fig. 10.2. The Monod relationship is so widely used more because of its mathematically

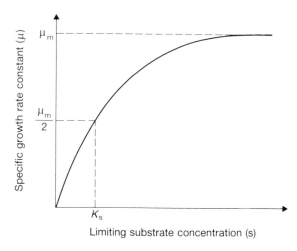

Fig. 10.2. The Monod relationship, which describes the effect of limiting substrate concentration on the specific growth rate constant.

tractability than because it provides a universal description of the growth kinetics of microorganisms. In order to evaluate μ_m and K_s, which is inversely proportional to the affinity of the microorganisms for the particular substrate, equation 14 can be rearranged as

$$\frac{1}{\mu} = \frac{K_s}{s\mu_m} + \frac{1}{\mu_m} \tag{15}$$

Hence, a plot of $1/\mu$ against $1/s$, i.e. a Lineweaver–Burk plot, will give a straight line with an intercept on the abscissa at $-1/K_s$ and an intercept on the ordinate at $1/\mu_m$. Only few data are available with respect to values of the saturation constant, K_s. For carbon energy substrates and for mineral nutrients, K_s values are usually $c.\ 10^{-5}$ M and, for oxygen, are usually between 10^{-6} and 10^{-5} M.

As far as the production of microbial biomass is concerned, the three equations that must be linked with equations describing the flow characteristics of microbial reactors are:

$$\frac{dx}{dt} = \mu x \tag{1}$$

$$\mu = \frac{\mu_m s}{(s + K_s)} \tag{14}$$

$$(x - x_0) = Y_{x/s}\,(s_0 - s) \tag{7}$$

In the case of growth associated product formation, equation 9, i.e.

$$(p - p_0) = Y_{p/s}\,(s_0 - s) \tag{9}$$

must also be used. However, by no means all microbial product
formation is growth associated and it is, therefore, essential to define
first the actual relationship between product formation and growth,
before application of the kinetic equations to microbial reactors.

Probably the most useful classification with respect to microbial
product formation in batch culture systems was that proposed some
twenty-five years ago by Gaden (1959). Essentially, this classification
divides microbial product production processes into three groups,
designated Types I to III for convenience.

Type I processes are those in which the main product appears as a
result of primary energy metabolism, where an essentially constant
stoichiometric relationship between substrate consumption and product formation exists, where metabolic routes are primarily serial and
where each rate process exhibits a single maximum which is essentially
coincident with that of other rate processes. Typical examples of Type I
processes include those for the production of ethanol, gluconic acid and
lactic acid from glucose and, of course, microbial biomass production in
general.

Type II processes are those in which the main product is formed
indirectly from reactions of energy metabolism and where the product
results from either some side reactions or subsequent interaction
between direct metabolic products. Reaction rates are complex and the
various rate processes do not exhibit coincident maxima. Typical
examples of Type II processes include those for the production of citric
and itaconic acids and for some amino acids.

Type III processes are those in which the main product is independently elaborated by the microbes and does not arise directly from
energy metabolism. In such processes, growth and metabolic activity
reach a maximum early in the batch process cycle and it is not until a
later stage, when oxidative activity is low, that maximum desired
product formation occurs. Products produced by such processes are
described as secondary metabolites and examples include the antibiotics, penicillin and streptomycin.

The three process types are shown diagrammatically in Fig. 10.3.
Obviously, adoption of such a classification is only for technical
convenience and classification in this way can be neither perfect nor
comprehensive. Particular processes may exhibit widely different
behaviour patterns with major changes in medium composition and
process conditions. However, strain variations seem to have little effect
on the general rate patterns. Exceptions are found, especially in Type
III processes, suggesting that, with increasing complexity of the

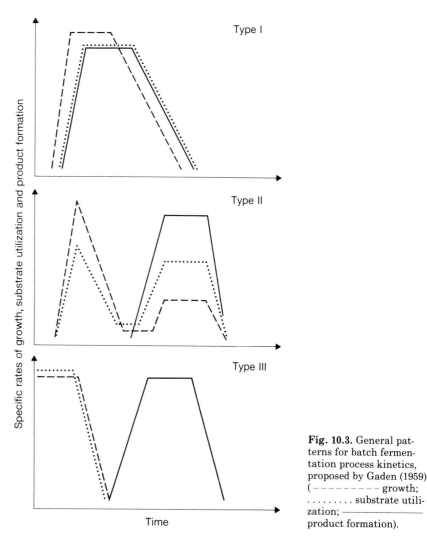

Fig. 10.3. General patterns for batch fermentation process kinetics, proposed by Gaden (1959) (– – – – – – – growth; . . . . substrate utilization; ————— product formation).

biosynthetic reactions involved, there is justification to subdivide this general type further.

Turning to continuous flow microbiological processes, kinetic information is essential to allow process operation under optimized steady-state conditions. For processes exhibiting Type I kinetics during batch operation, continuous flow operating conditions can be theoretically determined directly from batch process data. Essentially, processes of this type are best performed under steady-state conditions, in a single-stage, completely mixed bioreactor. However, those microbial processes that exhibit either Type II or Type III kinetic behaviour for batch operation are probably best performed in multiple bioreactors, operating in series, when the process is operated under continuous flow conditions. In such systems, process conditions can be manipulated so as to allow maximum growth and associated energy metabolism in one stage and maximum product formation in a subsequent stage. Prior to the conversion of such processes from batch to continuous flow operation, elucidation of the relationships between the various process rates and major controllable process variables is essential.

At present, several processes exhibiting Type I kinetic behaviour for batch operation are operated in the continuous flow mode, but few processes exhibiting either Type II or Type III kinetics for batch operation are operated continuously, although fed-batch type operation has become increasingly common for these latter types of processes in recent years.

10.2.3   *Substrate affinity and microbial growth*

The questions of residual feedstock (substrate) concentrations, of the competitive ability of microbes and of growth on mixed carbon energy substrates are all significant as far as culture performance in biotechnological processes is concerned. When the first question is considered, two important aspects emerge: the economic question of complete substrate utilization, and product quality contamination/safety questions concerned with the association of residual substrate with the product. Essentially, it is the affinity of the microbes for the particular substrate that controls the completeness of utilization under conditions where that particular substrate is limiting growth, either in batch or during continuous flow culture.

Probably the best known model for enzyme kinetics is the Michaelis–Menten model, which represents the reaction of an enzyme with a substrate as

$$E + S \underset{k_{-1}}{\overset{k_{+1}}{\rightleftharpoons}} E\text{-}S \overset{k_2}{\rightarrow} E + P \tag{16}$$

where E is the enzyme, S the substrate, E–S the enzyme–substrate complex, P the product, k_{+1} the forward reaction rate constant for the first reaction, k_{-1} the reverse reaction rate constant for the first reaction, and k_2 the reaction rate constant for the second reaction, which is considered irreversible. The rate of change of the enzyme–substrate concentration can be expressed as

$$\frac{dc}{dt} = k_{+1}(e - c)s - k_{-1}c - k_2c \tag{17}$$

where e is the enzyme concentration, c the enzyme–substrate complex concentration, and s the substrate concentration. If it is assumed that $s \gg e$, for a steady state,

$$c = \frac{es}{\left(\dfrac{k_{-1}+k_2}{k_{+1}}\right) + s} \tag{18}$$

The rate of product formation, v, for the enzymic reaction shown as equation 16 is

$$\mathrm{v} = k_2 c = \frac{k_2 es}{[(k_{-1}+k_2)/k_{+1}] + s}$$

$$= \frac{\mathrm{v_{max}}s}{K_\mathrm{s} + (k_2/k_{+1}) + s} = \frac{\mathrm{v_{max}}s}{K_\mathrm{m} + s} \tag{19}$$

where $\mathrm{v_{max}}$ is the maximum rate of production, i.e. when all the enzyme forms the enzyme–substrate complex, K_s is the equilibrium constant for the dissociation of the enzyme–substrate complex and $K_\mathrm{m} = K_\mathrm{s} + k_2/k_{+1}$, i.e. the Michaelis–Menten constant. When the rate of product formation is controlled by k_2, i.e.

$$k_2 \ll k_{+1'}$$

$$K_\mathrm{m} = K_\mathrm{s} \tag{20}$$

When the condition expressed in equation 20 applies, the value of K_m is inversely proportional to the affinity of the enzyme for substrate. The smaller the value of K_m, the greater will be the affinity. In the Monod relationship, which is derived from and is analogous with the Michaelis–Menten type enzyme kinetics expression, μ, μ_m and K_s are substituted for v, $\mathrm{v_{max}}$ and K_m, respectively, in equation 19, giving equation 14, which was introduced earlier. The form of this relationship was shown in Fig. 10.2. Microbes that exhibit a very steep curve in their μ–s diagram have a high affinity for a growth limiting substrate, whilst those that exhibit a less steep curve in their μ–s diagram have a lower affinity, and higher residual substrate concentrations can be predicted.

In a chemostat type continuous flow culture system, it is generally the affinity, rather than the maximum specific growth rate constant, μ_m, that controls competition between two microbial species for a single growth limiting substrate. For steady-state operation of a simple chemostat, with no biomass recycle, the limiting substrate concentration, $\bar{s}$, is given by

$$\bar{s} = \frac{K_\mathrm{s}D}{(\mu_\mathrm{m} - D)} \tag{21}$$

where D is the overall dilution rate. It is clear from equation 21 that the steady-state limiting substrate concentration, $\bar{s}$, is dependent on the growth rate that corresponds directly to the operating dilution rate and on the growth characteristics of the particular microbe. However, the growth characteristics of various microbes growing on the same substrate are most unlikely to be identical. In fact, as far as binary microbial cultures are concerned, two distinct possibilities exist when the specific growth rate constant, μ, is plotted against limiting substrate concentration, $\bar{s}$, and these are illustrated in Fig. 10.4, (i) and (ii), where the two microbial species comprising the binary culture are designated A and B. For the situation illustrated in Fig. 10.4 (i), species A will, at any given dilution rate (growth rate), always maintain a lower limiting substrate concentration than that maintained by species B at the same dilution rate (growth rate). Hence, in a chemostat where such a binary microbial culture is growing under limitation by the same substrate, species A will always outcompete species B, because species B is unable to maintain a sufficiently high growth rate at the substrate concentration maintained by species A, even though the maximum specific growth rate for species B has not been exceeded. In contrast, if the characteristics of the species A and B follow the pattern depicted in Fig. 10.4 (ii), the relative competitiveness of the two species, when they comprise a binary mixture in a chemostat with a common limiting substrate, will be dilution rate dependent. Such an analysis of competitive growth in chemostats can be extended to multi-component mixed microbial populations, but the wisdom of doing this is questionable in view of the several other interactions, which are likely to occur in complex mixed cultures. Further, where mixed cultures are used in microbial processes, such cultures usually comprise only one species capable of utilizing the principal substrate, whilst the several other species present modify the environment for the optimum growth of the primary species, particularly under conditions where the primary species tends to be somewhat fastidious.

In general, any particular heterotrophic microbe is able to grow on a diversity of carbon energy substrates. Most carbohydrates are utilized by a wide spectrum of microbes, hydrocarbons are utilized by a more restricted spectrum, and some synthetic organic compounds are utilized by a very restricted spectrum, whilst others do not even support microbial growth. One of the major questions concerning the utilization of multiple carbon energy substrates by an individual heterotrophic strain of microbe is whether mixed carbon energy substrates are utilized either concurrently or sequentially.

When a pure culture of microbes is grown, in batch mode, on a

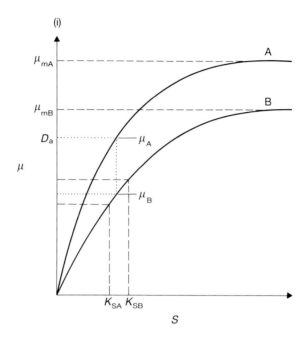

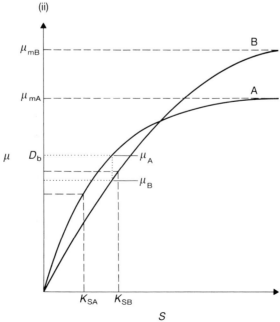

Fig. 10.4. The Monod relationships for a binary culture comprising microbes A and B and a single limiting substrate where: (i) $\mu_{mA} > \mu_{mB}$ and $K_{sA} < K_{sB}$; (ii) $\mu_{mA} < \mu_{mB}$ and $K_{sA} < K_{sB}$. In chemostat culture, at both dilution rates, D_a and D_b, microbe A will be retained and microbe B will be washed out (redrawn from Harder *et al.*, 1977).

mixture of two carbon energy substrates, the best known response is that of diauxic growth, i.e. the growth curve exhibits two distinct exponential phases, separated by a phase during which the growth rate passes through a minimum, caused by the fact that the two substrates are utilized sequentially. Although diauxic growth was first reported for sugars by Monod (1942), this pattern of growth is not restricted to sugars, and other carbon energy substrates have been shown to support diauxic growth. Diauxic growth is characterized by a failure on the part of microbes to synthesize the enzymes necessary for the metabolism of the second carbon substrate available for utilization, i.e. induction is prevented even in the presence of the inducer. This phenomenon is described as catabolite repression.

Other responses with respect to binary carbon energy substrate utilization in batch cultures can also be observed. Substrates do not necessarily interfere with the utilization of each other; when this is the case, they can be metabolized simultaneously and even growth stimulation is possible under such circumstances. In other situations, adaptation to the second substrate is not prevented, but its utilization is inhibited, resulting in a sequential pattern of utilization but not, strictly speaking, diauxic growth. For binary substrate mixtures that are used sequentially, it is usual for the more rapidly utilized substrate to be used first, although exceptions to this pattern do exist. However, for a particular binary substrate mixture, the more rapidly utilized substrate can vary with respect to the microbial species growing on the binary mixture.

In the context of mixed substrates, it is important to note that many culture media used for commercial-scale production of bio-technological products contain complex mixtures of substrates whose composition affects process performance. Behaviour with mixed substrates in continuous flow culture is not necessarily evident from the results of batch experiments with the same substrate mixture and microbe, and simultaneous utilization of the components of binary substrate mixtures can occur during continuous flow operation, even though the same binary mixture is utilized sequentially by the same microbe in batch culture. Particularly interesting are observations in chemostat culture that the critical dilution rate for complete utilization of the member of a binary pair of substrates supporting the lower growth rate in batch culture can be significantly extended beyond normal wash-out when present in a binary substrate mixture. In addition, distinct possibilities exist, with binary substrate mixtures in continuous culture, for manipulating fluxes of carbon over the various available metabolic pathways and to channel substrates into energy producing, assimilative or product-forming routes by the selection of either appropriate operating conditions or substrate mixture compositions.

10.2.4 *Fastidiousness and culture stability*

The overall suitability of a selected culture for application in a commercial process is a feature that is frequently ignored and, hence, is a factor of critical importance when the time comes for process scale-up. Obviously, in the overall context, genetic stability is paramount, although outside the scope of this chapter. Here, physical, chemical and nutritional factors that affect culture stability and the fastidiousness of microbes will be briefly considered. Earlier, the essential requirements for growth were listed, but the question always arises as to the suitability of the selected conditions for optimum growth and/or product formation, and effects that changes in growth conditions might promote, particularly with respect to process productivity.

Heat and mass transfer in bioreactors are frequently enhanced by the dissipation of energy in the system, usually as a result of intense mechanical agitation by rotating impellers. The shearing action of rotating impellers is twofold: to disperse air in the form of fine bubbles throughout the growth medium and to promote adequate mixing such that concentration gradients are minimized, reactor cooling is effective and bubble escape minimized. In addition to these essential requirements, the shearing action of impellers also tends to damage the cells employed, and such damage is particularly pronounced in the case of filamentous moulds, actinomycetes, bacteria with appendages such as flagella and fimbriae and, in the extreme case, where fragile tissue cells are grown in suspension culture.

Mould morphology can be important with respect to specific product formation and, in such cases, strains where morphology is significantly affected by culture conditions are obviously unsuitable for commercial process application. In addition, process conditions that injure mould mycelium are also best avoided.

As far as damage to bacterial cells is concerned, impeller tip speeds have always been considered to correlate with damage. However, the most plausible hypotheses, with respect to the cause of damage, suggest that cavitational type phenomena in the vortexes immediately behind the blades of rotating impellers are responsible. The question of damage to bacterial cells in intensely agitated culture systems remains a vexed one. If one considers the flow characteristics of discretely dispersed bacterial cell suspensions from the theoretical point of view, using the concept of equivalent spherical diameter for determination of the hydrodynamic characteristics of an individual bacterium, effects of sufficient magnitude to cause mechanical damage to the bacterium cannot be verified. However, such an approach does not take into consideration hydrodynamic effects on either fimbriae or flagella, which, particularly in the case of flagella, can have lengths many times the major dimension of the bacterium to which they belong, and it is entirely plausible that the stripping of such appendages and subsequent leakage of cell contents represents the adverse action of intense shear on bacterial cultures. It is interesting to note that a significant portion of genetic engineering research is designed to transfer specific characteristics from other bacteria into *Escherichia coli*, a bacterium which suffers with respect to its suitability for high intensity processes for the reasons discussed above, i.e. the presence of fimbriae and flagella and, hence, susceptibility to mechanical damage.

Whereas in the case of moulds and bacteria it is probably possible to select more robust microbes that are better suited to the conditions encountered in intense processes, such an option is unavailable with respect to tissue cell culture, particularly mammalian cell culture, where appropriate bioreactors have to be designed to accommodate the intrinsic fragility of the cells.

As far as other environmental conditions are concerned, microbes that exhibit broad plateaux with respect to their pH, dissolved oxygen and temperature optima are obviously more suited for process application than microbes that are fastidious with respect to such conditions, although improved process instrumentation and control potentially reduces the disadvantages of fastidious microbes in these respects. However, some newer designs of commercial-scale bioreactors, particularly column and pressure-cycle systems, subject the process microbes to significant gradients in both physical parameters and concentrations of nutrients, and if the advantages of such systems over more conventional systems, with respect to the non-biological aspects of process economics are sufficiently great, robust microbes adapted to high levels of performance under rapidly changing physical and chemical conditions are essential if good overall process economics are to be achieved.

Finally, it should be pointed out that the application of defined mixed cultures, with their inherent lack of fastidiousness, for commercial biotechnological processes, is an option that has frequently been ignored in recent times. Even so, with mixed culture processes, gross changes in process conditions can create imbalances with respect to the proportions of the constituent microbial species comprising mixed populations.

10.3   Process engineering factors affecting process performance and economics

Any microbial process, irrespective of its mode of operation, can be classified according to its purpose under one of two headings:

(1) production of microorganisms;

(2) achieving some desired chemical transformation.

As far as the latter category is concerned it can be further sub-divided:

(i) desired end-product formation;

(ii) decomposition of feedstock.

All chemical transformation processes have one major feature in common that sharply differentiates them from processes where the objective is to produce microbial biomass. This feature is that in chemical transformation processes, attendant growth of the microbes that carry out the particular transformation is not an absolute requirement. The importance of this latter fact is frequently underestimated, even though it has been common knowledge for many years that chemical transformations can be effected by microbial cells that are neither growing nor even viable. The transformations result from the capacities of the enzymes present in the cells and, obviously, the most controlled way of achieving many transformations and that giving potentially the highest yields is the use of either immobilized resting cells or immobilized enzymes extracted from the cells. Such systems are proving to be important for some industrial-scale transformations but, in many cases, problems still exist with respect to the active life of such biocatalytic systems, cofactor regeneration, and effective means for overcoming the intrinsic diffusional resistance that many proposed systems exhibit. Chemical transformation by growing microbial cultures is still the most commonly applied technology at the industrial scale. An important biochemical distinction in microbial processes is that between catabolic processes, where complex compounds are broken down into simpler compounds, and biosynthetic processes, where simple compounds are built up to form complex compounds.

Although the most essential feature of all microbially mediated processes is either a reaction or a series of reactions, it is unusual for the kinetics of either the reaction or the series of reactions concerned to be directly responsible for the rate-controlling step in the process, when the process is carried out on an industrial scale. Generally, it is some other, apparently less significant feature, which is rate controlling and, hence, critical with respect to any commercial exploitation of the process concerned.

In a bioreactor, productivity and conversion are the key factors that affect process economics but, in themselves, they are both in turn controlled by those physical phenomena occurring in the bioreactor that affect both mass and heat transfer capacities. Such phenomena include the following:

    (1) hydrodynamic properties of suspended microbes;
    (2) rheological properties of culture fluids;
    (3) electrokinetic properties of microbes;
    (4) pressure effects;
    (5) surface, wall and interfacial effects;
    (6) multi-phase flow effects;
    (7) effects in foam, froth and spray environments;
    (8) flotation, sedimentation and segregation effects.

It is interesting to speculate how little process oriented microbiological research provides realistic information on any of these phenomena and/or their effects on either microbial growth or product formation.

10.3.1   *Bioreactor classification and productivity*

Chemical reactions can be classified in many ways, but probably the most useful form of classification as far as the satisfactory design of reactors is concerned, is that based on whether the reaction occurs in either a single-phase or a multiphase environment. Reactions occurring in a single phase are termed homogeneous reactions, whilst those occurring in multiple phases are described as heterogeneous reactions. As far as most enzymic and many microbial reactions are concerned, such a classification is by no means clear-cut, as such reactions fall within a grey area between homogeneous and heterogeneous reactions. In order to resolve this lack of distinction in the case of biologically mediated reactions, a convention has been adopted that describes reaction systems where only micro-scale concentration gradients are evident as essentially homogeneous, and reaction systems where macro-scale concentration gradients exist as heterogeneous. Hence, for biological reactions, the concepts of homogeneity and heterogeneity are inconsistent with the way in which these same concepts are usually applied to chemical reactions.

Any system for which boundaries can be defined, and in which biochemical reactions occur, can be described as a bioreactor. As far as industrial-scale biological processes are concerned, the bioreactor is the process vessel in which growth and/or transformations occur.

Industrial-scale bioreactors can be operated in the batch, fed-batch, semi-continuous (semi-batch) and continuous flow modes. Historically, industry has favoured the batch mode of operation for transformations and the semi-continuous mode of operation for microbial biomass production. More recently, transformations have frequently been carried out in fed-batch bioreactors, whilst continuous flow operation is now favoured for microbial biomass production. Traditionally, the only major large-scale uses of bioreactors operating in the continuous flow mode were for aerobic wastewater and sewage treatment, processes with the greatest throughput of all processing operations, and vinegar manufacture. With the exception of these two examples, the biological industries have been remarkably conservative with respect to the adoption of continuous flow process technology, frequently citing unsubstantiated reasons as the basis for their decisions.

Detailed examination of the several potential modes of bioreactor operation clearly indicates the potential superiority of the continuous flow mode of operation. Essentially batch operation involves placing all the necessary materials, with the exception of oxygen for aerobic processes, in the reactor at the start of the operation and then allowing the processes to proceed until completion, when the product is harvested. Fed-batch operation allows either intermittent or continuous addition of a substrate to the reactor, but with no provision for product withdrawal, so that harvesting again occurs only at the end of process operation. Semi-continuous operation, which is also described as either semi-batch or fill-and-draw operation, involves an operating procedure where, at the completion of an initial batch operation, the reactor is half emptied, to allow partial product harvesting, and then refilled with an equal charge of fresh medium, and the process allowed to go to completion, when the reactor is again subjected to the same sequence of operations, so that an approach is made towards better process plant utilization.

An interesting comparison between the overall rate of production, i.e. the productivity, in batch and completely mixed continuous flow operation can be made for the same bioreactor operating in each of the two modes for microbial biomass production. Batch operation comprises a series of time periods for preparation, growth lag, exponential growth and product harvesting, of which only the exponential growth period is productive as far as the bioreactor is concerned. The time required for the entire batch cycle, t_c, can be expressed as the sum of the productive time, t, and the total non-productive time, t_n, i.e.

$$t_c = t + t_n \tag{22}$$

From equation 3, where for exponential growth $\mu = \mu_m$,

$$t = \frac{1}{\mu_m} \ln\left(\frac{x}{x_0}\right) \tag{23}$$

so that

$$t_c = \frac{1}{\mu_m} \ln\left(\frac{x}{x_0}\right) + t_n \tag{24}$$

For batch operation, overall productivity of microbial biomass is defined as the total production divided by the total cycle time. Total production is expressed by equation 7 but, as far as such batch processes are concerned, the residual substrate concentration, s, is virtually zero. Therefore, total production can be expressed as

$$(x - x_0) = Y_{x/s}s_0 \tag{25}$$

Hence, the productivity of the batch production process, P_b, is

$$P_b = \frac{Y_{x/s}\, s_0\, \mu_m}{\ln\left(\dfrac{x}{x_0}\right) + t_n\, \mu_m} \tag{26}$$

By definition, the productivity of microbial biomass in a completely mixed continuous flow bioreactor, P_c, can be expressed as

$$P_c = D\bar{x} \tag{27}$$

where $\bar{x}$ is the steady-state biomass concentration in the bioreactor effluent. Assuming that both the yield coefficient, $Y_{x/s}$, and the substrate concentration in the fresh medium, s_0, are identical for both batch and continuous operation, and $\bar{s}$ is the residual limiting substrate concentration, equation 27 becomes

$$P_c = D\, Y_{x/s}\, (s_0 - \bar{s}) \tag{28}$$

which, provided $\bar{s} \ll s_0$, reduces to

$$P_c = D\, Y_{x/s}\, s_0 \tag{29}$$

Therefore, the ratio of productivities is

$$\frac{P_c}{P_b} = \frac{D}{\mu_m}\ [\ln(x/x_0) + t_n\, \mu_m] \tag{30}$$

In a completely mixed continuous flow bioreactor, the maximum value of the dilution rate occurs when $D = \mu_m$. However, to prevent problems of instability in continuous flow bioreactors, it is accepted practice to operate at a dilution rate value of $c.\ 0.8\ \mu_m$. On the basis of this assumption,

$$\frac{P_c}{P_b} = 0.8\ [\ln(x/x_0) + t_n\mu_m] \tag{31}$$

The comparison between productivities in batch and in completely mixed continuous culture is illustrated in graphical form, for the parameters listed, in Fig. 10.5.

The operating characteristics of continuous flow bioreactors are best evaluated on the basis of material balance calculations with respect to microbial biomass production, growth limiting substrate utilization and product formation. As far as any gross classification of continuous flow bioreactors employing suspended microbes is concerned, essentially two extreme types exist, i.e. completely mixed stirred tanks and plug flow type bioreactors. Completely mixed stirred tank type bioreactors can be operated either in a chemostatic or in a turbidostatic manner. Essentially, chemostatic operation involves maintenance of the microbial culture density by exhaustion of either a

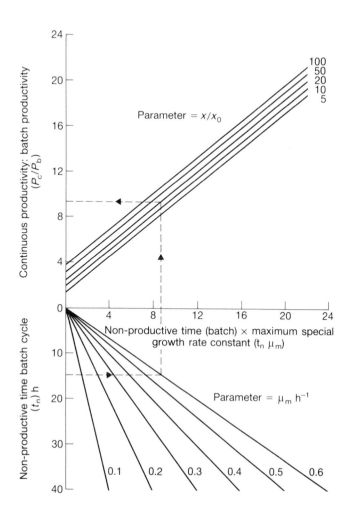

Fig. 10.5. The relationship between microbial biomass productivity in continuous flow and batch systems for various inoculum sizes, maximum specific growth rate constants and non-productive batch cycle times, when $D = 0.8 \mu_m$ in the continuous flow system (adapted from Aiba et al., 1973).

limiting substrate or a limiting nutrient, whilst turbidostatic operation involves maintenance of a constant microbial culture density by optical density measurement control of the medium in the reactor. As far as industrial application is concerned, chemostatic operation is preferred and turbidostatic operation will not be further discussed.

Any analysis of bioreactor performance involves both an understanding of the flow characteristics of the reactor and the kinetics for the growth and transformation processes occurring in the reactor. The flow characteristics of continuous flow reactors are most conveniently described by the residence time distribution of material passing though the reactor. In the case of a perfect plug flow reactor, all material passing through the reactor does so in an orderly manner such that no mixing occurs between any given element of fluid and material that entered the reactor either before or after the element under consideration, i.e. no axial mixing occurs. Thus, for the steady-state operation of a plug flow reactor, all fluid entering the reactor has an identical residence time within the reactor. The mean residence time is given by

$$\tau = \frac{V}{F} = \frac{1}{D} \tag{32}$$

where τ is the mean residence time, V the volume of fluid in the reactor, and F the volumetric flow of fluid into and out of the reactor. In contrast, the continuous flow stirred tank type of reactor is sufficiently intensely agitated so that the reactor contents are entirely uniform, with the consequence that the outflow from the reactor is of identical composition to the fluid within the reactor. Because the feed to the reactor is perfectly mixed with the reactor contents immediately on entering the reactor, it passes through the reactor with a range of residence times. However, the mean residence time, τ, for fluid in the reactor, is also given by equation 32, as it is for plug flow reactors, even though the residence time distribution curves for the two reactor types are entirely different.

The wash-out curves for the two reactor types are also very different from each other. If we consider each type of reactor to be filled with fluid in which inert, non-segregating particles are both discretely and uniformly distributed prior to the initiation of a flow of clear fluid into the reactor and an identical volumetric flow out of each reactor, the quantity of particles in each reactor will be

$$Q_0 = Vx_0 \tag{33}$$

where Q_0 is the quantity, V, the fluid volume and x_0 the initial concentration. Upon initiation of the flow of clear fluid into the reactor, the effluent that maintains a constant fluid volume in the reactor will comprise fluid and particles, the concentration of which will be different for the two different reactor types. In addition, the quantity, Q, of particles remaining in the reactors will steadily reduce.

For an ideal plug flow reactor, the concentration of particles in the reactor effluent, x, will be equal to x_0 until the elapsed time, t, is equal to the residence time, τ, but immediately $t = \tau$, x will become zero. The quantity, Q, of particles remaining in the reactor will decrease linearly with time, until the elapse time $t = \tau$, when no particles will remain in the reactor. For real plug flow reactors, axial mixing and diffusion modify the patterns described.

For a completely mixed stirred tank type reactor, the concentration of particles in the reactor effluent will always be equal to the concentration of particles, x, in the fluid in the reactor. The wash-out curve for such reactors will be of the negative exponential form, as is the residence time distribution curve, and the rates of change of particle concentration and quantity in the reactor can be expressed as

$$-\frac{dx}{dt} = -\frac{dQ}{dt} = Dx \tag{34}$$

Using the wash-out relationships discussed above, and combining them with equations for microbial growth, product formation and yield coefficients, allows the development of material balance equations that describe ideal continuous flow bioreactor operation. Any expression of a material balance for a continuous flow bioreactor involves the establishment of an expression where the accumulation of the particular material under consideration in the bioreactor is expressed in terms of an increase due to inflow in the feed stream, a decrease due to outflow in the effluent stream, an increase due to either growth or production in the reactor, a decrease due to utilization in the reactor and an increase due to recycle, as applicable.

The material balance expressions for a continuous flow stirred tank bioreactor with no recycle, i.e. a simple chemostat, as shown in Fig. 10.6, where microbial biomass is the only product, are as follows.

For microbial biomass,

Accumulation = Growth – Output

or, expressed in symbols for an infinitely short time interval, dt,

$$(V)dx = (V\mu x)dt - (Fx)dt \tag{35}$$

For limiting substrate,

Accumulation = Input – Output – Utilization

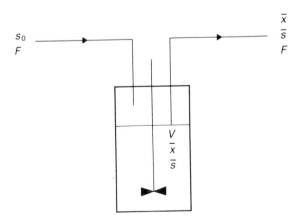

Fig. 10.6. Continuous flow, completely mixed, stirred tank bioreactor (simple chemostat) showing flows and concentrations (symbols are defined in the text).

or for an infinitely short time interval, dt,

$$(V)ds = (Fs_0)dt - (Fs)dt - \left(\frac{V\mu x}{Y_{x/s}}\right)dt \qquad (36)$$

where V is the constant liquid volume in the bioreactor, F the flow rate into and out of the bioreactor, $Y_{x/s}$ the yield coefficient for microbial biomass from the limiting substrate, x the microbial biomass concentration in the bioreactor, s_0 and s the concentration of limiting substrate in the feed to the bioreactor and in the bioreactor, respectively, and μ the specific growth rate constant.

Dividing equations 35 and 36 by $(V)dt$:

$$\frac{dx}{dt} = (\mu - D)x \qquad (37)$$

and

$$\frac{ds}{dt} = D(s_0 - s) - \frac{\mu x}{Y_{x/s}} \qquad (38)$$

For steady-state operation, both $dx/dt = 0$ and $ds/dt = 0$. Therefore,

$$(\mu - D)\bar{x} = 0 \qquad (39)$$

or

$$\mu = D \qquad (40)$$

and

$$D(s_0 - \bar{s}) - \frac{\mu \bar{x}}{Y_{x/s}} = 0 \qquad (41)$$

where $\bar{x}$ and $\bar{s}$ are the steady-state concentrations of microbial biomass and growth limiting substrate, respectively.

The Monod relationship can be rewritten as

$$\mu = \frac{\mu_m \bar{s}}{(\bar{s} + K_s)} \tag{42}$$

The steady-state limiting substrate concentration, $\bar{s}$, may be obtained by substituting D for μ in equation 42, so that

$$\bar{s} = \frac{K_s D}{(\mu_m - D)} \tag{21}$$

and the steady-state microbial biomass concentration, $\bar{x}$, may be obtained by first substituting D for μ in equation 41 and then substituting for $\bar{s}$ from equation 21, so that

$$\bar{x} = Y_{x/s}\left[s_0 - \frac{K_s D}{(\mu_m - D)} \right] \tag{43}$$

The steady-state expression for microbial biomass productivity, P_c, for the system under consideration is

$$P_c = D\bar{x} \tag{27}$$

or, substituting for $\bar{x}$,

$$P_c = DY_{x/s}\left[s_0 - \frac{K_s D}{(\mu_m - D)} \right] \tag{44}$$

The relationships between $\bar{x}$, $\bar{s}$ and P_c and dilution rate D, according to equations 43, 21 and 27 are shown in Fig. 10.7, for the values of μ_m, $Y_{x/s}$ and s_0 given.

The second system for which material balance expressions will be developed comprises a completely mixed continuous flow bioreactor, producing only microbial biomass, to which a part of a concentrated stream of microbial biomass is recycled from a biomass concentration device on the bioreactor outlet, such as a sedimentation tank, as used in wastewater treatment, a centrifuge or an ultra-filtration system, in order to enhance process intensity in the bioreactor. In such a system, illustrated in Fig. 10.8, the overall dilution rate, D, for the system is equal to F/V where F is the liquid flow rate through the entire system, and V is the constant liquid volume in the bioreactor. The outflow from the bioreactor, but not from the whole system, can be expressed as

$$F_s = F + aF_s \tag{45}$$

or

$$F_s = \frac{F}{(1 - a)} \tag{46}$$

where a is the fraction of the bioreactor outflow that is recycled and F_s is the total outflow from the bioreactor. If it is assumed that the microbial biomass is concentrated by a factor, g, the microbial biomass recycled to the bioreactor will be $aF_s\,gx$.

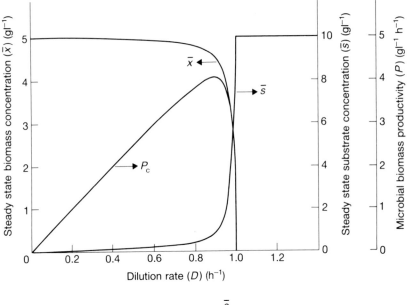

Fig. 10.7. Theoretical relationships between the steady-state microbial biomass concentration, the steady-state limiting substrate concentration, the microbial biomass productivity and the dilution rate for a simple chemostat ($\mu_m = 1.0\ \mathrm{h}^{-1}$, $K_s = 0.1\ \mathrm{gl}^{-1}$, $Y_{x/s} = 0.5$, and $s_0 = 10\ \mathrm{gl}^{-1}$)

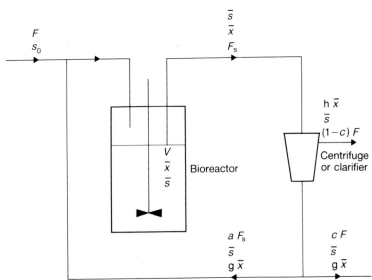

Fig. 10.8. Continuous flow, completely mixed, stirred tank bioreactor with a centrifuge or clarifier and recycle of a concentrated microbial biomass (symbols are defined in the text).

The bioreactor material balance for biomass is

$$Accumulation = Growth - Output + Recycle$$

or, in symbols, for an infinitely short time interval, dt,

$$(V)dx = (V\mu x)dt - (F_s x)dt + (aF_s gx)dt \tag{47}$$

The bioreactor material balance for limiting substrate is

$$Accumulation = Input + Recycle - Output - Utilization$$

or, for an infinitely short time interval, dt,

$$(V)ds = (Fs_0)dt + (aF_s)sdt - (F_s s)dt - \left(\frac{V\mu x}{Y_{x/s}}\right)dt \tag{48}$$

After substituting for F_s from equation 46 and dividing by $(V)dt$, equation 47 can be written, for steady-state conditions where $dx/dt = 0$, as

$$\left[\mu - \frac{(1 - ag)D}{(1 - a)}\right]\bar{x} = 0 \tag{49}$$

or

$$\mu = \frac{(1 - ag)D}{(1 - a)} \tag{50}$$

Assuming Monod type kinetics,

$$\frac{\mu_m \bar{s}}{(\bar{s} - K_s)} = \frac{(1 - ag)D}{(1 - a)} \tag{51}$$

or

$$\bar{s} = \frac{DK_s (1 - ag)}{\mu_m (1 - a) - D (1 - ag)} \tag{52}$$

After substituting for F_s from equation 46 and dividing by $(V)dt$, equation 48 reduces, for steady-state conditions where $ds/dt = 0$, to the equations for a simple chemostat, i.e. equation 41. Substituting equation 52 into equation 41,

$$\bar{x} = \frac{Y_{x/s} (s_0 - s) (1 - a)}{(1 - ag)} \tag{53}$$

In order to determine the concentration of microbial biomass, hx, in the clarified stream leaving the separator employed in the above system, it is necessary to construct a material balance for microbial biomass around the separator. Assuming steady-state operation, i.e. no overall accumulation of biomass in the separator, the material balance for biomass can be written as

$$Input = Recycle + Concentrated\ waste + Clarified\ waste$$

or, in symbols,

$$F_s\bar{x} = aF_sg\bar{x} + cFg\bar{x} + (1 - c)\ Fh\bar{x} \tag{54}$$

where c is the fraction of flow F leaving the system as the concentrated waste stream, and h the concentration factor for microbial biomass in the clarified waste stream leaving the separator.

Substituting for F_s from equation 46,

$$h = \frac{(1 - ag) - cg(1 - a)}{(1 - a)(1 - c)} \tag{55}$$

For conditions where the bioreactor is operating under steady-state conditions, the biomass productivity for the bioreactor (not the system), P'_c, will be

$$P'_c = \frac{F_s\bar{x}}{V} - \frac{aF_sg\bar{x}}{V} \tag{56}$$

or, substituting for F_s,

$$P'_c = \frac{D\bar{x}(1 - ag)}{(1 - a)} \tag{57}$$

The relationships between both the productivity, P'_c, the steady-state microbial biomass concentration, $\bar{x}$, the steady-state limiting substrate concentration, $\bar{s}$, and dilution rate are shown, for the parameters indicated, in Fig. 10.9, and clearly demonstrate the potential extension of operating dilution rate that results from either recycle, as discussed here, or active biomass retention.

Turning to the analysis of ideal plug flow bioreactors, we have the situation in which essentially exponential growth ($\mu = \mu_m$) occurs in each small element of fluid, dv, passing through the bioreactor, provided the first limiting substrate concentration significantly exceeds the value of the saturation constant, K_s. The expression for growth in an ideal plug flow bioreactor with no biomass recycle is

$$\ln x = \ln x_0 + \mu_m t \tag{58}$$

where x_0 is the microbial biomass concentration in the fluid entering the bioreactor and x is the concentration after time, t. If v is the volume of culture fluid leaving the bioreactor in time, t, then $t = v/F = v/VD$. Substituting for t in equation 57 gives

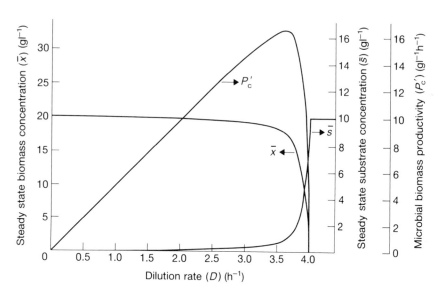

Fig. 10.9. Theoretical relations between the steady-state microbial biomass concentration, the steady-state limiting substrate concentration, the bioreactor microbial biomass productivity and the dilution rate for a chemostat with microbial biomass concentration and recycle ($\mu_m = 1.0$ h^{-1}, $K_s = 0.1$ gl^{-1}, $Y_{x/s} = 0.5$, $s_0 = 10$ gl^{-1}, $a = 0.2$ and g = 4).

$$\ln x = \ln x_0 + \frac{v\mu_m}{DV} \tag{59}$$

When microbial biomass recycle is incorporated with an ideal plug flow bioreactor, no external source of inoculum is required. In such a system the total flow out of the bioreactor, but not the system, is given by equation 46, as for the previously discussed bioreactor with recycle. If the microbial biomass is concentrated in the separator by a factor, g, then the fraction of the microbial biomass recycled to the bioreactor will be ag. Under steady-state conditions, where x_w is the microbial biomass concentration leaving the bioreactor, $x_0 = agx_w$. The time, t, for a volume of culture, v, to flow out of the bioreactor is given by

$$t = \frac{v(1-a)}{F} \tag{60}$$

By putting $v = V_e$, where exhaustion of the growth limiting substrate occurs, and substituting for x and t in equation 59, for steady-state operation,

$$V_e = \frac{F}{(1-a)\mu_m} \left(\ln \frac{1}{ag}\right) \tag{61}$$

When $V_e = V$, the bioreactor liquid volume, equation 61 becomes

$$\frac{1}{D} = \frac{1}{(1-a)\mu_m} \left(\ln \frac{1}{ag}\right) \tag{62}$$

The maximum value of x_w, x_m, is given by

$$x_m = Ys_0 + x_0 \tag{63}$$

As far as suspended growth bioreactors are concerned, ideal plug flow systems are rarely realized in practice. The vast majority of bioreactors more closely approach completely mixed systems, certainly as far as the liquid phase is concerned, although the assumption of complete mixing should never be made until possibilities of segregation, in the form of wall growth, etc., have been fully examined. Probably the nearest approach to ideal plug flow that can be achieved for suspended growth bioreactors is a cascade of completely mixed stirred tank bioreactors in series which have no intermediate feed addition points. However, only when the number of bioreactors comprising the cascade is infinite, is ideal plug flow behaviour achieved. Non-ideal plug flow behaviour is achieved with more than six bioreactors in series.

10.3.2 *Dimensional analysis and scale-up*

Dimensional analysis and modelling with the aid of dimensionless groups play an important role in chemical engineering and the use of such techniques is equally important for bioreactor design and scale-up. Few microbiologists are familiar with either dimensional analysis or dimensionless groups. Dimensional analysis is primarily a method by which partial information may be obtained about the relationships which apply between the variables that characterize definite physical systems. The advantage of such an approach is that it can be applied when only partial knowledge concerning a system is available, and the partial information obtained is frequently of great value in restricting the experiments needed to generate complete information.

When one considers, for example, the physical absorption of oxygen in a sparged bioreactor, it is clear that if one could establish complicated hydrodynamic equations that govern the motion of the gaseous and liquid phases and diffusion in the liquid phase, a complete description of the system, from the physical point of view, could be obtained. The amount of oxygen absorbed would obviously be a highly complicated function of the gas velocity and of all the other variables which enter the equations by which motion is determined. The determination of the exact form of the relationship between velocity and the other variables is hopelessly complicated, but dimensionless analysis can be used to define certain important connections that must be satisfied by any such relationship. This is possible because scientific systems of measurement satisfy the requirement that the number which expresses the ratio of the measurements of two objects is independent of the size of the fundamental units, and systems of scientific equations are expressed in a form that makes them independent of the size of the particular system of units employed.

When approaching specific processing problems, chemical engineers frequently start with a broad-brush analysis, leaving out as much of the fine structure of the problem as is considered to be allowable. Such analyses frequently result in the situation where the primary character-istics of a reactor system can be described in terms of a simple relationship between either two or three dimensionless groups and appropriate constants of an empirical nature. Nearly three hundred dimensionless groups of potential interest to process engineering exist, but most biotechnologists will only encounter some twenty on a regular basis.

The most widely used dimensionless group is the Reynolds number, N_{Re}, which is used in the description of fluid flow and is

$$N_{Re} = \frac{L \, u \, \rho}{\eta} \tag{64}$$

where L is a characteristic dimension, u a velocity term, ρ the fluid density and η the fluid viscosity, all in consistent units. In bioreactor design, Reynolds numbers for flow in pipes, for drops, particles and bubbles, and for rotating impellers are encountered. For fluid flow in a pipe,

$$N_{Re} = \frac{d \, u \, \rho}{\eta} \tag{65}$$

where d is the pipe diameter, and u the flow velocity. For particles, bubbles or drops, either rising or sinking in a fluid,

$$N_{Re} = \frac{d_e U \rho_c}{\eta_c} \tag{66}$$

where d_e is the equivalent spherical diameter of the particle, bubble or drop, i.e. the diameter of a sphere of equal volume to that of the particle, bubble or drop, U either the free rising or sinking (falling) velocity of the particle, bubble or drop, ρ_c the density of the continuous phase, and η_c the viscosity of the continuous phase. For rotating impellers,

$$N_{Re} = \frac{D_i^2 N \rho_c}{\eta_c} \tag{67}$$

where D_i is the impeller diameter, and N the rotational speed of the impeller.

Some other dimensionless groups commonly used by chemical engineers are, the Nusselt number, N_{Nu}, the Prandtl number, N_{Pr}, the Grashof number, N_{Gr}, the Froude number, N_{Fr}, the Péclet number, N_{Pe}, the Stanton number, N_{St}, the Schmidt number, N_{Sc}, the Sherwood number, N_{Sh}, the Weber number, N_{We} and the power number, N_P. Just as with the Reynolds number, some of these other numbers, e.g. the Weber number, have multiple definitions. Others, such as the Péclet, Prandtl and Nusselt numbers have analogous forms that apply either to heat or to mass transfer. It should also be noted that some dimensionless groups such as, for example, the Stanton number, N_{St}, are combinations of others and some have more than one name.

To provide an example of the use of dimensionless groups, one can look at the question of heat transfer. For conditions where only natural convection occurs, the velocity is dependent solely on buoyancy effects, represented by the Grashof number, and the Reynolds number can be omitted from dimensionless equations describing the transfer process. However, for conditions where forced convection occurs, the effects of buoyancy effects are usually negligible and the Grashof number can be omitted. Thus the type of equation used to describe natural convection is of the form

$$N_{Nu} = f(N_{Gr}, N_{Pr}) \qquad (68)$$

and for forced convection, of the form

$$N_{Nu} = f(N_{Re}, N_{Pr}) \qquad (69)$$

Scale-up concerns problems associated with transferring data obtained in laboratory and pilot-plant scale equipment to large-scale commercial production equipment. Equipment for biotechnological industries can only be designed and operated effectively if scale-up technology is fully appreciated. Obviously the question of scale-up applies to all items of process plant equipment used in biotechnological processes, including heat exchangers, centrifuges, evaporators, dryers, etc. but here emphasis will be placed only on bioreactors.

Material objects and physical systems are characterized by three qualities: size, shape and composition. All three of these qualities are independently variable, so that two objects may either differ in size but have the same shape and composition, or be alike only in shape and have entirely different sizes and compositions. The principle of similarity concerns the relationships between physical systems of different sizes and is fundamental to either scaling up or scaling down of processes and equipment. This principle states that the spatial and temporal configuration of a system is determined by ratios of magnitudes within the systems and is independent of both the size and the nature of the units in which the magnitudes are measured. The principle of similarity is usually coupled, and frequently confused, with the technique of dimensional analysis, even though they are logically quite distinct. The former is a general principle of nature, whilst the latter is one technique by which the former can be applied to specific cases.

The biotechnologist is concerned with complex systems comprising solid bodies and fluids in which the transfer of both energy and matter takes place and reactions occur; for such systems, the principle of similarity is particularly concerned with the concept of shape. When the concept of shape is applied to complex biotechnological systems, it involves, in addition to geometrical proportions, factors such as fluid-flow patterns, temperature gradients, time–concentration profiles, etc. Systems exhibiting the same configuration in either one or more of these respects can be considered to be similar.

Similarity can be defined in two ways, by specifying the ratios either of different measurements in the same body, i.e. intrinsic proportions, or of corresponding measurements in different bodies, i.e. scale ratios. Geometrical similarity is best defined in terms of correspondence and scale ratios, but similarity with respect to variables such as velocity, force or temperature are usually best defined by a single intrinsic ratio for each system. These intrinsic ratios are the dimensionless groups which define similarity under different conditions.

The four similarity states that are important in biotechnological process systems are (1) geometrical similarity, (2) mechanical similarity, (3) thermal similarity, and (4) chemical similarity.

Each of the states is dependent on the previous states listed. Complete chemical similarity requires geometrical, mechanical and thermal similarity, but it is frequently necessary to accept an approximation to chemical similarity when, for example, substantial divergences in mechanical similarity occur. All actual cases of similarity involve an element of approximation because disturbing factors, which prevent ideal similarity from being attained, are always encountered in real systems. As far as fluid flow is concerned in systems fabricated to have geometrically similar dimensions, it is essentially impossible to ensure that the dimensions of surface roughness are geometrically similar, and such factors exhibit significant affects on flow character-istics.

Considering the four states of similarity listed above, geometrical similarity exists between two bodies when for every point in the one body a corresponding point exists in the other. However, it is also possible for each point in the first body to have more than one corresponding point in the second body and it is also not necessary for the scale ratio to be the same for all axes; the relation between two bodies where such differences in scale ratio exist is described as one of distorted similarity.

Mechanical similarity comprises static, kinematic and dynamic similarity, and each of these can be regarded as an extension of the concept of geometrical similarity to either stationary or moving systems subjected to forces. Static similarity is primarily concerned with the deformation of structures and is of little direct interest to biotechnologists. However, both kinematic similarity and dynamic similarity are of direct interest and concern systems in motion. Whereas geometrical similarity involves rectangular coordinates, kinematic similarity introduces an additional dimension—time. Geometrically similar moving systems are described as being kinematically similar when corresponding particles trace out geometrically similar paths in corresponding time intervals. When two geometrically similar fluid systems are kinematically similar, then flow patterns are geometrically similar and mass and heat transfer rates in the two systems bear a simple relationship to each other. Kinematic similarity in fluids entails both geometrically similar eddy systems and geometrically similar streamline boundary films. Dynamic similarity concerns forces (gravitational, centrifugal, etc.) which either accelerate or retard moving masses in dynamic systems. In either fluid systems or systems composed of discretely dispersed particles, kinematic similarity implies dynamic similarity, since the motions of such systems are functions of the applied forces. Dynamically similar systems are defined as geometrically similar moving systems where the ratios of all corresponding forces are equal. In fluid flow, dynamic similarity is of direct importance when it is required to predict either pressure drops or power consumption, but in mass and heat transfer, it is only of indirect importance as a means of establishing kinematic similarity.

Thermal similarity is concerned with process systems where there is a flow of heat and, hence, introduces the dimension of temperature to those of length, force and time in the concept of similarity. Heat can flow from one point in a system to another by radiation, conduction, convection and bulk movement of matter through the action of a pressure gradient. For the first three processes to occur, a temperature difference is necessary, but the fourth process depends on flow patterns in the system where such a transfer process is occurring. Thermal similarity occurs in geometrically similar systems when corresponding temperature differences bear a constant ratio to each other and, when the systems are in motion, they are also kinematically similar. Thermal similarity is of obvious importance in the sterilization of bioreactors, particularly with respect to heat damage of labile material, and in questions of bioreactor cooling.

Chemical similarity is concerned with reacting systems in which either composition varies from point to point or in the case of batch and cyclic processes, with respect to time. No new fundamental dimension is introduced into the similarity concept, but there are either one or more concentration parameters, depending upon the number of independently variable chemical constituents in respect of which similarity is to be established. Chemical similarity occurs in geometrically and thermally similar systems when corresponding concentration differences bear a constant ratio to each other and when the systems, if moving, are kinematically similar. The intrinsic criteria which define chemical similarity, in addition to those required for kinematic and thermal similarity, are the rate of product formation relative to either the rate of bulk flow or to the rate of molecular diffusion.

The mechanical, thermal or chemical similarity between geometrically similar systems is specified in terms of criteria which are intrinsic ratios of measurements, forces or rates within the systems. Since these criteria are ratios of like quantities, they are dimensionless, and can be derived by dimensional analysis.

Although dimensionless similarity criteria are ratios of physical quantities which are functions of either the various forces or resistances which control rates, in real systems there are frequently multiple controlling factors. For example, resistance to fluid motion may result from viscous drag, gravitational forces or surface tension for which the corresponding dimensionless groups are the Reynolds, Froude and Weber numbers, respectively. For the simplest case of homologous systems of different absolute magnitudes, these three criteria are incompatible, since each one requires the fluid velocity to vary as a different function of the linear dimensions, i.e. for equal Reynolds numbers, velocity $\propto$ length^{-1}, for equal Froude numbers, velocity $\propto$ length$^{0.5}$ and for equal Weber numbers, velocity $\propto$ length$^{-0.5}$. In scaling up a complex biotechnological process, it would be advantageous to select conditions such that the rate of the overall process depends predominantly upon one dimensionless criterion, but in such systems, at least two dimensionless criteria are always involved.

For impeller agitated bioreactors, where air is being dispersed throughout a microbial growth medium, it is frequently the power consumption for mixing at different scales of operation that is the key economic factor. The generalized dimensionless equation for two-phase fluid flow in impeller agitated vessels is

$$N_P = f(N_{Re}, N_{Fr}, N_{We}) \tag{70}$$

or

$$\frac{P}{\rho_c N^3 D^5_i} = f\left(\frac{D^2_i N \rho_c}{\eta_c}, \; \frac{D_i N^2}{g}, \; \frac{D_i^3 N^2 \rho_c}{\sigma}\right) \tag{71}$$

where P is the power consumption, ρ_c the continuous phase density, N the rotational rate of the impeller, D_i the impeller diameter, η_c the viscosity of the continuous phase, g the gravitational constant and σ the surface tension.

For gas–liquid systems the effects of the Reynolds and Weber numbers in the fluid-flow equation can, with fair accuracy, be represented by power functions of these numbers, although the exponents are not constant over the entire operating range. Further, the Froude number only plays a role when vortexing occurs at the liquid surface, and in impeller agitated bioreactors, vortexing is largely eliminated because of baffles in the reactor. Hence, for impeller agitated, aerated bioreactors,

$$N_p = f(N_{Re})^\alpha (N_{We})^\gamma \tag{72}$$

where the exponents α and γ vary according to operating conditions, particularly the flow regime.

Dimensional analysis of the mass transfer coefficient in aerated, agitated, geometrically similar bioreactors gives

$$N_{Sh} = f(N_{Re})^\alpha (N_{Sc})^\gamma \tag{73}$$

or

$$\frac{k_L D_i}{\mathscr{D}} = f\left(\frac{D^2_i N \rho_c}{\eta_c}\right)^\alpha \left(\frac{\eta_c}{\rho_c D}\right)^\gamma \tag{74}$$

where k_L is the mass transfer coefficient, and $\mathscr{D}$ the molecular diffusivity for oxygen in the liquid (continuous) phase. It does not seem unreasonable to accept that equations 72 and 73 are applicable to impeller agitated bioreactors but, even so, it must be emphasized that difficulties in assessing the effects that agitation has on metabolic activity are by no means adequately understood, and these might well invalidate the approach discussed. The impeller tip velocity, $\pi N D_i$, determines the maximum shear rate in impeller agitated, aerated bioreactors and, without doubt, influences potential damage to viable cells and, therefore, particularly in the case of processes employing either filamentous or flocculating microbes, is obviously of critical importance. In scaling up biotechnological processes, there remains a need for infinitely greater understanding of the relationships between mechanical and physical factors on the one hand, and biological factors on the other. Until this need is satisfied, a high degree of uncertainty with respect to bioreactor scale-up will remain.

10.3.3   *Oxygen transfer*

The efficient transfer of oxygen in commercial- and technical-scale aerobic microbiological processes represents one of the major challenges to chemical engineers concerned with bioreactor design. Irrespective of whether an aerobic microbiological process is operated in the batch, semi-continuous or continuous flow mode, oxygen must be continuously supplied to the process if acceptable productivities are to be achieved. The oxygen requirement of an aerobic culture is controlled by the concentration of microbes in the reactor, their rate of growth, and the appropriate yield coefficient. Other considerations, such as a requirement to strip carbon dioxide from the culture, can also become important in specific situations, particularly in processes where oxygen has been substituted for air as the source of oxygen. However, in the vast majority of aerobic microbiological processes, air is used as the source of oxygen. In the typical temperature range for microbiological process operation, 10–40 °C, and at atmospheric pressure, equilibrium dissolved oxygen concentrations, which result from contacting aqueous microbiological growth medium with air, are extremely low when compared with dissolved concentrations of almost any other major nutrient or substrate unless it is the nutrient or substrate that is limiting growth in continuous flow chemostatic operation. Hence, in aerobic microbiological processes, oxygen limitation frequently occurs and reductions in productivity and either off-specification product or by-product formation result. In most aerobic microbiological processes, the costs incurred for oxygen transfer comprise a significant fraction of total operating expenditure.

The solubility of a gas in a liquid has, when equilibrium is attained, a limited and definite value which depends on the nature of both the gas and the liquid involved and on the temperature and the pressure of the system. In the case of gas mixtures, such as air, it is the partial pressure of each component of the gas mixture which affects their individual solubilities. When other factors remain constant, an increase in temperature results in a decrease in solubility, whilst an increase in pressure (or partial pressure) results in an increase in solubility. The quantitative relationship between solubility and pressure is known as Henry's Law, which states that, at constant temperature, the mass of gas dissolved by a given volume of liquid is proportional to the pressure of the gas with which the liquid is in equilibrium. Most microbiological growth media are dilute aqueous solutions and, for simplicity, can be considered to be similar to water. The concentration of oxygen dissolved in water in equilibrium with air, i.e. at saturation, at constant temperature and constant total pressure, is

$$x_{O_2} = \frac{p_{O_2}}{H_{O_2}} \tag{75}$$

where x_{O_2} is the dissolved oxygen concentration, p_{O_2} is the partial pressure of oxygen in air, and H_{O_2}, the Henry's Law constant for the system at the appropriate temperature. Henry's Law constants are listed in physical and chemical data books, but care should be exercised to ensure the units used are appropriate.

Three hydrodynamic models that describe the transfer (absorption) of gases in liquids are in common use:

(1) the two-film model of Lewis & Whitman;
(2) the systematic surface renewal penetration model of Higbie;
(3) the random surface renewal penetration model of Dankwerts.

The Lewis & Whitman model is the most widely used model for describing physical gas absorption in aerobic microbial culture systems and, hence, is the only one that will be described in detail. Oxygen transfer from the bulk liquid medium to the microbes is also frequently considered in an analogous manner to the two-film model.

The Lewis & Whitman model assumes that at a gas-liquid interface both the gas and the liquid that comprise the thin layer (film) immediately adjacent to the interface are essentially stagnant, whilst both the bulk gas and the bulk liquid phases are in turbulent motion. Absorption results from steady-state molecular diffusion processes in the two stagnant films. Instantaneous equilibrium is assumed at the actual gas–liquid interface. The rate of absorption is controlled by the relative rates of diffusion in the two films on either side of the interface. For absorption from gas bubbles dispersed in a liquid, the radius of curvature of the bubbles is orders of magnitude larger than the hypothetical film thickness and, hence, oxygen transfer from an air bubble to a liquid can be illustrated by Fig. 10.10. An oxygen partial pressure gradient exists in the gas film from p_{O_2} the partial pressure in the bulk gas phase, to p_{O_2i}, the partial pressure in the gas film immediately adjacent to the interface, and a dissolved oxygen concentration gradient exists in the liquid film from c_{O_2i}, the equilibrium

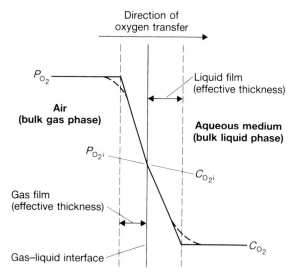

Fig. 10.10. The Whitman two (laminar) film model for the physical absorption (transfer) of oxygen from air into an aqueous liquid medium (symbols are defined in the text).

saturation concentration corresponding to p_{O_2i}, immediately adjacent to the interface, to c_{O_2}, the bulk liquid concentration.

The rate of oxygen transfer per unit area, N_{O_2}, in both the gas and liquid films will be equal, as oxygen accumulation cannot occur at the interface, and is

$$N_{O_2} = k_G \, (p_{O_2} - p_{O_2i}) = k_L \, (c_{O_2i} - c_{O_2}) \tag{76}$$

where k_G and k_L are the mass transfer coefficients for the gas film and the liquid film, respectively. For air and other gas mixtures containing oxygen, p_{O_2i} and c_{O_2i} cannot be determined, and it is necessary to introduce the concept of overall transfer coefficients defined on the basis of both the gas, K_G, and the liquid, K_L, as shown below:

$$N_{O_2} = K_G \, (p_{O_2} - p_{O_2}^*) = K_L \, (c_{O_2}^* - c_{O_2}) \tag{77}$$

where $p_{O_2}^*$ is the equilibrium partial pressure corresponding to c_{O_2}, and $c_{O_2}^*$ is the equilibrium concentration corresponding to the partial pressure p_{O_2}, the bulk liquid-phase dissolved oxygen concentration and the bulk gas-phase oxygen partial pressure, respectively. Applying Henry's Law (equation 75) to relate partial pressures and concentrations, one obtains

$$p_{O_2} = H_{O_2} \, c_{O_2}^* \tag{78}$$

$$p_{O_2}^* = H_{O_2} \, c_{O_2} \tag{79}$$

and

$$p_{O_2i} = H_{O_2} \, c_{O_2i} \tag{80}$$

Substitution into equations 75 and 76 gives

$$\frac{1}{K_G} = \frac{1}{k_G} + \frac{H_{O2}}{k_L} \tag{81}$$

and

$$\frac{1}{K_L} = \frac{1}{H_{O_2}k_G} + \frac{1}{k_L} \tag{82}$$

The reciprocals of the coefficients are known as transfer resistances, so that $1/k_G$ and $1/k_L$ are the gas and liquid film resistances, respectively, and $1/K_G$ and $1/K_L$ are the overall gas and liquid side resistances, respectively. The relative magnitudes of the resistances depend on the solubility of the gas undergoing transfer. For the absorption of a very soluble gas, such as ammonia in water, the Henry's Law constant, H_{NH_3}, is very small, and the last term in the analogous equation to equation 81 is very small and can be neglected, so that the overall gas-phase transfer coefficient, K_G, is virtually equal to the gas film coefficient, k_G, and the absorption process is described as gas film controlled. For the absorption (transfer) of oxygen, which is only slightly soluble in water, the Henry's Law constant, H_{O_2}, is very large, so that the term, $1/H_{O_2}k_G$ in equation 82 is very small and can be neglected. The overall liquid-phase transfer coefficient, K_L, is virtually equal to the liquid film transfer coefficient, k_L, and oxygen transfer is described as a liquid film controlled absorption process. For the case where the gas phase comprises oxygen, $p_{O_2i} = p_{O_2}$ and k_G is infinite, so that the transfer process is also liquid film controlled.

For most aerobic bioreactors, oxygen is transferred from air bubbles dispersed throughout the liquid medium in which the microbes are either growing or respiring, and it becomes necessary to assess the rate of oxygen transfer on a unit volume basis. Therefore equation 77 must be modified accordingly:

$$R_{O_2} = K_L a \, (c_{O_2}^* - c_{O_2}) \tag{83}$$

where R_{O_2} is the rate of oxygen transfer per unit liquid volume, and a is the specific gas–liquid interfacial area, i.e. the interfacial area per unit liquid volume. At saturation, $c_{O_2}^* = c_{O_2}$, and no oxygen transfer occurs. However, either growing or respiring aerobic microbes will utilize oxygen dissolved in the liquid medium such that $c_{O_2}^* > c_{O_2}$ and transfer occurs. In the case of the two-film model K_L is defined as

$$K_L = \frac{\mathscr{D}}{\delta} \tag{84}$$

where $\mathscr{D}$ is the diffusivity, and δ the liquid film thickness. For the other gas absorption models K_L is defined differently. In the Higbie model:

$$K_L = 2\sqrt{\mathscr{D}/\pi t^*} \qquad\qquad (85)$$

and in the Dankwerts model:

$$K_L = \sqrt{\mathscr{D}/t_D} \qquad\qquad (86)$$

where t^* is defined as the residence time of a fluid element at the interface, and t_D as the equivalent diffusion time.

The transfer of oxygen from a bubble to the liquid medium is only one part of the overall process of oxygen transfer to the individual microorganisms. As has been seen for oxygen transfer from a bubble to a liquid, the transfer process is subject to transfer resistances, one of which is controlling. The concept of transfer resistances can be extended to the further steps in the overall process. Once oxygen is transferred to the bulk liquid, it has then to be transferred to the sites of utilization in the microorganisms. Conceptually, the additional transfer resistances are the bulk liquid transfer resistance, $1/k_B$, which, for non-viscous, Newtonian fluids under conditions of fully developed turbulence, is negligible, the liquid film resistance adjacent to the microbial cell, floc, pellet, or mycelial surface, $1/k_M$, and either an intracellular resistance in the case of discretely dispersed cells or an intercellular resistance (including an intracellular resistance) in the case of flocs, pellets and mycelium, $1/k_c$. The overall transfer process resistances are illustrated in Fig. 10.11. In the case of discretely dispersed cells, the bubble liquid film resistance, $1/k_L$, invariably controls the overall oxygen transfer process, whilst in cases where the microorganisms grow as flocs, as pellets or as a mycelial mass, the intra-/intercellular or clump resistance $1/k_c$ can, in some cases, take over from the bubble liquid film resistance as the controlling resistance in the overall oxygen transfer process.

In view of the obvious significance of bubble–liquid oxygen transfer, it is important to examine equation 83 in order to evaluate possibilities for enhancing the rate of oxygen transfer. Potentially, the three component terms, K_L, a and $(c^*_{O_2} - c_{O_2})$, the driving force term, are amenable to modification. However, K_L is essentially a property of the system and, as such, is not considered amenable to independent modification and, therefore, is usually considered to be constant, even though surface active agents can and do modify it in microbiological process systems. Both the specific gas–liquid interfacial area, a, and the driving force term can be readily increased by controlled changes in

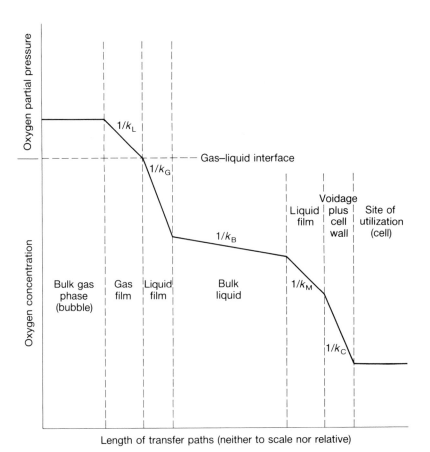

Fig. 10.11. Resistances to oxygen transfer from a bubble to a microbe, based on an extension on the Whitman model (symbols are defined in the text).

process operating conditions; usually power dissipation for the former and either overall or partial pressure increases for the latter.

The gas–liquid interfacial area, a, can be defined as

$$a = \frac{6\varepsilon}{\bar{D}_b} \tag{87}$$

where ε is the gas hold-up per unit volume of dispersion, and $\bar{D}_b$ a mean value for the bubble diameter of a swarm of non-uniformly sized bubbles. Usually, a Sauter mean bubble diameter is employed and this is defined as

$$\bar{D}_b = \frac{\sum\limits_1^n n\, D_b^3}{\sum\limits_1^n n\, D_b^2} \tag{88}$$

where n is the number of bubbles, and D_b the equivalent spherical diameter of the bubbles. Obviously, to enhance oxygen transfer by increasing a, it is necessary either to increase gas hold-up, ε, or to reduce mean bubble diameter. Both possibilities are potentially realistic process engineering options.

The dispersion of gas bubbles in liquids is a complex process and, irrespective of the technique used to produce a bubble swarm, a size distribution of bubbles will occur in the swarm. The size of any individual bubble results from equilibrium being reached between the two major stresses that act on bubbles in dispersions, i.e. the shear stress and the interfacial tension stress. The presence of both dissolved inorganic and dissolved organic matter in aqueous systems markedly affects bubble size either by influencing the coalescence characteristics or by changing the interfacial tension.

In bioreactors, neither coalescence characteristics nor interfacial tension can be varied independently at will, although undoubtedly they can and do change during the time course of a process and also as a result of the addition of foam control agents. Reduction in the mean bubble size in gas–liquid dispersions theoretically results from increased turbulence, brought about by increasing the power dissipated. However, at most practical levels of power dissipation, the effect of changes on mean bubble size is relatively minor, whilst the effect on gas hold-up, the other component of a, is major, because of increased liquid recirculation, which in turn hinders the escape of the gas bubbles from the system. The mechanisms of bubble break-up in isotopic turbulent flow will not be discussed but, under non-coalescing conditions, such considerations have resulted in an expression that successfully correlates much of the available laboratory data for very high power input systems:

$$\bar{D}_b = J\left[\frac{\sigma^{0.6}}{\rho_1^{0.2}(P_g/V)^{0.4}}\right] \tag{89}$$

where J is a constant, σ the interfacial tension, ρ_1 the density of the liquid, and P_g/V the power input per unit volume under gassed conditions. To account for coalescence, equation 89 can be modified:

$$\bar{D}_b = J\left[\frac{\sigma^{0.6}}{\rho_1^{0.2}(P_g/V)^{0.4}}\right]\varepsilon^\alpha \tag{90}$$

where the power α to which the hold-up, ε, is raised, is the coalescence tendency.

As is obvious from above, the hold-up and the mean bubble size are intimately related. Hold-up depends on both the power dissipation rate and the volumetric gas flow rate, and empirical relationships that correlate the effects of these factors on gas hold-up, for conditions where fluid properties remain constant, are of the following type:

$$\varepsilon = \mathbf{J'} \, (P_g/V)^\beta \, (V_s)^\gamma \qquad (91)$$

where $\mathbf{J'}$, β and γ are system and equipment dependent constants, and V_s is the superficial gas velocity, i.e. the quotient of the volumetric gas flow rate and the cross-section area of the bioreactor. In high power input, impeller agitated bioreactors, considerable gas-phase recirculation occurs, and the influence of the superficial gas velocity, V_s, on gas hold-up is correspondingly reduced, whilst in non-impeller agitated, sparged bioreactors V_s becomes relatively much more important.

In a recent evaluation of gas–liquid oxygen transfer in microbiological reactors, Andrew (1982) has suggested that on the industrial scale all aerobic bioreactors, including large impeller agitated, tank reactors, can, for design purposes, be regarded as free bubble rise type systems of the bubble column type. The inclusion of large, impeller agitated, tank bioreactors in the bubble column type bioreactor category is a consequence of the definition of the category in terms of mass transfer phenomena, and the failure of the impeller to create, at typical industrial-scale power inputs, sufficient liquid recirculation so as to re-entrain bubbles escaping from the liquid surface.

Returning to the question of altering the oxygen transfer rate by modification of the driving force term in equation 83, i.e. $(c_{O_2}^* - c_{O_2})$, one finds that several possibilities to achieve this exist. The most fundamental of these is to minimize the operating dissolved oxygen concentration in the medium, c_{O_2}. However, once this is reduced to essentially zero, no further increase in the driving force term is possible using this approach. The maximum, but constant, rate of oxygen transfer, $R_{O_{2m}}$, achievable is

$$R_{O_{2m}} = \mathbf{K}_L a \, c_{O_2}^* \qquad (92)$$

Caution must be exercised with respect to reducing the dissolved oxygen of the growth medium to zero, because the culture will now be growing under oxygen limited conditions, which might not be optimal with respect to growth and/or product formation. In cases of the growth of microorganisms that exhibit floc, pellet or mycelial type morphologies and where $1/k_c$ has become the controlling resistance in the overall oxygen transfer process, c_{O_2} must be maintained at a value significantly higher than zero, if aerobic metabolism is to occur in a significant fraction of the microbial biomass present in the reactor.

Prior to discussion of modifications to $c_{O_2}^*$, it is first necessary to identify which oxygen partial pressure it is that corresponds to $c_{O_2}^*$ in particular bioreactor types. For high intensity, impeller agitated, tank bioreactors of the type used in the laboratory, it has been shown that both the gas and the liquid phases are completely mixed and, hence, the partial pressure of oxygen in the dispersed bubbles is equal to the partial pressure of oxygen in the gas escaping from the liquid, i.e. the gas in the bioreactor head space. For bubble column type bioreactors, including industrial-scale stirred tank bioreactors operating at typical power inputs, it can be assumed, as a first approximation, that the liquid phase is completely mixed, but that the gas phase is in plug flow, i.e. an oxygen partial pressure gradient exists from the bottom to the top of the bioreactor. For these two extreme cases, equation 83 can be modified thus:

$$R_{O_2} = \frac{K_L a P}{H_{O_2}} \; (y_2 - y_3) \tag{93}$$

for the completely mixed case and

$$R_{O_2} = \frac{K_L a P_{av}}{H_{O_2}} \left[\frac{(y_1 - y_2)}{\ln \dfrac{y_1 - y_3}{y_2 - y_3}} \right] \tag{94}$$

for the plug flow case, where P is the head-space pressure, P_{av} the mean reactor operating pressure (head-space + hydrostatic), and y_1, y_2 and y_3 the mole fractions of oxygen in the inlet gas, the outlet gas and the gas mixture corresponding to the equilibrium dissolved oxygen concentration, respectively.

It is evident from equations 93 and 94 that increases in the bioreactor head-space pressure enhance the rate of oxygen transfer in both cases, whilst increasing the partial pressure of oxygen, by either enriching or replacing the air sparged to the bioreactor with oxygen, would enhance the rate of oxygen transfer in the plug flow case. For completely mixed bioreactors, reducing oxygen conversion would increase the oxygen partial pressure in the bubbles and, hence, the oxygen transfer rate, but such a procedure would certainly be economically questionable.

In view of the tendency of microbes to concentrate at air–liquid interfaces, it is probably realistic to modify equation 82 by the introduction of an enhancement factor, E′, so that the equation for oxygen transfer becomes

$$R_{O_2} = E' K_L a \, (c_{O_2}^* - c_{O_2}) \tag{95}$$

In the absence of microbial accumulation in the liquid film adjacent to the interface, E' will obviously be equal to one. However, particularly in the case of bacteria whose dimensions are considerably less than the typical film thickness of *c.* 20 μm, substantially higher bacterial concentrations can be expected in the liquid film than in the bulk liquid; the effect on oxygen transfer will probably be somewhat similar to effects occurring in gas absorption with simultaneous chemical reaction, although direct equivalence should be avoided because of both a non-linear bacterial concentration profile within the film and restriction in film fluidity from the physical presence of the bacteria.

10.3.4   *Heat transfer and process cooling*

One of the most neglected factors in biotechnology is the question of the various heat transfer processes involved in controlling the temperature of the individual microorganisms in aerated culture medium at a level coincident with the temperature optimum for their anabolic and catabolic processes. The optimum temperature range for either growth of or product formation by most microorganisms is narrow, in many cases only a few degrees. Below the optimum temperature for growth, the growth rate of any particular microorganism increases relatively slowly with increasing temperature, whilst at temperatures only a few degrees in excess of the optimum for growth, a precipitous decline in the growth rate, to zero, occurs. The vast majority of microorganisms used in biotechnological processes are mesophilic, and their optimum temperature for growth is somewhere between 15 and 40 °C. Thermophilic microorganisms have temperature optima considerably in excess of 40 °C and obligate psychrophilic microorganisms have temperature optima below 10 °C.

If one considers the production of any bulk microbiological product by an aerobic route, process cooling is a very significant technological problem and a major operating cost, when the temperature optimum for maximum yield and productivity is below 40 °C. Currently, it is general practice to use cooling water rather than a refrigerant as the cooling medium for industrial-scale biotechnological processes. The maximum temperature of the cooling water is of critical importance for sizing the cooling surfaces needed to maintain the temperature of any particular process and the operating costs for cooling systems are governed by the average cooling water temperature at the particular process location. It must be stressed that most bulk product refining and manufacturing processes require cooling to temperatures appreciably higher than those typically required for biological processes. However, some of the newer physical separation processes that will find industrial application in the future will be operated at near ambient temperature, and may allow more possibilities for the integration of biotechnological processes into manufacturing locations.

Cooling water temperatures are dependent on a number of factors which include the origin of cooling water, the climatic conditions and variations at the plant location, and the type of cooling water system employed, which can be either a once-through system or a recycle system employing a cooling tower. The probable future trend for process cooling will be an increasing use of recycle systems in order to eliminate both heat and possible chemical pollution as a result of cooling water discharge from once-through systems. In temperate regions, the use of mesophilic microorganisms for biotechnological processes will mean that only relatively small temperature differences will exist between the optimum process temperature and the temperature of the cooling water in large, integrated manufacturing locations.

Comparatively few studies of heat production during microbial growth have been undertaken, in spite of the fact that calorimetric techniques for measuring heat production during the growth and respiration of microorganisms have been available for many years. Where heat production has been measured, it has usually been as a means of assessing the rate and/or the amount of growth. In aerobic microbiological processes, the rate of heat production is directly proportional to the rate of oxygen consumption, which is, in turn, related to growth by an appropriate yield coefficient. The empirical correlation between the rate of oxygen consumption and the rate of heat production is considered sufficiently reliable to allow the heat production rate to be calculated and heat production rates are only rarely measured. The correlation is

$$Q_n = k\,R_{O_2} \tag{96}$$

where Q_n is the rate of heat generation by the aerobically growing microbes per unit volume, R_{O_2} the rate of oxygen consumption per unit volume (equal to the rate of oxygen transfer), and k a constant, which has the numerical value 124 when Q_n is expressed in kcal m^{-3} h^{-1} and R_{O_2} in mol m^{-3} h^{-1}. The empirical equation (96) is reasonably consistent with calculated heat production on the basis of heats of combustion data for both substrates and microbial biomass of appropriate composition.

Most bioreactors used for industrial-scale processes are provided with a cooling jacket and/or internal cooling coils. Recently, for higher intensity processes, external cooling circuits have been introduced for bioreactors both of conventional design and of more innovative configurations. In some of the latter types, the provision of internal cooling would so disrupt essential hydrodynamic operating characteristics as to make it impracticable. With increasing reactor scale, the provision of adequate internal cooling surface becomes increasingly difficult. For geometrically similar vessels, volume increases in proportion to the cube of the linear dimension, whilst area increases only in proportion to the square of the linear dimension. The main factors that affect the effective cooling of aerobic bioreactors are as follows:

(1) the transfer of heat from the individual microorganisms to the bulk liquid medium;

(2) the transfer of heat from the bulk liquid medium, containing dispersed air bubbles and microorganisms, to the cooling surfaces;

(3) the addition of heat by the mechanical energy dissipated during agitation of the medium for aeration;

(4) evaporative cooling, temperature equilibration and expansion work resulting from the passage of air through the medium;

(5) the effect, on overall heat transfer coefficients, of fouling of the cooling surfaces by accumulation and growth of microorganisms on these surfaces.

For each process system, it is important to examine the relative magnitudes and the potential significance of these factors.

In cases where the microorganisms are discretely dispersed throughout the liquid medium, as is common in processes where bacteria and yeasts are cultivated, it is the transfer of heat from the bulk liquid medium to the cooling surfaces that controls the overall rate of the heat transfer process.

The contents of an aerobic bioreactor, operating at a high level of process intensity, can be described as a bubble bed, but with the added complication that the liquid phase also contains suspended microorganisms, which can affect the viscosity of the medium, although this depends either on their size, morphology and concentration or on their ability to produce extracellular polymeric products.

One major objective in the aeration of bioreactor is to produce small bubbles, in order to maximize the area available for gas–liquid oxygen transfer. Largely because of the nature of liquid growth media, bubbles produced in such media are spherical and subject to comparatively little distortion. Essentially they are rigid spheres with hydrodynamic properties similar to those of solid spherical particles. Obvious differences exist between the density and the thermal conductivity of solids and bubbles, but the behaviour of bubble beds is, in some respects, analogous to the behaviour of liquid fluidized beds, although in the latter the dispersed-phase hold-up is generally higher.

The densities of microorganisms are very similar to those of aqueous microbial growth media and, for cultures of discretely dispersed bacteria and yeasts, which usually have equivalent spherical diameters of less than 10 μm the culture medium can, for heat transfer purposes, be considered as a single phase. In aerated bioreactors, the transfer of heat from the culture medium to the cooling surfaces results from forced convection. In order to evaluate the potential effectiveness of cooling, it is necessary to assess the compatibility between the characteristics of gas–liquid dispersions that result in optimum oxygen transfer and those factors that enhance heat transfer from the dispersion to the surface.

The good heat transfer properties of fluidized bed reactors have resulted in their adoption for reactions where close control of temperature is required. Fluidized particles enhance the heat transfer coefficient between the bed and the heat transfer surfaces. Three main mechanisms are considered to enhance heat transfer coefficients in fluidized systems:

(1) the high heat capacity per unit volume of the particles which enables them to act as heat-transferring agents;

(2) the erosion of the laminar sub-layer at the heat transfer surface by the particles, and the consequent reduction in the effective thickness of the sub-layer;

(3) an unsteady-state heat transfer process resulting from 'packets' of particles moving to and from the heat transfer surface.

On examination of the application of these mechanisms to gas–liquid dispersions, it can be concluded that if the first mechanism plays a significant role in gas–liquid dispersions, the heat transfer coefficient between the bubble bed and the heat transfer surface will be effectively decreased as a result of the low heat capacity of the bubbles. If the second mechanism occurs, the heat transfer coefficient will be enhanced, whilst if the third mechanism occurs, excessive coalescence and slug formation might result. In liquid-fluidized systems, the hold-up of solid particles is often as high as 0.4, whilst in conventional bioreactors with typical power inputs, the gas hold-up is unlikely to exceed 0.12. At hold-ups in excess of 0.2, slugs of gas start to form.

In sparged columns, the bubbling of gas through the liquid induces two types of motion:

(1) a general circulation motion provoked by the ensemble of bubbles;

(2) local micro-motions provoked by the individual bubbles. Experimental data for the overall heat transfer coefficient when gas–liquid dispersions are cooled by heat transfer to the column wall can be correlated by

$$h = \mathrm{B}k_1 \left(\frac{g\varepsilon\rho_1^2}{\eta_1^2}\right)^{0.33} \mathrm{N}_{Pr}^{\;0.33} (\eta_\mathrm{w}/\eta_1)^{-0.14} \tag{97}$$

and where heat transfer is to a cooling coil, the data can be correlated by

$$h = \mathrm{B}'k_1 \left(\frac{g\varepsilon\rho_1^2}{\eta_1^2}\right)^{0.33} \mathrm{N}_{Pr}^{\;0.33} (\eta_\mathrm{w}/\eta_1)^{-0.14} (d_\mathrm{c}/d_\mathrm{p})^{0.33} \tag{98}$$

where h is the overall heat transfer coefficient, B and B′ are system dependent constants, k_1 is the thermal conductivity of the liquid, g the gravitational constant, ε the gas hold-up, ρ_1 the liquid density, η_1 the liquid viscosity, N_{Pr} the Prandtl number, η_w the liquid viscosity at the temperature of the heat transfer surface, d_c the column diameter, and d_p the diameter of the tube forming the cooling coil. Neither the correlation for transfer to a wall nor that for transfer to a cooling coil can be applied directly to intensely agitated stirred tank type bioreactors. Because of the significant recirculation in such reactors, it is necessary to modify the values of the exponents in the correlations.

As far as oxygen transfer is concerned, aeration systems in bioreactors are designed to produce small bubbles. However, for bubble swarms, large bubbles enhance heat transfer, and bubble size reduction is contrary to enhanced heat transfer. However, heat transfer is enhanced by increased gas hold-up, which suggests that systems that exhibit enhanced oxygen transfer rates will also exhibit enhanced heat transfer rates, other factors being equal.

10.3.5   *Conversion and recycle*

As has been discussed in a previous section, the principal engineering factor that affects process economics in microbiological processes is the bioreactor productivity. However, in the case of either essentially insoluble or immiscible substrates and non-growth limiting nutrients, conversion also becomes an important economic factor. The fractional conversion of either a substrate or a nutrient can be defined as the weight utilized per unit weight of that supplied. In microbiological processes, the fractional conversions of a completely soluble, growth limiting substrate will be virtually one, with only very small residual concentrations depending on the affinity of the process culture for the limiting substrate. For either essentially insoluble or immiscible substrates and nutrients and for non-growth limiting nutrients, the fractional conversion achieved in microbial processes will frequently be significantly less than one. For such substrates and nutrients, the conversion coefficient, defined as the weight of either dry microbial biomass or product produced per unit weight of either substrate or nutrient supplied, will be less than the corresponding yield coefficient. The conversion of oxygen and gaseous carbon energy substrates, of water immiscible and insoluble carbon energy substrates, and of soluble non-growth limiting nutrients will, in some cases, be affected by the mode of operation of the bioreactor whilst, in others, mass transfer characteristics, bioreactor geometry and operating variables will be the most important factors.

Although aerobic microbiological processes do not incur raw material costs by using air as their oxygen source, considerable capital and operating costs accrue to the process because of the need to compress and sterilize the air prior to its use in the process. For industrial-scale processes, it is important to minimize such cost by achieving maximum conversion of the oxygen supplied, provided the resultant costs for maximization do not result in an overall de-optimization of the process economics.

The fractional conversion of oxygen achieved in high intensity laboratory bioreactors rarely exceeds 0.2, and in conventional industrial-scale bioreactors is typically about 0.1, when air is used as the source of oxygen. Even at these low fractional conversions, oxygen is frequently the growth limiting nutrient for the process, because the rate of oxygen transfer fails to meet the potential rate of demand for oxygen by the culture. Examination of the equation that describes the physical absorption of oxygen in microbial growth media (equation 83) indicates that, in systems with a completely mixed gas phase, high fractional oxygen conversions are inconsistent with the maximization of the oxygen transfer rate, and obviously a technical and economic optimum must be sought between these two important, but conflicting, factors. As far as bioreactors where the gas phase is in plug flow are concerned, significantly higher fractional conversions at particular oxygen transfer rates, when compared with completely mixed systems, are predictable.

One of the most common means of enhancing fractional conversions in process systems is the introduction of recycle. Such an approach is equally applicable to both the gas and liquid phases in microbial processes, but is not without potential ramifications. These can be examined, for gas-phase recycle, on the basis of the two flow schemes shown in Fig. 10.12, for single-pass and recycle gas-phase operation, making the assumption that the gas-phase component undergoing transfer is a minor component, as is oxygen in air, and hence the gas

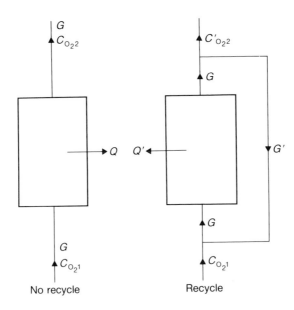

No recycle   Recycle

Fig. 10.12. Gas-phase recycle for enhanced fractional conversion of oxygen at low conversions (symbols are defined in the text).

flows into and out of the bioreactor are virtually equal. The fractional conversions of oxygen in air without recycle, X_{O_2}, can be expressed as

$$X_{O_2} = \frac{C_{O_2}1 - C_{O_2}2}{C_{O_21}} = \frac{Q_{O_2}}{G \, C_{O_21}} \qquad (99)$$

and with recycle, X'_{O_2}, can be expressed as

$$X'_{O_2} = \frac{C_{O_21} - C'_{O_2}2}{C_{O_21}} = \frac{Q'_{O_2}}{(G - G')C_{O_21}} \qquad (100)$$

where C_{O_21} is the concentration of oxygen in the inlet gas flow to the overall systems, C_{O_22} and C'_{O_22} are the concentrations of oxygen in the outlet gas flow from the system for non-recycle and recycle operation, respectively, Q_{O_2} and Q'_{O_2} the oxygen utilization rates of the total culture for non-recycle and recycle operation, respectively, and G and G' the gas flow rate through the bioreactor and in the recycle, respectively. When oxygen is not growth limiting, i.e. $Q_{O_2} = Q'_{O_2}$, the ratio of the fractional conversions with recycle and without recycle is given by

$$\frac{X'_{O_2}}{X_{O_2}} = \frac{G}{(G - G')} \qquad (101)$$

However, for the case where oxygen is the growth limiting nutrient, $Q_{O_2} > Q'_{O_2}$ because of a difference in the oxygen transfer driving force

term and, on the basis of a completely mixed gas phase, the expression
for the ratios of the fractional conversions becomes

$$\frac{X'_{O_2}}{X_{O_2}} = \frac{Q'G}{Q(G - G')} \tag{102}$$

Irrespective of any enhancement of the fractional conversion of oxygen,
bioreactor productivity will inevitably fall as a result of gas-phase
recycle due to a reduction in the oxygen transfer driving force,
assuming that the operating conditions are maintained constant.
Obviously, conversion should not be optimized independently of productivity.

For those processes where either gaseous or volatile carbon energy
substrates are supplied to bioreactors as either gases or vapours, very
different considerations apply with respect to conversion. Such substrates, unlike oxygen in air, are expensive raw materials and processes
using them must be designed either to achieve very high fractional
conversions, or, through process integration, to allow the effluent gas to
be utilized in another unit operation in the overall process system.
Probably the best known gaseous carbon energy substrate is methane,
although other gaseous alkanes are not without potential as commercial substrates. Provided methane can be used, in the form of natural
gas, as a microbiological processes feedstock, the fuel value of uncon-
verted methane can probably be recovered in heat-requiring processes
such as drying, as long as relatively high fractional conversions of
0.7–0.8 can be achieved in the bioreactor. Clearly, such fractional
conversions would require gas-phase plug flow, elevated pressure and,
possibly, gaseous phase recycle in bioreactors where realistic productivities are to be maintained. Alternatively, the employment of multiple,
high pressure, completely mixed bioreactors in series to achieve
respectable fractional conversions of gaseous substrates and nutrients
might offer an optimum solution.

Turning our attention to the liquid phase and to potential liquid-
phase recycle, it should be emphasized that all microbiological processes are characterized by relatively low concentrations of microbes,
substrate, nutrients and products in the growth medium and, therefore,
employ large volumes of water in which the microbes, substrate,
nutrients and products are either dispersed or dissolved. The concentration of microbes is usually limited by factors such as inhibition by
substrate, nutrients or products and, in aerobic processes, oxygen
transfer rates. However, very high microbe concentrations are possible
in process systems employing microbe recycle.

For the production of either high or medium value products by microbiological processes, the incentive to introduce either medium recycle or recovery is limited to processes in which partially utilized, very high value feedstocks are employed, but in the case of bulk products manufacture, the incentive to introduce medium recycle is considerable, both with respect to economizing on the quantities of good quality process water and of nutrients required, and reducing the volume of liquid effluent requiring treatment.

In any process water/medium recycle scheme, it is of critical importance to establish a comprehensive understanding of the interactions between the various unit operations involved in any recycle loop. The problems pertinent to medium recycle can be summarized:

(1) the build-up of inhibitory products in the medium because of either metabolic activity that results in by-product formation or lysis during recycling;

(2) the build-up of inhibitory products by metabolic activity, lysis or cell disruption in the bioreactor;

(3) the build-up of undesirable, unutilizable impurities in feedstocks;

(4) the recycle of unused substrates and nutrients such as to cause physical and chemical imbalances and dynamic instability in the bioreactor.

In the case of mixed culture processes, recycle could also result in some of the microbial components of the culture being selectively returned to the bioreactor, thereby causing population imbalances in the bioreactor.

Microbial biomass recycle is a means of enhancing the intensity of microbial processes, rather than a means for improving conversion. In contrast to the few examples of gaseous and liquid-phase recycle in industrial-scale microbial processes, microbial biomass recycle has been an established operating practice for nearly seventy years in activated sludge wastewater treatment processes.

Continuous flow microbial processes incorporating microbe recycle can be directly equated, in mathematical terms, to continuous flow microbial processes in which microbe retention occurs either by design or by accident, and it is probable that microbe retention will find extensive industrial application in the future. Microbe recycle and microbe retention offer important advantages when the process microbes are either slow growing or are producing a non-growth associated product by their enzymic capacity. However, when these practices are used to achieve process intensification, no increase in productivity will be possible unless heat and mass transfer capacities of the bioreactor employed are adequate to permit such increases in productivity. If either the heat or the mass transfer capacity of a bioreactor is limiting under operating conditions without either microbe recycle or microbe retention, process intensity can only be enhanced when the limiting capacity is also enhanced by some means.

10.3.6 *Characteristics of microbial and mycelial suspensions*

Although the analyses of microbes are generally given as the percentage of carbon, hydrogen, oxygen, nitrogen and ash, on a dry weight basis, microbes are, in their functional form, very largely composed of water: in fact, about 70% on a weight basis.

The shapes and sizes of various species of microbes are listed in the taxonomic literature, but it should be remembered that growth conditions frequently affect both size and shape. For an understanding of mass and heat transfer problems in bioreactors and microbe/liquid separation processes during product harvesting, the use of an equivalent size (equivalent spherical diameter) term is frequently introduced for cultures where the cells are growing in discretely dispersed suspension. The equivalent size concept is an essential simplifying assumption which enables the dynamic behaviour of cells in suspension to be described. In addition, the density of microbes is also of obvious importance. Densities generally fall within the range 1.07–1.09 g cm^{-3}, although values as low as 1.03 g cm^{-3} have been quoted. The density of aqueous culture fluids is usually slightly higher than that of water.

The only case for which the drag on an immersed sinking spherical particle has been calculated from purely theoretical considerations is that of a particle sinking at low velocity in an infinite volume of liquid. For this, Stokes obtained the relationship

$$F = 3(\pi \eta u d) \tag{103}$$

where F is the drag force, η the viscosity of the liquid, d the diameter of the sphere, and u the velocity of the fluid relative to the spherical particle. In practice, conditions under which Stokes' Law is applicable are rare, and in most practical systems, particles are invariably undergoing either acceleration or retardation, whilst even under laminar flow conditions, particles are rarely isolated from each other and particle–particle interactions occur.

According to Stokes' Law, the terminal sinking velocity of a particle, u_o is obtained by equating the viscous drag force to the effective gravitational force, i.e.

$$3(\pi \eta u_o d) = \frac{\pi d^2}{6} (\rho_s - \rho_l)g \tag{104}$$

where ρ_s and ρ_l are the particle and liquid densities, respectively. Hence

$$u_o = \frac{d^2(\rho_s - \rho_l)g}{18\eta} \tag{105}$$

Modifications to Stokes' Law for predicting the settling rates, u_c, of suspensions of fine uniformly sized particles have been proposed, i.e.

$$u_c \propto \frac{d^2\,(\rho_s - \rho_c)g}{\eta_c} \tag{106}$$

where ρ_c is the average density of the suspension, and η_c the apparent viscosity of the suspension. Here, account has been taken of the effect of the particle concentration on both the viscosity and density of the fluid. In addition, to take into account variations in particle shape and size, a function of the particle voidage can be included in equation 106, when non-uniform particles are considered.

As far as biological process systems are concerned, the viscosity of process fluids is a most important parameter. Viscosity is defined as a measure of the internal friction of a fluid. Under laminar flow conditions, i.e. flow at low Reynolds numbers, the ratio of the shear stress to the corresponding rate of shear is equal to the viscosity of the fluid, i.e.

$$T = \eta\frac{du}{dy} \tag{107}$$

where T is the shear stress, η the viscosity of the fluid, and du/dy the rate of shear expressed as a velocity gradient in the fluid.

The dimensions of viscosity can, with equal propriety, be considered as either those of the product of force, time and the reciprocal of length squared, or the product of mass and the reciprocals of both time and length. In all absolute systems of units, the numerical value of the absolute viscosity is the same on either basis. In engineering literature, viscosity is frequently expressed in terms of the kinematic viscosity, which is the product of the absolute viscosity and the reciprocal of the fluid density.

Any fluid in which the viscosity is constant and independent of the rate of shear, at constant temperature and pressure, is described as Newtonian. All gases and many homogeneous liquids exhibit Newtonian characteristics. However, most liquids that are not essentially homogeneous, i.e. many microbiological culture fluids, do not exhibit a constant viscosity, independent of the rate of shear, and are described as non-Newtonian fluids. For non-Newtonian fluids, the ratio of the shear stress to the shear rate is known as the apparent viscosity, which, of course, varies with the shear rate. Non-Newtonian liquids can be classified according to the form of relationship between shear stress and shear rate that is exhibited.

The great majority of bacterial and yeast culture fluids are essentially Newtonian in character, even at very high cell densities, unless significant production of extracellular biopolymer, which can result in the development of non-Newtonian characteristics, occurs. Culture fluids containing either moulds or actinomycetes are frequently non-Newtonian. The degree of departure from Newtonian behaviour depends on both the concentration of mycelium and on the morphology. For most mould and actinomycetes culture fluids, removal of the mycelium results in a fluid that is essentially Newtonian.

Many mould and actinomycetes culture fluids exhibit pseudo-plastic behaviour, i.e. the apparent viscosity decreases initially as the rate of shear increases, when the shear stress is plotted against the shear rate, for a pseudo-plastic fluid, the relationship initially takes the form of a curve, but subsequently it becomes linear. Extrapolation of the linear portion of the relationship to the shear stress axis gives an intercept at a point described as the yield stress. For fluids that exhibit Bingham plastic characteristics, the yield stress corresponds to the yield point. The several relationships between shear stress and shear rate are shown in Fig. 10.13. Occasionally mycelial culture fluids have been found to be thixotropic, i.e. besides the apparent viscosity decreasing as the shear

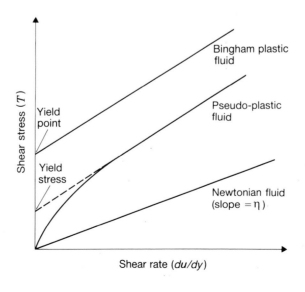

Fig. 10.13. The relationships between the shear stress and the shear rate for fluids with different rheological properties.

rate increases, the apparent viscosity also decreases to a minimum value with the time of stirring at a given shear rate.

In general, the relationship between shear stress and shear rate for non-Newtonian fluids is of the form

$$T = K(\frac{du}{dy})^n \qquad\qquad (108)$$

where K and n are constants. A pseudo-plastic fluid can be approximately characterized by specification of the yield stress and the plastic viscosity, i.e. the slope of the straight portion of the shear stress/shear rate linear coordinate plot. However, when the shear stress is plotted against the shear rate on logarithmic coordinates, a straight line of slope n and an intercept of K results for the initial part of the relationship. In truly Bingham plastic fluids, until a shear stress in excess of the yield point is reached, no fluid motion occurs, but in pseudo-plastic fluids, even at low shear stresses, some fluid movement occurs. Pseudo-plastic fluids differ markedly from Newtonian fluids with respect to gas hold-up, bubble size distribution, and mixing times for a given power input in vessels with identical geometries.

10.4   Ancillary unit operations in biotechnological processes

Much of the previous part of this chapter has been concerned with aspects of the central feature of biotechnological processes, the bioreactor. However, in any process it is impossible for the central production step to exist in isolation. For the production of high value, fine chemical type products, using batch processes, it is sometimes possible to carry out several steps, in addition to the actual production step, in the bioreactor. Medium preparation, medium sterilization, product concentration and product extraction are all operations that can, conceivably, be carried out, not necessarily optimally, in a batch bioreactor. The concept of extensive utilization of a single item of process plant for multiple process steps, although very valuable for some particular products and operating environments, is, with the exception of medium preparation and medium sterilization, largely an approach of the past, and with increasing frequency, specific purpose designed equipment for each process step is being introduced so as to improve the ease and effectiveness of overall process operation, in spite of possible elevation in the capital cost of processes.

For bulk biotechnological products, irrespective of either the feedstocks used or the process route employed, the production process will, generally, be operated on a commercial scale in a continuous flow mode in order to satisfy economic criteria. Such production processes will comprise an ordered and integrated series of unit processes that are sufficiently reliable for a product that meets specification to be produced for between 300 and 330 d/a. The fundamental objective is one of optimization of resource utilization but, unfortunately, the integration of unit processes in this way usually adversely affects overall process operability and can cause problems in maintaining high on-stream factors. A typical list of steps involved in an overall continuous flow biotechnological production process is as follows:

(1) feedstock storage, pretreatment and blending;
(2) contamination control;
(3) production;
(4) product concentration;
(5) product separation;
(6) product finishing;
(7) product storage.

With respect to both capital and operating expenditure, it is unlikely that the key production step will require more than about 30–40% of the expenditure in either cost category, thus indicating the very real importance of the ancillary process steps in overall process economics.

Obviously not all the process steps listed above are exclusive to biotechnological processes and the unit operations that comprise these steps find wide application in the process industry in general. However, some steps such as medium and air sterilization for contamination control are virtually specific to biotechnology and still others, such as concentration, drying and storage, must, when used for biotechnological product production, take account of the special properties of biological matter.

10.4.1   *Medium and air sterilization*

In many biotechnological processes, aseptic operation is an absolute process requirement and, hence, both medium and, in the case of aerobic processes, air sterilization are of critical importance. Even so, remarkably little effort has been devoted to the development and optimization of effective sterilization procedures for either medium or air. As far as biotechnological processes are concerned, liquid media are exclusively sterilized by heat, and air is virtually exclusively sterilized by filtration; only these two techniques for sterilization will be discussed here.

The rational design of sterilizers for liquid media is based on the death kinetics of microbes. The inactivation of microbes by heat involves a loss of viability, but not physical destruction. At a specific temperature, the rate of inactivation of microbes in a microbial population is described by

$$\frac{dN}{dt} = -k_d N \tag{109}$$

where k_d is a temperature dependent reaction rate constant, N the number of potentially viable microbes, and t time. It is important to note that the basis of equation 109 is cell and spore numbers, rather than cell weight as used earlier in growth equations. As far as sterilization is concerned, it is cell and spore numbers, as potential point centres of infection, that are critical.

As far as vegetative cells are concerned, the logarithmic rate of death indicated by equation 109 is a good approximation of reality, although many interfering influences can affect death kinetics. However, in the case of spores, the rate of death frequently deviates from the logarithmic pattern suggested by equation 109. For both vegetative cells and spores, the effect of temperature on the reaction rate constant k_d, in equation 109, is given by the Arrhenius expression, i.e.

$$k_d = \alpha e^{-E/RT} \tag{110}$$

where α is an empirical constant, E the activation energy, R the gas constant, and T the absolute temperature. Rearranging equation 109 and substituting for k_d from equation 110 gives

$$\frac{dN}{N} = \alpha e^{-E/RT} dt \tag{111}$$

or, in integral form, for $t = 0$ when $N = N_0$ and $t = t$ when $N = N$,

$$\ln \frac{N}{N_0} = -\alpha \int_0^t e^{-E/RT} dt \tag{112}$$

In all sterilization processes, the objective is to reduce the number of surviving potentially viable cells and spores to a number less than one. Hence, such processes are best considered from the probabilistic viewpoint. Medium sterilization processes can be operated as either batch or continuous flow processes. In batch processes, over-exposure of the medium to heat and resultant adverse affects on medium quality are inherent in the process, but continuous processes can be designed to overcome such deficiencies and thereby achieve improved bioreactor performance.

Essentially a batch medium sterilization cycle involves three distinct periods of time: a heating period, a holding period, usually at 121 °C, and a cooling period. The inactivation of cells and spores occurs during all three time periods; so, if the heating period extends for time t_1, the holding period for time t_2, the cooling period for time t_3, where t is the total cycle time, then

$$\ln\frac{N_0}{N} = \ln\frac{N_0}{N_1}+\ln\frac{N_1}{N_2}+\ln\frac{N_2}{N} = \alpha\left[\int_0^{t_1} e^{-E/RT}dt + \int_0^{t_2} e^{-E/RT}dt + \int_0^{t_3} e^{-E/RT}dt\right]$$
(113)

where N_1 and N_2 are the total number of surviving cells and spores after the heating and the holding periods, respectively.

In batch sterilization, the heating and cooling periods are frequently extensive, whilst the holding period usually lasts for 0.3–0.5 h at a temperature of 121 °C.

For continuous medium sterilization, two distinct types of system exist; those employing direct steam injection, which necessitates extra clean steam, and those employing indirect heating via heat exchangers. Obviously, both types of system operate under pressure. Direct injection systems comprise a steam injector, a holding section and an expansion valve and flash cooler. Indirectly heated systems usually comprise medium preheating with sterilized medium to recover waste heat, a high-rate heat exchanger which allows the medium to be heated to 135 °C in <1 min, a holding section, in which the medium is maintained at 135 °C for 3–5 min, and a rapid cooling section where first the sterilized medium is cooled with raw medium and subsequently with cooling water. Either plate or tube heat exchangers can be used, but the former are frequently preferred, as they permit easy cleaning of the heat exchange surfaces, which become easily fouled during the sterilization of many media.

In continuous medium sterilization, it is important to realize that velocity gradients will exist in the medium flowing through the heat exchangers and that perfect piston flow will not occur. In view of this, it is necessary, depending on the flow regime in the heat exchangers, to consider the appropriate residence time distribution function in assessing the probability of cell and spore survival.

Just as medium sterilization is critically important in many biotechnological processes, it is equally important to sterilize the air used for aeration in such processes. Air sterilization processes are required to remove microbes, with dimensions as small as 0.5 μm, from the vast volumes of air frequently required by aerobic processes. Obviously, heat generation during air compression is of considerable value in killing airborne microbes, but it is usually also necessary to filter the air using fibrous, sintered or acetylated polyvinyl alcohol (PVA) sheet filters. Essentially, air filters, with the exception of PVA filters, do not seek to exclude airborne microbes by their very small pore sizes, but by the tortuous path that the air is forced to take through the filter and, hence, the high probability of entrained microbes impacting with the expanded surface of the filter. In such filters, pressure drops are frequently considerable and the filter media used must permit steam sterilization of the filter and ancillary pipe work. Air filters must be kept dry for effective performance to be maintained.

10.4.2   *Product separation and finishing operations*

A wide range of product separation and finishing operations are used in the biotechnological industries. However, these same operations, with the exception of processes designed to break cells, are widely used in other process industries; therefore, individual unit process operations will not be discussed here. However, in connection with the application of such operations in biotechnological production processes, two requirements must always be stressed: first, the need to prevent product denaturation, which can result from a lack of appropriate process cooling, excessive overheating, chemical and biological contamination, or degradation, through biological action, when product is retained in an unstable state; second, the potential health and explosive hazard associated with proteinaceous dusts that arise with dried biological products. In effective process design, both problems must be avoided.

10.5   Future developments in industrial biotechnological processes

In the previous sections of this chapter, relatively detailed discussion of key aspects of the application of the chemical engineering to biotechnological, predominantly microbiological, processes has been presented, together with discussion of the philosophy of industrial-scale process systems and their economics. What is most evident with respect to data required for biotechnological process development is the incredible deficiency that exists with respect to such data. In fact, it would be entirely fair to state that no adequate data base exists concerning the physical behaviour of microbes, i.e. physical microbiology. In comparison with the degree of development of hydrodynamics, even in the last century, of surface and interfacial chemistry and of molecular biology, all subjects with key impacts on biotechnological process development, such a situation is entirely unacceptable and, furthermore, unnecessary, as many of the techniques required for the generation of appropriate physical microbiological data are available and are simply being ignored.

In contrast, the kinetics of biotechnological processes are often relatively well understood, and in recent years have been extensively applied in process development. Even so, the fundamental situation that still has to be overcome in the use of laboratory-generated data for industrial-scale process evaluation and process design is the vexed one of data variability and the range of appropriate values applicable to any situation other than the system in which the data was generated. For example, a 15% variation in either the yield or the conversion coefficients can make the difference between considerable commercial success and total commercial failure, but few process oriented microbiologists are prepared to describe either coefficient within closer limits than this when pressed to do so for either process evaluation or process design. Another typical example, in a similar vein, is the question of pressure and dissolved carbon dioxide effects, and still another concerns the effects of growth conditions on the size of microbes.

Far too much microbiological data concerns taxonomical aspects, where test conditions under which the information has been generated are far removed from those pertaining in process environments. Extrapolation of such data has frequently resulted in what can only be described as the mystic and folklore of microbiology. What is essential for the future development of effective and efficient biotechnological processes is both imaginative and realistic hypothesis construction, thorough testing under appropriate conditions to either confirm or deny hypotheses and, ultimately, the development of immutable theories and laws that clearly describe microbial behaviour relative to the pertinent environment in which they are used. Only then can it be expected that biotechnology will be capable of success in the many commercial opportunities that will become available for exploitation in future decades.

10.6   Recommended reading

AIBA S., HUMPHREY A.E. & MILLIS N.F. (1973) *Biochemical Engineering*, (2nd edn.), p. 434. Academic Press, New York.

ANDREW S.P.S. (1982) Gas–liquid mass-transfer in microbiological reactors. *Trans. Inst. Chem. Engrs*, **60**, 3–13.

ATKINSON B. & SAINTER P. (1982) Development of downstream processing. *J. Chem. Tech. Biotechnol.* **32**, 100–108.

BAILEY J.E. & OLLIS D.F. (1977) *Biochemical Engineering Fundamentals*, p. 753. McGraw-Hill, Kogakusha, Tokyo.

CALDERBANK P.H. (1967) Gas absorption from bubbles. *Chem. Engr, Lond.* No. **212** Oct, 209–233.

COONEY C.L., WANG D.I.C. & MATELES R.I. (1968) Measurement of heat evolution and correlation with oxygen consumption during microbial growth. *Biotechnol. Bioengng*, **11**, 269–281.

GADEN E.L. (1959) Fermentation process kinetics. *J. biochem. microbial Technol. Engng*, **1**, 413–429.

HARDER W., KUENEN J.G. & MATIN A. (1977) Microbial selection in continuous culture. *J. appl. Bacteriol.* **43**, 1–24.

HERBERT D. (1961) A theoretical analysis of continuous culture systems. In *Continuous Culture of Micro-organisms, S.C.I. Monograph No. 12*, pp. 21–53. Soc. Chem. Industry, London.

JOHNSTONE R.E. & THRING M.W. (1957) *Pilot Plants, Models, and Scale-up Methods in Chemical Engineering*, p. 307. McGraw-Hill, New York.

MONOD J. (1958) *Recherches sur la croissance des cultures bactériennes, 2nd edn.*, p. 210. Hermann, Paris.

PAYNE W.J. (1970) Energy yields and growth of heterotrophs. *Ann. Rev. Microbiol.* **24**, 17–52.

PIRT S.J. (1975) *Principles of Microbe and Cell Cultivation*, p. 274. Blackwell Scientific Publications, Oxford.

RUDD D.F. & WATSON C.C. (1968) *Strategy of Process Engineering*, p. 466. John Wiley & Sons, New York.

RICHARDS J.W. (1961) Studies in aeration and agitation. *Prog. ind. Microbiol.* **3**, 141–172.

Index

Genera of microorganisms which have wide application in biotechnological fields are sub-divided by the processes in which they are involved, rather than by species.